Handbook of Corrosion Engineering

Handbook of Corrosion Engineering

Modern Theory, Fundamentals, and Practical Applications

Chaudhery Mustansar Hussain

Department of Chemistry and Environmental Sciences, New Jersey Institute of Technology (NJIT), Newark, NJ, United States

Chandrabhan Verma

Department of Chemical Engineering, Khalifa University of Science and Technology, Abu Dhabi, United Arab Emirates

Jeenat Aslam

Department of Chemistry, College of Science, Taibah University, Yanbu, Saudi Arabia

Ruby Aslam

Department of Applied Chemistry, Aligarh Muslim University, Aligarh, India

Saman Zehra

Department of Applied Chemistry, Aligarh Muslim University, Aligarh, India

ELSEVIER

Elsevier
Radarweg 29, PO Box 211, 1000 AE Amsterdam, Netherlands
The Boulevard, Langford Lane, Kidlington, Oxford OX5 1GB, United Kingdom
50 Hampshire Street, 5th Floor, Cambridge, MA 02139, United States

Notices
Knowledge and best practice in this field are constantly changing. As new research and experience broaden our understanding, changes in research methods, professional practices, or medical treatment may become necessary.

Practitioners and researchers must always rely on their own experience and knowledge in evaluating and using any information, methods, compounds, or experiments described herein. In using such information or methods they should be mindful of their own safety and the safety of others, including parties for whom they have a professional responsibility.

To the fullest extent of the law, neither the Publisher nor the authors, contributors, or editors, assume any liability for any injury and/or damage to persons or property as a matter of products liability, negligence or otherwise, or from any use or operation of any methods, products, instructions, or ideas contained in the material herein.

ISBN: 978-0-323-95185-2

For Information on all Elsevier publications
visit our website at https://www.elsevier.com/books-and-journals

Publisher: Candice G. Janco
Acquisitions Editor: Anita Koch
Editorial Project Manager: Emerald Li
Production Project Manager: Kumar Anbazhagan
Cover Designer: Matthew Limbert

Typeset by MPS Limited, Chennai, India

Contents

Part 8 Corrosion protection in industries

Part 9 Sustainable corrosion management by computational modelling

Part 10 Emerging trends in corrosion protection

About the authors

Prof. Chaudhery Mustansar Hussain is an adjunct professor and director of laboratories in the Department of Chemistry and Environmental Science at New Jersey Institute of Technology, Newark, New Jersey, United States. His research is focused on the applications of nanotechnology and advanced materials, environmental management, analytical chemistry, smart materials and technologies, and other various industries. Dr. Hussain is the author of numerous papers in peer-reviewed journals as well as a prolific author and editor of around 150 books, including scientific monographs and handbooks in his research areas. He has published with Elsevier, American Chemical Society (ACS), Royal Society of Chemistry, Springer, John Wiley & Sons, and CRC Press.

Dr. Chandrabhan Verma works at the Department of Chemical Engineering, Khalifa University of Science and Technology, Abu Dhabi, United Arab Emirates. He obtained his PhD degree in materials science/chemistry from the Indian Institute of Technology (Banaras Hindu University) Varanasi, India. He is an ACS member and serves as a reviewer and editorial board member for various internationally recognized ACS, RSC, Elsevier, Wiley, and Springer platform. Dr. Verma is an associate editor-in-chief of Organic Chemistry Plus Journal. He is the author of several research and review articles published by ACS, Elsevier, RSC, Wiley, Springer, etc. He has a total citation of more than 9388 with an h-index of 54 and i-10 index of 143. He has edited many books for the ACS, Elsevier, RSC, and Wiley. He received several awards for his academic achievements.

Dr. Jeenat Aslam is currently working as an associate professor at the Department of Chemistry, College of Science, Taibah University, Yanbu, Al-Madina, Saudi Arabia. She earned her PhD degree in chemistry from the Aligarh Muslim University, Aligarh, India. Materials & corrosion, nanotechnology, and surface chemistry are the primary areas of her research. Dr. Jeenat has published a number of research and review articles in peer-reviewed international journals such as ACS, Wiley, Elsevier, Springer, Taylor & Francis, and Bentham Science. She has authored over 30 book chapters and edited more than 20 books for the American Chemical Society, Elsevier, Springer, Wiley, De-Gruyter, and Taylor & Francis.

Dr. Ruby Aslam is currently a research associate under CSIR-HRDG, New Delhi, in the Department of Applied Chemistry, Aligarh Muslim University, Aligarh, India. She received her MSc, MPhil, and PhD degrees from the same university. Her main areas of interest in research include the development of stimuli-responsive smart coatings for corrosion detection and protection as well as the assessment of environment-friendly corrosion inhibitors. She has authored/coauthored several research papers in international peer-reviewed journals of wide readership, including critical reviews and book chapters. She has edited many books for ACS, Elsevier, Springer, Wiley, De-Gruyter, and Taylor & Francis.

Dr. Saman Zehra is working as scientist on the Women Scientist (WOS-A) program of DST, New Delhi, in the Department of Applied Chemistry of Aligarh Muslim University, Aligarh, India. She received her PhD and MSc degrees from the same university. Her research interest lies broadly in developing stimuli-responsive smart coating materials for corrosion sensing and protection and evaluating green corrosion inhibitors. She has published many research articles in international peer-reviewed journals, including reviews and book chapters. She has also edited one book for Springer. Dr. Zehra presented her research work at various conferences held nationally and internationally.

PART 1

Fundamentals of corrosion & forms of corrosion

1

Basics of corrosion and its impact

As iron is eaten with rust, so are the envious consumed by envy.

(Antistenes, c.455 BCE to c.365 BCE).

1.1 Introduction

Corrosion is a natural process that causes slow degradation of metallic and nonmetallic materials by the reaction of the environment. Corrosion is the process through which fabricated metals regress to their original oxidation state. According to Fontana, "corrosion is the deterioration of materials due to a reaction with their environment." According to Uhlig, "corrosion is the destructive attack on a metal by a chemical or electrochemical reaction with the environment." According to NACE International, corrosion is "the destruction of a substance (usually a metal) or its properties because of a reaction with its environment." In this reduction–oxidation reaction, the metal is oxidized by its surroundings, generally oxygen in the air. These reactions are preferred both electrochemically and spontaneously. Corrosion is a natural occurrence, like water running downhill. In the real world, metallic materials are subjected to corrosion in both atmospheric and aquatic conditions. The corrosion of metals is one of the challenges that we have frequently experienced in our industrialized world; as a result, it has been extensively investigated since the late 18th century's industrial revolution. The local cell model described by Evans (1937) and the corrosion potential model given by Wagner (1938) unveiled modern corrosion science in the early 20th century. The modern electrochemical theory of corrosion combines these two approaches and describes corrosion of metals as a coupled electrochemical reaction which involves anodic metal dissolution as well as cathodic oxygen reduction. The theory may be used to explain wet metal corrosion at room temperature and dry metal oxidation at high temperatures (Wagner, 1974).

Corrosion, as well as its mitigation, has a long and fascinating history. Material resources will deteriorate because of their interactions with their environments. Corrosion affects numerous materials, including ceramics, polymers, and composites, even though it is primarily concerned with metallic materials. It is one of the most common types of material deterioration, posing significant issues in various industries and potentially affecting our health. The difficult dilemma of degradation threatens to cost society significantly, in addition to its impact on the introduction of modern and important technology. As a result, corrosion, like mortality and taxation, must be avoided at all costs. The primary cause of corrosion is a reduction in a system's Gibbs energy. Almost all metals provide energy to the

Handbook of Corrosion Engineering. DOI: https://doi.org/10.1016/B978-0-323-95185-2.00001-0

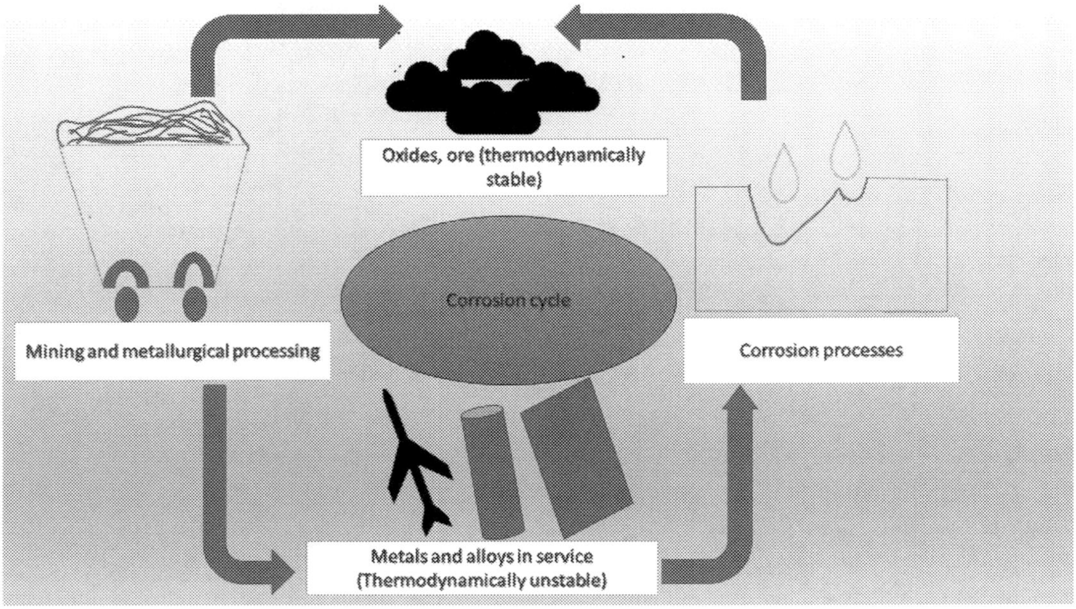

FIGURE 1–1 Corrosion cycle.

system during manufacturing. The metal demonstrates a strong tendency to revert to its low-energy oxide form, that is, corrosion, due to this strenuous thermodynamic conflict (Fig. 1–1).

Even though it is unavoidable, potential substitutes (corrosion management methods) are employed to decelerate the process toward equilibrium. As a result, the rate at which equilibrium is achieved, which is determined by the nature of the metal surface and the surrounding environment, is always a matter of concern.

1.2 Impact of corrosion

The topic of corrosion may pique the interest of various people, including pure scientists, engineers, architects, economics, health officials, and environmentalists because it has numerous vital consequences for our society. The following are the three most severe outcomes of corrosion:

1. costs of repairing damage caused by corrosion
2. devastative corrosion accidents
3. environmental threats

1.2.1 Costs of repairing damage caused by corrosion

To design strategies that alter the costs of material corrosion, the thorough knowledge of corrosion processes from both a scientific and technological aspect is essential. Corrosion can be considered as more than an economic issue when viewed in this light. Hadfield made the first attempt to assess the cost of corrosion in 1922. This method relies solely on a hypothetical calculation of the global annual rate of corrosion of iron and steel. Table 1–1 (Al-Kharafi, Al-Hashem, & Matrouk, 1995; Behrens, 1967; Bennett, Kruger, Parker, Passaglia, & Reimann, 1978; Bhaskaran et al., 2014; Bhaskaran et al., n.d.; Cherry & Skerry, 1983; Committee on Corrosion Loss in Japan: Boshoku Gijutsu, 1977; Committee on Corrosion Loss in Japan: Zairyo-to-Kankyo, 2001; Hoar, 1976; Hou et al., 2017; Koch et al., 2016; Linderborg, 1967; Payer & Koch, 2002; Payer, Dippod, Boyd, Berry, & Brooman, 1978; Potter, 1972; Revie & Uhlig, 1974; Sastri, Elboujdaini, & Perumareddi, 2003; Tems & Al Zahrani, 2006; Tradgaidh, 1965; Zhu, 1986) summarizes the results of corrosion cost studies conducted in many countries commencing in 1949, with the year the results were published. Prof. Uhlig (1950) of the United States deserves credit for being the first to identify and analyze the cost of corrosion correctly. Uhlig took a cautious approach, and his corrosion cost estimate was likely to be conservative.

1.2.1.1 Percentage of GDP

The United Kingdom Committee employed a different procedure than Uhlig, who obtained cost data from the manufacturer. Dr. T.P. Hoar chaired the Committee on the Cost of Corrosion, established in 1966 and primarily based on direct interaction with companies. The Hoar technique also allows to calculate every individual industrial sector's potential percentage contribution to a country's national GDP. In 1978, NBS-BCL utilized an input–output method to forecast the cost of corrosion in the United States, which had some inherent quantification issues. There was also the problem of getting current information from a single source. The use of life cycle cost and the net present value method, which was first presented by Dillon in 1966 and later endorsed by the Hoar Committee in its entirety, is a more systematic approach to analyze the cost of corrosion. As a result, many governments have conducted corrosion expenditure evaluations to estimate the cost of corrosion. Corrosion is estimated to cost businesses, municipalities, and militaries billions of dollars each year. The overall cost of corrosion is divided into two main types: direct and indirect costs. The annual expenditures on maintenance and repairs, as well as corrosion control procedures such as coatings, linings, chemical treatment, cathodic protection (CP), corrosion-resistant alloys, and nonmetallic materials, are all included in the direct cost of corrosion. Corrosion costs also include the expenditures of inspection and testing. Indirect expenses associated with plant shutdowns, excess capacity, redundant equipment, product loss, equipment efficiency loss due to corrosion, product contamination, and overdesign. Indirect costs associated with "lost or deferred output" can be many times the cost of replacing or repairing corroded equipment, yet they are notoriously difficult to calculate.

The review of published studies will be discussed in the section below:

1.2.2 The Uhlig report

This approach (Uhlig, 1950) was first used in the United States in 1949 and in Japan in 1974. The method is also known as the "materials and services" method. It adds direct

Table 1–1 The cost of corrosion studies undertaken by specific countries.

Author	Year of issue	Country	%GNP	Cost	References
Uhlig	1949	United States	2.1	\$5.5 billion	Uhlig (1950)
Worner	1955	Australia	–	\$240 million	Worner (1955)
Rajgopalan	1962	India	–	\$320 billion	Bhaskaran et al. (n.d.)
Tradgaidh	1965	Sweden	–	–	Tradgaidh (1965)
Linderborg	1967	Finland	–	\$54 billion	Linderborg (1967)
Behrens	1969	West Germany	3	\$6 billion	Behrens (1967)
Hoar	1971	United Kingdom	3.5	Stg. £1365 billion	Hoar (1976)
Potter	1972	Australia	3.5	\$900 million	Potter (1972)
Revie and Uhlig	1974	Australia	1.5	\$470 million	Revie and Uhlig (1974)
Committee on corrosion loss in Japan	1977	Japan	1.8	\$9.2 billion	Committee on Corrosion Loss in Japan: Boshoku Gijutsu (1977)
Battelle	1978	United States	4.2	\$70 billion	Payer et al. (1978), Bennett et al. (1978)
Zhu	1986	China	–	–	Zhu (1986)
Cherry and Skerry	1983	Australia	1.5	\$2 billion	Cherry and Skerry (1983)
Al-Kharafi et al.	1995	Kuwait	5.2	\$1 billion	Al-Kharafi et al. (1995), (n.d.)
illion http://www.Corrosion-Club.com	1996	Switzerland	3–5	–	http://www.Corrosion-Club.com
JSCE	2001	Japan (Uhlig method)	0.77	3938 billion yen	Committee on Corrosion Loss in Japan: Zairyo-to-Kankyo (2001)
	2001	Japan (Hoar method)	1.02	5258 billion yen	
http://www.Corrosion-Club.com17.	2001	Australia	2	–	http://www.Corrosion-Club.com
Payer and Koch	2001	United States	3.1	US\$276 billion	Payer and Koch (2002)
Sastri et al.	2003	Canada	–	US\$46.4 billion	Sastri et al. (2003)
Tems et al.	2006	Saudi Arabia	–	US\$900 million	Tems and Al Zahrani (2006)
R Bhaskaran et al.	2012	India	2.4	US\$26.1 billion	Bhaskaran et al. (2014)
Koch et al.	2016	United States	3.4	US\$2.5 trillion	Koch et al. (2016)
Baorong Hou	2017	China	3.3	2127.8 billion RMB (~310 billion USD)	Hou et al. (2017)

and indirect costs for the owner/operator and consumers of corroding components to compute overall corrosion costs. Direct expenses are calculated by adding cost estimates for corrosion control products and services, cathodic protection, coatings, corrosion-resistant metals, and other things. Indirect costs are expenses incurred due to things like replacing a residential water heater, repairing an automobile's internal combustion engine, or replacing automotive silencers, among other things. Uhlig calculated the annual direct cost of corrosion in the United States to be $5,427 million US dollars by combining anti-corrosion products with corrosion-induced maintenance and replacement expenses (2.1% of gross national product (GNP) at the time).

1.2.3 The Hoar report (1971)

The Hoar method (Hoar, 1976), also known as "industry sector analyses." This technique calculates corrosion costs for specific industrial sectors. The study includes both direct corrosion costs and prevention and mitigation expenditures. The capital cost, in addition to operational expenditures, can be included. The United Kingdom and Japan were the first countries to use this method to estimate corrosion costs in 1970 and 1974, respectively. The Hoar method was used to analyze corrosion costs in 10 different industrial sectors in the United Kingdom, resulting in a total cost of 1365 million pounds per year (3.5% of the UK GNP at the time). Hoar also predicted that effective corrosion control might save 310 million pounds per year on this expense.

1.2.4 The input/output economic report

Wassily Leontief devised the input/output (IO) analysis (Payer, Boyd, Lippold, & Fisher, 1980). In 1973 he was given the Nobel Prize for this work. The Battelle study was the first to bring together corrosion and economics professionals to assess the monetary influence of corrosion in the United States. The overall cost was further separated into avoidable and unavoidable expenditures in the study. The former refers to costs that can be reduced by implementing the best corrosion mitigation measures. Unavoidable costs are those that cannot be consolidated with currently available technologies. IO is a general equilibrium economic model that shows how much one sector needs other sectors' inputs to produce its output and how much each sector sells to the other. The IO model represents the income activity growth in every other sector required to boost net production in one sector. An economic IO model was run on three alternative situations:

1. a real-life scenario involving corrosion (universe I),
2. a hypothetical world devoid of corrosion (universe II), and
3. an ideal world in which everyone, at all times, exercised the most cost-effective corrosion prevention. (universe III)

After calculating the GDP for each universe, the corrosion cost and avoided corrosion cost are determined using the following equations:

$$\text{Corrosion cost} = \text{GDP(UniverseII)} - \text{GDP(UniverseI)} \quad (1-1)$$

$$\text{Avoidable corrosion cost} = \text{GDP(UniverseIII)} - \text{GDP(UniverseI)} \quad (1-2)$$

By comparing the different "universes," the corrosion cost and avoided corrosion cost were computed. In the 1970s, the Battelle study used this approach to account for the direct impacts of corrosion on specific sectors and their relationships. The purpose of this model is to organize corrosion-related capital and operational costs into a logical framework so that their implications can be evaluated. Reduced annual capital equipment replacement or purchase would result in direct and indirect labor, energy, material, and other capital requirements. Corrosion makes the real world spend more on input to generate the same result as the world without corrosion. According to a study conducted by Battelle NBS, the entire cost of corrosion was 70 billion dollars per year or 4.5% of GNP at the time, and 14% of corrosion costs could be reduced directly using existing anticorrosion technologies. The inaccuracy of this approach, according to the NBS, is expected to be 30%. Australia (Cherry & Skerry, 1983), Kuwait (Al-Kharafi et al., 1995), and Japan (n.d.) later adopted this IO approach in 1983, 1995, and 1999, respectively.

1.2.5 US FHWA study (2002)

The Federal Highway Administration (FHWA) funded this study (Koch, Brongers, Thompson, Virmani, & Payer, 2002), which was carried out in collaboration with NACE International by C&C Technologies, Inc. The Uhlig and Hoar methodologies were combined in this project, along with extensive expert knowledge input. The gross yearly cost of direct corrosion was estimated to be US$276 billion, or about 3.1% of the US GDP in 1998. This cost was computed by extrapolating findings from 26 industrial areas where corrosion is known to occur to a countrywide estimate. Infrastructure, utilities, transportation, production and manufacturing, and governance were split into five key groups. Corrosion's indirect cost was tentatively evaluated to be equivalent to its direct cost. The key indirect cost examined was social, that is, public time and productivity lost due to interruptions and business disruption triggered by corrosion and corrosion management activities. The industries with the highest direct corrosion impact were drinking water and sewer systems (US$36 billion), motor vehicles (US$23.4 billion), and defense (US$20 billion). Every year, corrosion management methods and services cost US$121 billion. Corrosion costs are divided into two categories: direct and indirect. Corrosion-related direct costs include material selection, design, manufacture, and construction costs, additional material, that is, expansion of wall thickness, corrosion-prevention resources, that is, coatings, inhibitors, sealants, cathodic protection, and application which includes labor and equipment costs.

1.2.6 India (2011–12)

R. Bhaskaran of Lovely Professional University in Phagwara, Punjab, India and N.S. Rengaswamy of Central Electrochemical Research Institute in Karaikudi, Tamil Nadu, India led the cost of corrosion research for 2011–12 in India. The NBS IO economic model was employed to assess the cost of corrosion (Payer et al., 1978). This study included one of the most detailed sector breakdowns of any national corrosion costs assessment to date. The direct cost of corrosion for India was assessed to be US$26.1 billion, equivalent to 2.4% of India's GDP. Corrosion costs that may have been avoided totaled US$9.3 billion, or 35% of overall corrosion costs. Indirect corrosion cost India US$39.8 billion, which was equivalent to 3.6% of GDP.

1.2.7 US (2016) International Measures of Prevention, Application, and Economics of Corrosion Technologies study

NACE International launched the International Measures of Prevention, Application, and Economics of Corrosion Technologies (IMPACT) study (Koch et al., 2016) to investigate the current role of corrosion management in industry and government and adopt 10 best practices in Trends in Oil and Gas Corrosion Research and Technologies. The worldwide cost of corrosion was calculated using national corrosion costs available at the time of the study. The IMPACT cost of corrosion research comprised studies from India in 2011–12, the United States in 1998, Japan in 1997, Kuwait in 1987, and the United Kingdom in 1970. Each of these investigations yielded information that could be categorized into the three economic sectors mentioned earlier. Fig. 1–2 depicts the economic breakdown of the five nations studied in this study. With modern industrial and service economies, the United States, the United Kingdom, and Japan are relatively comparable. However, India and Kuwait have distinct profiles due to a substantial agricultural economy and oil industries. The global economy was divided into economic zones with similar economies to handle the economic sectors for different parts of the world. The countries under consideration include the United States, India, the European Region, the Arab World (as defined by the World Bank), China, Russia, Japan, the Four Asian Tigers Plus Macau, and the Rest of the World. Corrosion is predicted to cost the world US$2.5 trillion, or 3.4% of the global GDP (2013). It was projected that by employing available corrosion control measures, savings of between 15% and 35% of the cost of corrosion might be recognized, equating to between US$375 billion and US$875 billion annually on a global scale.

1.2.8 Devastative corrosion accidents

Corrosion damage may be considerably worsened by conditions. Many corrosion accidents have gone undiscovered throughout history, either due to liability or because the

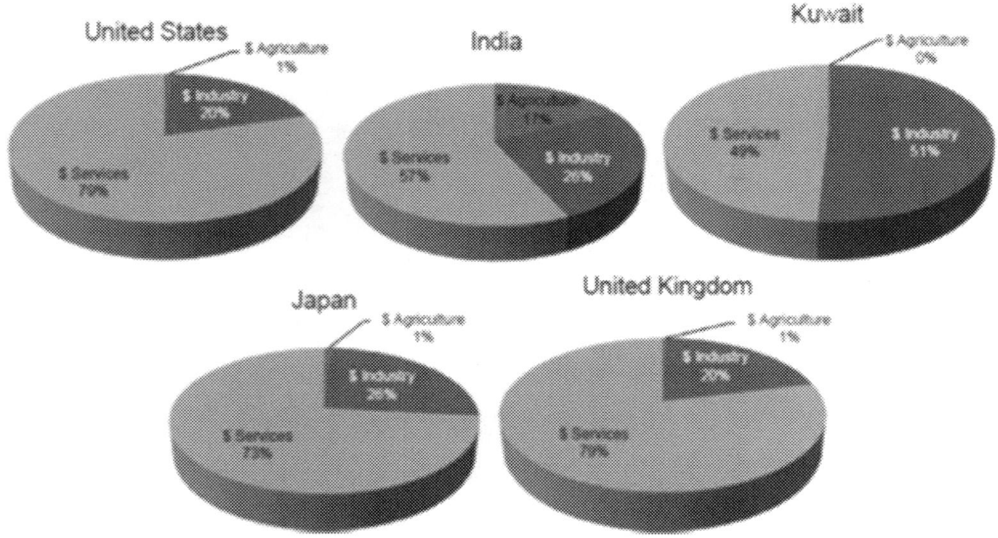

FIGURE 1–2 Economic sectors for the five countries used in the global cost of corrosion study. *From (n.d.). http://impact.nace.org/economic-impact.aspx.*

proof was lost after the disastrous event; others have made the news. The following are a few well-known incidents:

1. Wing lift strut assembly failure and collision with terrain
   With one pilot and one passenger on board, a privately registered Piper J3C-65 aircraft on skis was conducting a VFR fly from Minaki, Northwestern Ontario to Snowshoe Lake, Northwest Ontario on March 30, 2019. The pilot lost control of the plane during a low pass before landing, and it collided with the lake's frozen surface (Fig. 1–3). Both individuals died as a result of their injuries. Heavy corrosion and thinning of the load-bearing wall inside the wing lift strut, as well as fatigue and subsequent overload failure, induced the detachment of the left main spar wing lift strut assembly (Fig. 1–4) found in the lower fork end attachment, leading to two casualties.
2. Eurocopter EC135T1 accident
   The Royal Cayman Islands Police Service's Airbus Helicopter EC135T1 VP-CPS was damaged on take-off at Owen Roberts International Airport in the Caribbean Cayman Islands on February 26, 2019. The main rotor actuator's longitudinal axis had collapsed, according to a subsequent investigation of the helicopter. Pitting corrosion on a tie bar within the actuator led to intercrystalline corrosion and cracking, resulting in overload failure of the remaining material. It is possible that moisture got into the actuator and corroded the tie bar. Corrosion may have been aided by the Cayman Islands' environmental conditions.

FIGURE 1–3 Wreckage site.

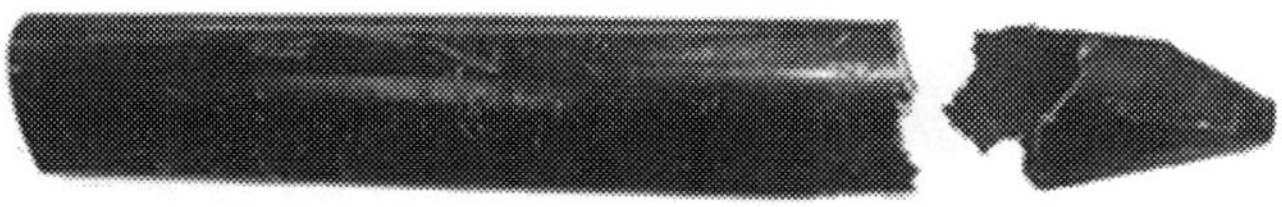

FIGURE 1–4 Damage on the upper half of the left main spar wing lift strut assembly. *From Wing lift strut assembly failure Piper J3, C-FLDQ. (2019). https://www.news.scudrunners.com/wing-lift-strut-assembly-failure-piper-j3-c-fldq/.*

3. Pipeline rupture and fire

A 36-inch pipeline operated by Enbridge Inc. that was delivering sweet natural gas burst approximately 13 km northeast of Prince George, British Columbia, on October 9, 2018. The natural gas ignited after the rupture, resulting in a fire. As a precaution, 125 persons living within a 2-km radius of the incident were evacuated. The blast did not result in any injuries, but it did prompt the evacuation of approximately 125 individuals within a 2-km radius. On the exterior surface of the investigated pipe sections, magnetic particle examination revealed several signs compatible with colonies of stress corrosion cracking (SCC). Due to a near-neutral pH environment and operating stress, the fractures expanded over time and eventually consolidated into a larger single crack. This decreased the pipeline steel's load-bearing capacity until the critical crack size was exceeded, resulting in an instantaneous overstress fracture of the pipe at normal operating pressures. The crack patterns were similar with SCC with a pH of near-neutral.

4. In-flight engine shutdown

The incident took place on April 20, 2017 when a Qantas Airbus A380-800, registration VH-OQG with 480 people on board, stepped climbing from FL320 to FL340 about 800 nm west southwest of Los Angeles. The flight crew was forced to shut the engine down as the outboard right-hand engine (Trent 972) emanated a bang and streaks of flames and sparks. The plane dropped to FL290 and landed safely in Los Angeles around 4:15 hours after taking off. Internal corrosion of low-pressure turbine stage 2 blades caused fatigue failure, blade debris separation, and downstream engine damage were among the reasons for incidents as stated by the Australian Transportation

Safety Board. And the blade deterioration was triggered by chemical residue leftover from the previous engine service's cleaning procedure.

5. Rupture of a natural gas transmission pipeline

   The incident took place on April 29, 2016 in Pennsylvania, United States. Due to corrosion, a natural gas transmission pipeline with a 30-inch diameter and two circumferential welds collapsed. One property was destroyed, three residences were partially damaged, and numerous homes were evacuated due to this incident.

6. Landing of a Jetstream away from the runway

   The left main landing gear of a Jetstream 31 plane collapsed after it landed at Doncaster's Robin Hood Airport overnight on August 15, 2014. The single passenger and the flight crew were unharmed when they exited the plane. According to the findings, the failure was caused by intergranular/SCC in the forward yoke pintle at the top of the left landing gear leg, which caused the leg to collapse.

7. Damage of a turbine engine

   The failure of the engine (June 24, 2013) on the Manchester trip was attributed to a blade falling off due to corrosion. Multiple crack start points caused by sulfidation corrosion were discovered during a laboratory investigation of the broken blade root.

8. Carlsbad pipeline explosion accident

   A natural gas transmission pipeline with 30-inch diameter operated by El Paso Natural Gas Company ruptured near Carlsbad, New Mexico, on August 19, 2000. The gas erupted and flared for 55 minutes after it was released. There were a total of 12 persons killed. Property and other damages or losses totaled US$998,296 according to El Paso Natural Gas Company. Investigators from the National Transportation Safety Board concluded that substantial internal corrosion caused the explosion. As an outcome, the pipe wall thickness decreased to the point where the remaining metal could no longer sustain the pressure inside the pipe. According to the report, corrosion was likely caused by a combination of microorganisms and contaminants such as moisture, chlorides, oxygen, carbon dioxide, and hydrogen sulfide.

9. Sinking of the Erika

   The oil tanker Erika split in two and drowned in the Bay of Biscay (Fig. 1–5) on December 12, 1999. There were no injuries in this incident. However, the oil leak wreaked havoc on the ecosystem. Significant deterioration of the vessel's interior structure was the primary cause of the accident.

10. Guadalajara sewer explosion

    Several gas explosions in the New Mexico city drainage system devastated 8 km of roadways on April 22, 1992. There were 252 people were dead, approximately 1500 people injured, and 15,000 people made homeless. Corroded water pipes caused water leaks, which led to rusty gas pipes and a gas leak into the sewer.

11. Aloha Boeing 737 airlines accident

    A huge section of the plane's upper fuselage was practically ripped off while flying at 8000 m above the earth, leaving a large hole in the passenger compartment on April 18, 1988 in the United States. The lap joints of the fuselage skin were separated, rusted, and

FIGURE 1–5 Sinking of the Erika. *From Environmental Catastrophe: Sinking of the Erika. (n.d.). https://www.corrosion-doctors.org/Pollution/erika.htm.*

cracked during the aircraft examination. There were 89 travelers and six crew members. One flight attendant was pulled overboard during the decompression and is presumed dead; seven passengers and one flight attendant sustained injuries.

12. Bhopal methyl isocyanate release accident

The tragedy occurred in India on December 2–3, 1984. It was India's first significant industrial catastrophe. Union Carbide India Limited's pesticide plant in Bhopal, Madhya Pradesh, India was the site of the incident (Fig. 1–6). Approximately 45 tons of the toxic chemical methyl isocyanate leaked from the plant, resulting in the deaths of tens of thousands of people in Bhopal. The death toll has been estimated to range between 3,800 to 16,000 individuals, while current government statistics placed the figure at 15,000. The heat of the reaction, along with elevated chloroform levels and the presence of an iron catalyst produced by corrosion of the stainless-steel tank wall, ended in response with such intensity that safety mechanisms could not control the gases expelled. Toxic waste has stayed in the environment for three decades. And those who were exposed to the gas gave birth to children who are still physically and mentally disabled. Survivors have been battling for decades for the site to be cleaned up.

13. Demolition of the silver bridge

On December 15, 1967, the Silver Bridge, which connects West Virginia and Ohio, collapsed (Fig. 1–7). As a result of the disaster, 46 people died, and 11 others were seriously injured. The catastrophe occurred due to stress corrosion of the bridge's building compound and chain.

1.2.9 Environmental threats

Infrastructure initiatives that are poorly built will lead to the need to rebuild a corroded structure, which is a waste of natural resources and goes against global sustainability concerns. Economic factors also have an impact on technology. Corrosion difficulties are

FIGURE 1–6 Union carbide factory site. *From (n.d.). https://www.corrosion-doctors.org/Pollution/bhopal.htm.*

FIGURE 1–7 Silver Bridge (A) before and (B) after collapse.

preventing much of the growth of modern technology since materials are required to withstand higher temperatures, higher pressures, and more intensely corrosive conditions in multiple scenarios at the same time. Degradation frequently contributes to environmental pollution due to leakage from perforated tanks, drums, and pipelines. More and more pollution of this type can be predicted as the number and length of oil, gas, water, and sewage pipelines grow. The impact of oil leaks caused by pipeline corrosion on wildlife is likewise horrifying. Because the chemical components of spilled oil are hazardous, it has the potential to kill living organisms. Oil can affect organisms when ingested or inhaled internally and externally through skin and eye irritation. Oil can suffocate small fish and crustaceans and coat feathers and hair, reducing birds' and mammals' power to manage their body temperatures. Corrosion-related pipeline leaking also puts deep-sea species in danger of being exposed to potentially toxic pollution.

1.3 Corrosion science vs. corrosion engineering

Corrosion science terms cover theories, laws, and explanations of phenomena confirmed by intersubjective observation or experiments, such as different types of corrosion, rate of corrosion, and mechanism of corrosion. Corrosion science "knows why," while corrosion engineering "knows how," because corrosion engineering applies principles evolved from corrosion science to minimize or prevent corrosion. Corrosion engineering requires the formulation of corrosion prevention strategies and the implementation of specific rules and practices. It offers the tools needed to conduct a proper corrosion evaluation and choose the best corrosion protection approach based on the harsh environment and the targeted service life.

1.4 Reasons to study corrosion

1. To comprehend how corrosion damages materials and how corrosion protection technology can be used to "protect material and treasured assets."
2. Because the deterioration and disintegration of critical structures, pipelines, and other productive assets have an impact on people's safety, industrial reliability, and the global economy.
3. Without knowing corrosion, engineering knowledge is inadequate; airplanes, ships, vehicles, and other structures cannot be designed without knowledge of the corrosion behavior of the materials utilized in these structures.
4. Corrosion is a concern to the environment, as corrosion products can contaminate water, making it unsafe to drink.
5. The design of artificial human implants necessitates a thorough understanding of corrosion science and engineering. Because of the corrosive nature of human blood, surgical implants must be highly corrosion resistant.

1.5 Corrosion education

Personnel educated and trained in corrosion engineering techniques, principles, and practices is required to increase competence in managing and reducing material degradation in manufactured products. In collaboration with university engineering leaders, the corrosion industry and government stakeholders are tasked with meeting the national demand for a robust, viable corrosion engineering workforce. Various studies conducted by major institutions such as the National Research Council (NRC) of the United States (Tems & Al Zahrani, 2006; Bennett, et al., 1978) and NACE International (Koch et al., 2016) concluded that the current generation of civil engineering professionals is not adequately trained to address societal, economic, and technological challenges. Undergraduate corrosion education was found to be nonexistent at many engineering institutions, according to the NRC study. This was attributed to a lack of corrosion research funding and insufficiently skilled, willing, and interested faculty. Surprisingly, the NRC research advocated for a knowledge and skills-based education. A group of corrosion experts is needed as instructors to educate knowledge-based education that will produce the next generation of "corrosion-aware" design engineers and experts and specialists who will usher in the next wave of fundamental knowledge. As a result, a recommended strategic goal would be to offer money for the government and industry fellowships, endowed chairs, or national institutes in corrosion excellence, all of which would help assure the survival of corrosion knowledge-based education programs. The committee proposed a 2-year master's degree and a 3–5-year Ph.D. program. Distance learning, short courses, and educational programs with a shortened duration can continue to train the workforce.

Imperial College London offered a Master of Science degree in "Engineering Materials Corrosion" in the 1990s. A similar program was also provided by the University of Manchester Institute of Science and Technology, which is now part of the University of Manchester, Leeds University in England, and the University of Rhode Island in the United States, offering programs on corrosion. The curricula include study materials on corrosion control and comprehension. Mars G Fontana, a well-known corrosion engineer, is honored at Ohio State University with a corrosion center. Akron University offered the sole bachelor's degree in corrosion engineering in the United States, which graduated its first class in 2015. This program had widespread support from the corrosion community and the US Department of Defense (DoD). Both parties were concerned that a shortage of corrosion professionals would be a significant roadblock to advancing corrosion engineering and management in the future. And in this corrosion engineering curriculum, the science and technology of corrosion processes, mitigation, and control, rather than corrosion management, was the primary focus. The Curtin Corrosion Centre has been performing cutting-edge scholarly and demand-driven corrosion research in petroleum, mining, defense, and chemical processing. NACE International provides the majority of professional development and vocational training for corrosion specialists.

1.6 Units of corrosion rates

The average corrosion rate can be estimated by employing the formula:

$$\text{Corrosion rate} = \frac{KW}{\rho AT}$$

where K = a constant, T = time of exposure in hours to the nearest 0.01 h, A = area in cm^2 to the nearest 0.01 cm^2, W = mass loss in g, to nearest 1 mg (corrected for any loss during cleaning), and ρ = density in g/cm^3 (ASTM G1, Standard Practice for Preparing, Cleaning, and Evaluation CorrosionTest Specimens).

Corrosion rates are expressed in a variety of different units. The corrosion rate may be computed in a number of units as discussed in Fig. 1–8 with the proper value of K using the aforementioned units for T, A, W, and ρ.

Corrosion Rate Units Desired	Constant (K) in Corrosion Rate Equation
mils per year (mpy)	3.45×10^6
inches per year (ipy)	3.45×10^3
inches per month (ipm)	2.87×10^2
millimetres per year (mm/y)	8.76×10^4
micrometres per year (μm/y)	8.76×10^7
picometres per second (pm/s)	2.78×10^6
grams per square metre per hour ($g/m^2 \cdot h$)	$1.00 \times 10^4 \times D^A$
milligrams per square decimetre per day (mdd)	$2.40 \times 10^6 \times D^A$
micrograms per square metre per second ($\mu g/m^2 \cdot s$)	$2.78 \times 10^6 \times D^A$

FIGURE 1–8 Conversion table of corrosion rates in different units. [A]The corrosion rate in these units may be determined without considering density. The density in the corrosion rate equation cancels out the density in the constant *K*.

Note 1—These constants can also be used to translate corrosion rates across different measurement systems, if needed. Multiplying by KY/KX will change the corrosion rate in unit X to a rate in unit Y.

$\mu A/cm^2$, mA/cm^2, and A/m^2 are common measures for the corrosion current density. When the mass loss is the empirically measured variable, other units have been utilized. These consist of mdd (milligrams per square decimeter per day) and g/cm^2 each day. The corrosion rate may occasionally be expressed as a uniform penetration rate. Inches per year, inches per month, and mils per year (where 1 mil is equal to 1 inch) are some of the units.

1.7 Corrosion cell

For corrosion to take place, the formation of a corrosion cell is essential. A corrosion cell is essentially comprised the following four components:

1. anode
2. cathode
3. electrolyte
4. metallic path

1.7.1 Anode

One of the two dissimilar metal electrodes in an electrolytic cell is represented as the cell's negative terminal. Electrons are released at the anode, which is the more reactive metal. Electrons are insoluble in aqueous solutions, and they only move through the wire connection into the cathode. For example, in a battery, zinc casing acts as the anode. Also in a Daniel cell, zinc is the anode as oxidation occurs on it and electrons are released (Fig. 1–9).

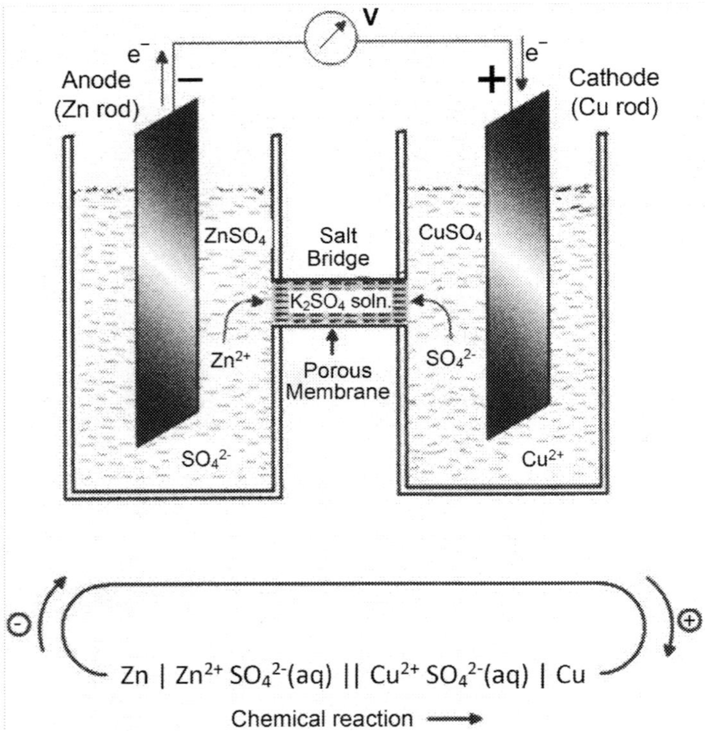

FIGURE 1–9 Daniel cell (A galvanic cell).

Corrosion nomenclature is the opposite of electroplating nomenclature, where an anode is positive, and the cathode is negative.

1.7.2 Cathode

One of the two electrodes in an electrolytic cell is represented as a positive terminal of a cell. Reduction takes place at the cathode and electrons are consumed. For example, carbon electrodes in a battery, and copper electrodes in a Daniel cell. Fig. 1–10 shows the reduction of hydrogen ions. The electron is always a reducing agent. Electrolyte: It is the electrically conductive solution (e.g., salt solution) that must be present for corrosion to occur. Note that pure water is a bad conductor of electricity. Positive electricity passes from anode to cathode through the electrolyte as cations, for example, Zn^{2+} ions dissolve from a zinc anode and thus carry positive current away from it, through the aqueous electrolyte.

1.7.3 Metallic path

The two electrodes are connected externally by a metallic conductor. In the metallic conductor, "conventional" current flows from (+) to (−) which is really electrons flowing from (−) to (+).

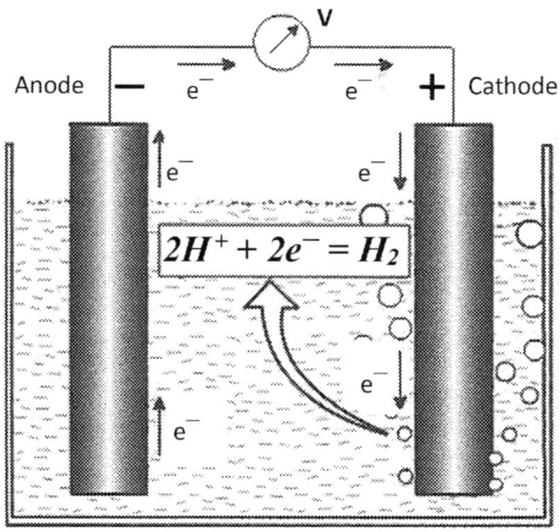

FIGURE 1–10 Reduction of hydrogen in acid medium.

Metals provide a path for the flow of conventional current which is actually the passage of electrons in the opposite direction.

1.7.4 Current flow

Conventional current flows from anode (−) to cathode (+) as Zn^{2+} ions through the solution. The current is carried by these positively charged ions. The circuit is completed by the passage of electrons from the anode (−) to the cathode (+) through the external metallic wire circuit (outer current). Electron Flow:

$$2H^+ + e^- \rightarrow H_2\uparrow$$

Although the anode (e.g., Fe or Zn) is the most negative of the two metals in the cell, this reaction does not occur there because its surface is emanating Fe^{2+} ions which repel H^+ ions from discharging there. The circuit is completed by negative ions (−) which migrate from the cathode (+), through the electrolyte, toward the anode (−). They form $Fe(OH)_2$ when they enter the cloud of Fe^{2+} ions coming from the anode.

Anions: Migrate toward the anode (OH^-) but precipitate as $Fe(OH)_2$ before reaching it. Cations: Migrate toward the cathode (Fe^{2+}). A galvanic cell (Daniel cell). Current flow in an electrochemical cell is shown in Fig. 1–11.

1.8 Types of corrosion cells

Three main types of cells take part in corrosion reactions. In engineering practice, cells responsible for corrosion may combine these three types.

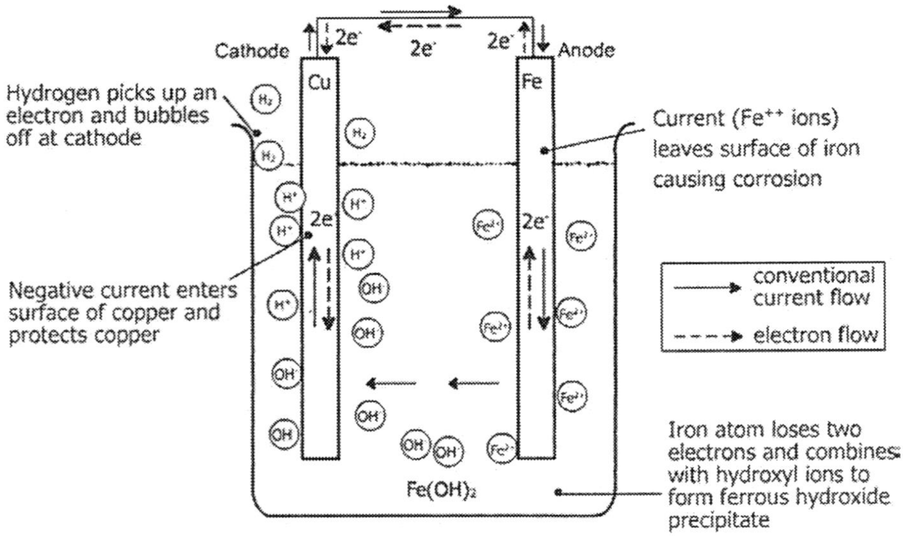

FIGURE 1–11 Current flow in an electrochemical cell.

1.8.1 Dissimilar electrode cells or galvanic

A copper pipe joined to an iron pipe, a bronze propeller in contact with a steel ship hull, and a metal-containing electrically conducting impurities on the surface as a separate phase are examples of dissimilar electrode cells. Additionally, grain-boundary metal in contact with grains, cold-worked metal in contact with the same metal annealed, and one metal crystal with a certain orientation in contact with another crystal with a different orientation are examples of dissimilar electrode cells.

1.8.2 Concentration cells

These are cells with two identical electrodes, each in contact with a solution of different composition. There are two kinds of concentration cells. The first is called a salt concentration cell. For example, if one copper electrode is exposed to a concentrated copper sulfate solution, and another to a dilute copper sulfate solution (Fig. 2–3), on short-circuiting the electrodes, copper dissolves (i.e., $Cu \rightarrow Cu^{2+} + 2e^-$) from the electrode in contact with the dilute solution (anode) and plates out (i.e., $Cu^{2+} + 2e^- \rightarrow Cu$) on the other electrode (cathode). The two solutions often reach the same concentration as a result of these reactions. Differential aeration cells are the second type of concentration cell, which is the most significant in actual use. The electrolyte around two iron electrodes in a diluted sodium chloride solution might be completely aerated around one electrode (the cathode), and deaerated around the other electrode (the anode), perhaps by bubbling nitrogen through the solution. The difference in oxygen concentration produces a potential difference. It makes the current move. Crevice corrosion, a form of visible deterioration at crevices, is caused by this sort of cell. Many engineering designs have crevices, such as the intersection of two linked pipes and threaded connectors. Crevices have a lower oxygen concentration than surrounding areas, and these lower oxygen concentration locations are anodic relative to surrounding areas with a greater oxygen concentration (outside crevices). Additionally, differential aeration cells may result in rust-related pitting damage. near the water line, or the point where water and air meet Less oxygen accesses areas of the metal that are coated in rust or other insoluble reaction products than areas where the permeable coating is thinner or nonexistent. Differential aeration cells may also cause crevice corrosion, which affects passive metals like stainless steel, aluminum, and nickel that are exposed to aqueous conditions like saltwater.

1.8.3 Electrolytic cell

This type of cell is produced when the system receives an external current. It might include galvanic cells, an external electrical energy source, and all the essential components of focus. In an electrolytic cell, the anode has (+) polarity, and the cathode has (−) polarity when an external current is applied. This specific cell type is installed to offer cathodic protection, which shields structures from electrical harm. The polarity of a galvanic (corrosion) cell is the inverse of that of an electrolytic cell.

1.8.4 Differential temperature cells

These cells consist of identical metal electrodes that operate at various temperatures and are each immersed in an electrolyte with a unique initial composition. Differential temperature cells' practical importance and underlying theory are less well-known than those of the previously covered cells. These cells are found in heat exchangers, boilers, immersion heaters, and other related equipment. The copper electrode in a copper sulfate solution that has been heated to a higher temperature is known as the cathode. The anode, in contrast, is the copper electrode that is heated to a lower temperature. When the cell is shorted, copper builds up on the hot electrode and dissolves from the cold electrode. Silver acts similarly to lead while having a different polarity. The more amazing metal of the same composition for iron submerged in diluted aerated sodium chloride solutions is anodic to the hot electrode. However, the polarity may change after a number of hours, depending on aeration, stirring speed, etc.

1.9 Electrochemical aspects of various types of electrodes

Fundamental electrochemical aspects of commonly used reference electrodes are illustrated below:

1.9.1 Silver–silver chloride reference electrode

The redox reaction at the electrode is:

$$AgCl + e^- \rightarrow Ag + Cl^-$$

The electrode consists of a silver wire coated with silver chloride immersed in a chloride solution (Fig. 1–12). Variations in chloride ion concentrations influence the potential.

$$E = E^{\circ} + RT/nF \log \frac{1}{Cl^{-}}$$

1.9.2 Saturated calomel electrode

A typical saturated calomel electrode (SCE) is represented in Fig. 1–13. The redox reaction for the electrode is:

$$Hg_2Cl_2 + 2e^{-} \rightarrow 2Hg + 2Cl^{-}$$

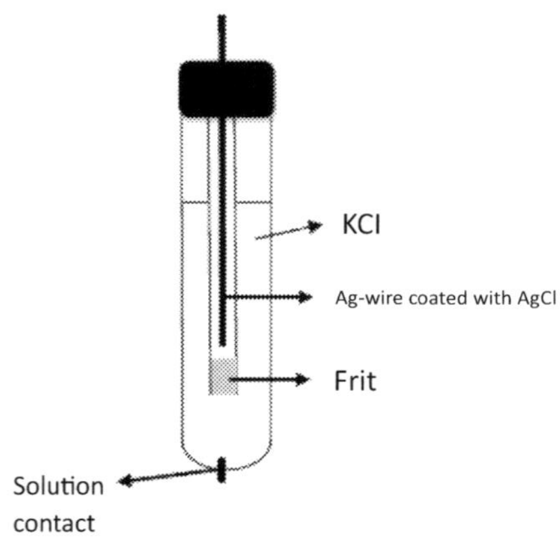

FIGURE 1–12 Silver–silver chloride reference electrode.

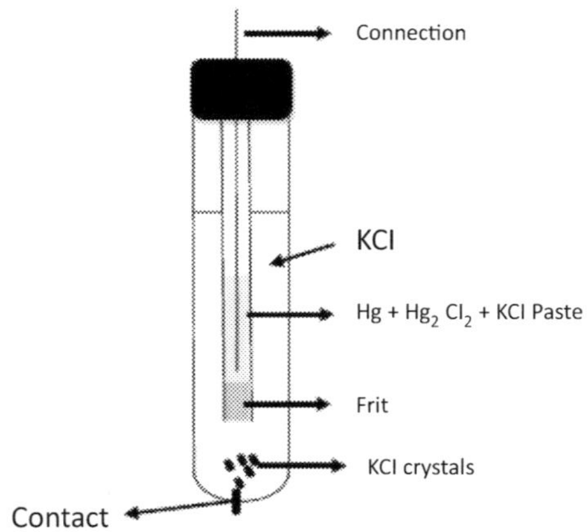

FIGURE 1–13 Saturated calomel electrode.

$$E = E^{\circ} + RT/nF \log \frac{1}{[\mathrm{Cl}^-]^2}$$

$$E = E^{\circ} + 0.059/2 \log \frac{1}{[\mathrm{Cl}^-]^2}$$

1.9.3 The hydrogen electrode

An acid solution with unit H^+ activity comes into touch with a platinum wire. At standard temperature, pure H_2 gas bubbles into a solution in one atmosphere.

$$2\mathrm{H}^+ + 2\mathrm{e}^- \rightarrow \mathrm{H}_2$$

$$E = E^{\circ} + \frac{0.059}{2} \log \left[H^+\right]^2$$

$$E = E^{\circ} - 0.059\mathrm{pH}$$

With a standard potential of $E^{\circ} = 0.00$ V, the standard hydrogen electrode (SHE) is also known as the normal hydrogen electrode. A typical SHE is represented in Fig. 1–14.

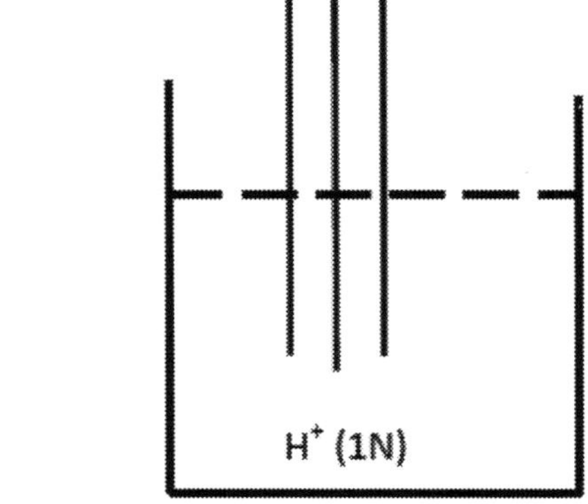

FIGURE 1–14 A typical standard hydrogen electrode.

1.9.4 Copper–copper sulfate electrode

The robust and stable reference electrode is often used in cathodic protection systems to measure pipe-to-soil potentials. Copper metal is placed in a solution of copper sulfate (saturated) (Fig. 1–15).

$$Cu^{2+} + 2e^- \rightarrow Cu$$

$$E_{\frac{Cu^{2+}}{Cu}} = 0.34 + \frac{0.059}{2}\log Cu^{2+}$$

The standard potential for the electrode is +0.316 V

1.9.5 Junction potentials

A slight potential difference at the two electrodes' interface (junction) arises due to different ionic compositions. Liquid junction corrections must be made to avoid interferences in measured electrode potentials.

1.10 EMF series

According to their standard electrode potentials, different metals are arranged in the EMF series in order of their electrochemical activity. It is a thermodynamic series in which the E° is determined from information about free energy. The most active metal in the series possesses the highest negative E° (anode), while the nobler metal has less negative or more

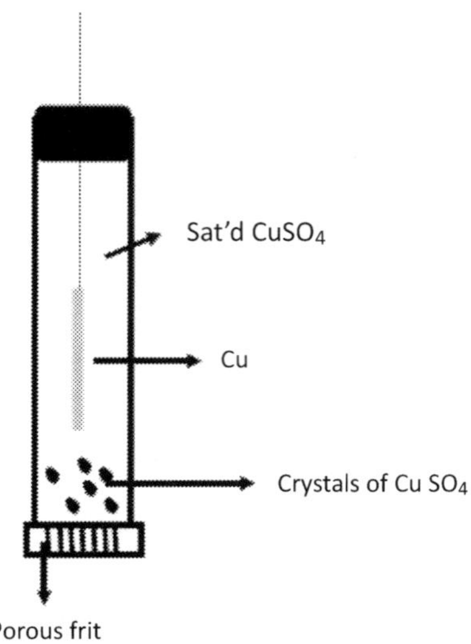

FIGURE 1–15 Copper–copper sulfate electrode.

positive electrode potential (cathode). There are exceptions to the thermodynamically predicted metal activities as listed in Fig. 1–16. Although this approach complies with the Stockholm Convention of 1953, corrosion scientists, primarily interested in the opposite (i.e., oxidation) process, find it somewhat irksome. (While specific half-cell reactions may have somewhat different values listed in various compilations of standard electrode potentials,

Metal	Metal / Formed Ion	Potential* (V)
Lithium	Li / Li +	−2.96
Rubidium	Rb Rb +	−2.93
Potassium	K / K +	−2.92
Strontium	Sr / Sr ++	−2.92
Barium	Ba / Ba ++	−2.90
Calcium	Ca / Ca ++	−2.87
Sodium	Na / Na +	−2.71
Aluminum	Al / Al +++	−1.70
Beryllium	Be / Be ++	−1.69
Manganese	Mn /Mn ++	−1.10
Zinc	Zn / Zn ++	−0.76
Chromium	Cr / Cr ++	−0.56
Iron (Ferrous)	Fe / Fe ++	−0.44
Cadmium	Cd / Cd ++	−0.40
Indium	In / In +++	−0.34
Thallium	Tl /Tl +	−0.33
Cobalt	Co / Co ++	−0.28
Nickel	Ni / Ni ++	−0.23
Tin	Sn / Sn ++	−0.14
Lead	Pb / Pb ++	−0.12
Iron (ferric)	Fe / Fe +++	−0.04
Hydrogen	H2/ H +	0.00
Antimony	Sb / Sb +++	+ 0.10
Bismuth	Bi / Bi +++	+ 0.23
Arsenic	As /As +++	+ 0.30
Copper (cupric)	Cu / Cu ++	+ 0.34
Copper (cuprous)	Cu /Cu +	+ 0.47
Tellurium	Te / Te ++++	+ 0.56
Silver	Ag / Ag +	+ 0.80
Mercury	Hg / Hg ++	+ 0.80
Platinum	Pt/ Pt ++++	+ 0.82
Palladium	Pd / Pd ++++	+ 0.86
Gold (auric)	Au / Au +++	+ 1.36
Gold (auras)	Au / Au +	+ 1.50

FIGURE 1–16 EMF series.

these differences are typical of the order of millivolts and should not be taken seriously.) The standard electrode potential in Fig. 1–16 has a sign that varies from positive to negative values, passing through the value 0.000, which is designated as such for the hydrogen reduction process. The most positive potential listed in Table 3–1 is for $Au^{3+} + 3e^{-} \rightarrow Au$ ($E^{\circ} = +1.498V$)

1.10.1 Applications of EMF series

1. In other media, the relative order of metals is not always the same as in the EMF series (and is typically not the same) (e.g., seawater, groundwater, sulfuric acid, and artificial perspiration).

$$Zn_2 + 2CuSO_4 \rightarrow 2ZnSO_4 + 2Cu$$

$$Fe + CuSO_4 \rightarrow FeSO_4 + Cu$$

Oxidation:

$$Fe \rightarrow Fe^{2+} + 2e^{-}$$

$$Zn \rightarrow Zn^{2+} + 2e^{-}$$

$$\text{Reduction } Cu^{2+} + 2e^{-} \rightarrow Cu$$

Electrode potentials show the tendency for corrosion and depositing. Metals (ions) in the series above hydrogen are more easily reduced. Metals with a heavier oxidation propensity are those below hydrogen.

2. One may anticipate using oxygen (or other oxidants) as an oxidizer for various metal ions and metals under various partial pressures as well as using hydrogen as a reducing agent under various pressures. When two metals are combined, the one with the nobler potential serves as the cathode, while the one with more active potential acts as the anode. Fe, for instance, will be cathodic to anodic Zn in the Fe-Zn pair.

1.10.2 Limitation of EMF series

1. Alloys and other composite materials are not included; only pure metals are.
2. The corrosion potentials of metals and alloys in a given corrosive environment are of practical utility and interest, not the thermodynamic electrode potentials for the different metal/metal ion concentrations.

1.11 Galvanic series

Metals and alloys are arranged in the galvanic series according to their susceptibility to environmental corrosion. The metals and alloys' noblest (positive) potentials are measured in the desirable conditions, while the most active potentials are measured at the bottom. A galvanic series of a few industrial metals and alloys in saltwater is shown in Table 1–2. A SCE is used to measure the potentials in seawater.

Table 1–2 Galvanic series in seawater at room temperature.

S. no.	Galvanic series	S. no.	Galvanic series
1.	Magnesium (LEAST NOBLE)	38.	Nickel–silver (18% Ni)
2.	Mg alloy AZ-3 IB	39.	Stainless steel 316L (active)
3.	MgalloyHK-31A	40.	Bronze 220Copper 110
4.	Mg alloy (hot-dip, die-cast, or plated)	41.	Red brass
5.	Beryllium (hot pressed)	42.	Stainless steel 347 (active)
6.	Al 7072 clad on 7075	43.	Molybdenum, commercial pure
7.	A12014-T3	44.	Copper-nickel 715
8.	A11160-H14	45.	Admiralty brass
9.	Al 7079-T6	46.	Stainless steel 202 (active)
10.	Cadmium (plated)	47.	Bronze, phosphor 534 (B-l)
11.	Uranium	48.	Monel 400
12.	Al 218 (die cast)	49.	Stainless steel 201 (active)
13.	Al 5052-0	50.	Carpenter 20 (active)
14.	Al 5052-H12	51.	Stainless steel 321 (active)
15.	Al 5456-0, H353	52.	Stainless steel 316 (active)
16.	Al 5052-H32	53.	Stainless steel 309 (active)
17.	Al 1100-0	54.	Stainless steel 17-7PH (passive)
18.	Al 3003-H25	55.	Silicone bronze 655
19.	A16061-T6	56.	Stainless steel 304 (passive)
20.	Al A360 (die cast)	57.	Stainless steel 301 (passive)
21.	A17075-T6	58.	Stainless steel 321 (passive)
22.	Al 6061-0	59.	Stainless steel 201 (passive)
23.	Indium	60.	Stainless steel 286 (passive)
24.	Al 2014-0	61.	Stainless steel 316L (passive)
25.	Al 2024-T4	62.	AM355 (active)
26.	A15052-H16	63.	Stainless steel 202 (passive)
27.	Tin (plated)	64.	Carpenter 20 (passive)
28.	Stainless steel 430 (active)	65.	AM355 (passive)
29.	Lead	66.	A286 (passive)
30.	Steel 1010	67.	Titanium 5A1, 2.5 Sn
31.	Iron (cast)	68.	Titanium 13 V, 11 Cr, 3 Al (annealed)
32.	Stainless steel 410 (active)	69.	Titanium 6 Al, 4 V (solution treated and aged)
33.	Copper (plated, cast, or wrought)	70.	Titanium 6 Al, 4 V (annealed)
34.	Nickel (plated)	71.	Titanium 8 Mn
35.	Chromium (plated)	72.	Titanium 13 V, 11 Cr, 3 Al (solution heat treated and aged)
36.	Tantalum	73.	Titanium 75 A AM350 (passive)
37.	AM350 (active)	74.	Silver gold graphite (MOST NOBLE)

1.11.1 Characteristics of the galvanic series

The general characteristics of the galvanic series are discussed below:

1. Instead of potentials, the relative locations of the metals and alloys are given in the galvanic series.
2. The series is based on actual measurements of equilibrium corrosion potential. By attaching the metal or alloy to the negative terminal of a voltmeter and the positive terminal to a reference electrode, often the calomel electrode, one may determine the potential of a corroding metal in a specific medium.
3. According to the galvanic series, alloys can be linked without corroding. For instance, it is safe to pair alloys near one another in the series. Monel may be connected to copper or bronze without the danger of galvanic corrosion, as illustrated in the Table 1–2. Tin and brass cannot be linked together, though, as their positions on the table are too dissimilar, and doing so might result in dangerous galvanic corrosion.
4. In the table, certain alloys are listed twice. For instance, steel (18/8) exists in its active and passive states. Steel in a passive state has an oxide coating covering it, changing the metal's potential from active to passive. Severe corrosion may result if the two kinds (active and passive) are combined. An example of stainless steel in the active state is if it is being continuously scraped.
5. Linking the metals and alloys in brackets is convenient without any corrosion danger.

1.11.2 Limitations of the galvanic series

1. Different galvanic series are needed for each environment; for instance, a static seawater galvanic series cannot be used to forecast galvanic corrosion in turbulently running salt water.
2. Along with their proximity in the galvanic series, the degree of polarization of the metals in alloys affects galvanic corrosion. Predictions may not sufficiently explain galvanic corrosion based on their place in the galvanic series. For instance, cathodic polarization is the dominant process in galvanic corrosion. For instance, in saltwater, titanium has the propensity to polarize cathodically. Due to titanium's cathodic polarization, any less corrosion-resistant metal linked to it will not experience corrosion, as one might assume. Although alloys rather than pure metals are employed in engineering applications, the EMF series only has a limited utility. Design engineers must read this series immediately.

1.12 Awards in the field of corrosion

Young ICorr Award
- Morris Cohen Graduate Student Award
- T.P. Hoar Award
- U.R. Evans Award
- T.J. Hull Award of NACE International

Galloway Award
Lionel Shrier Award
Outstanding achievement awards in corrosion science

Useful websites

https://www.corrosionpedia.com/the-8-most-common-forms-of-metal-corrosion/2/1680
https://www.electrochem.org/corrosion-science/
http://www.nace.org
https://fcc.osu.edu/
http://www.Corrosion-Club.com
http://www.iom3.org

The World Fact Book, GDPdComposition, by Sector of Origin (%). https://www.cia.gov/library/publications/the-world-factbook/fields/2012.html.

Short questions

Q.1. Define the term corrosion? Why is corrosion a problem?
Q.2. Explain how corrosion can be considered as extractive metallurgy?
Q.3. How is corrosion different from rusting?
Q.4. What are the economic impacts of corrosion?
Q.5. What are the factors that increase and decrease the cost of corrosion?
Q.6. Discuss the effect of corrosion on environment and human health.
Q.7. How the corrosion science is different from corrosion engineering?
Q.8. What are the dangers of corrosion?

References

Al-Kharafi, F., Al-Hashem, A., & Matrouk F. (1995). *Economic effects of metallic corrosion in the State of.*

Behrens, D. (1967). *British Corrosion Journal, 10.*

Bennett, L. H., Kruger, J., Parker, R. L., Passaglia, E., & Reimann C. (1978). *Economic effects of metallic corrosion in United States.*

Bhaskaran R., Bhalla L., Rahman A., Juneja S., Sonik U., Kaur S., . . . Rengaswamy N. S. (2014). An analysis of the updated cost of corrosion in India. Materials performance. 53 (8) 56–65. Available from: http://mp.epubxp.com/i/354215.India.

Bhaskaran, R., Palaniswamy, N., Rengaswamy, N. S., & Jayachandran M. (n.d.). *A review of differing approaches used to estimate the cost of corrosion (and their relevance in the development of modern corrosion prevention andp control strategies).* Available from: http://krc.cecri.res.in/ro_2005/002-2005.pdf.

Cherry, B. W. & Skerry, B. S. (1983). Corrosion in Australia. *Report of the Australian National Center of Corrosion Prevention and Control Feasibility Study.*

Committee on Corrosion Loss in Japan: Boshoku Gijutsu. (1977), 26, 401–512.

Committee on Corrosion Loss in Japan: Zairyo-to-Kankyo. (2001), 50, 490–512.

Evans, U. R. (1937). *Metallic corrosion and protection.*

Hoar, T. P. (1976). *Proceedings of the Royal Society, 348,* 1–18.

Hou, B., Li, X., Du, X.C., Zhang, D., Zheng, M., Xu, W., . . . Ma, F. (2017). *The cost of corrosion in China.*

Koch, G., Brongers, M., Thompson, N., Virmani, P., Payer, J. (2002). *Corrosion cost and preventive strategies in the United States.*

Koch, G., Varney, J., Thompson, N., Moghissi, O., Gould, M., & Payer, J. (2016). *International measures of prevention, application, and economics of corrosion technologies study.*

Linderborg, S. (1967). *Kemiam Teollusius, 24.* Available from http://urlm.co/http://www.corrosion-club.com.

Payer, J. H., Dippod, D. G., Boyd, W. K., Berry, W. E., Brooman, N. (1978). *Economic effects of metallic corrosion in United States.*

Payer, J. H., Koch G. H., (2002), Materials performance, pp. 2–11.

Payer, W. K., Boyd, D. G., Lippold, W. H., Fisher, (1980). NBS-Battelle cost of corrosion study ($70 billion!), Part 1–7, Materials performance.

Potter, E. C. (1972). *Australian corrosion engineering.*

Revie, R. W., & Uhlig, H. H. (1974). *Journal of the Institution of Engineers, Australia, 46,* 3–5.

Sastri, V. S., Elboujdaini, M., Perumareddi, J. R. (2003). Environmental degradation of materials and corrosion control in metals, in: Proceedings of the *international symposium on annual conference of metallurgists of CIM,* pp. 321–328.

Tems, R., & Al Zahrani, A. M. (2006). Cost of corrosion in oil production and refining. *Saudi Aramco Journal of Technology* (SUMMER), 2–14.

Tradgaidh, F. K. (1965). Tekn. *Tedskrift, 95,* 1191.

Uhlig, H. H. (1950). The cost of corrosion to The United States. *Corrosion, 6*(1), 29–33. Available from https://doi.org/10.5006/0010-9312-6.1.29.

Wagner, C. (1974). Corrosion in aqueous solutions and corrosion in gases at elevated temperatures - analogies and disparities. *Werkstoffe und Korrosion, 25*(3), 161–165.

Wagner, C. (1938). Traud Uber die Deutung von Korrosionsvorgangen durch Uberlagerung von elektrochemis- chen Teilvorgangen und uber die Potentialbildung an Mischelektroden. *Z. Elektrochem., 44,* 391–402.

Worner, H. K. (1955). *Committee report on the symposium on corrosion.*

Zhu, R. (1986). In *Proceedings of the corrosion 86 symposium on international approaches to reducing corrosion costs.*

2 Uniform or general corrosion

2.1 Occurrence

When evaluating the long-term performance of materials, an engineer must consider the impacts of corrosion as well as other properties such as strength and formability. An engineer must first understand the nature of corrosion, its various, and the various types of corrosion protection strategiesto select the appropriate materials. There are many forms of corrosion. The science and knowledge related to these processes are constantly changing. Here's a brief overview of some of the most common types (Table 2–1) (Sully & Taylor, 1987).

Damaged test specimens or failed equipment can frequently provide useful information for resolving corrosion problems. According to Fontana and Greene's standard textbook on corrosion engineering, first published in 1967 (Fontana, Greene, & Klerer, 1968), each type of corrosion is described by how it manifests. Its various forms are categorized according to how the corroded metal appears, with visual inspection distinguishing corrosion types (Fig. 2–1).

In most circumstances, the naked eye will suffice, however, magnification may be beneficial or necessary in other cases. Dillon (1982) categorized the types of corrosion into three categories shown below:

Group I: Corrosion determined by visual observation

1. uniform corrosion
2. pitting Corrosion
3. crevice corrosion
4. galvanic corrosion

Group II: Corrosion needing additional means of examination

1. erosion, cavitation, exfoliation
2. fretting
3. intergranular
4. dealloying

Group III: Corrosion requiring microscopic verification

1. stress corrosion cracking
2. environmental cracking
3. high-temperature corrosion (scaling, internal attack)
4. corrosion Fatigue

Handbook of Corrosion Engineering. DOI: https://doi.org/10.1016/B978-0-323-95185-2.00002-2

Table 2–1 ASM classifications of corrosion types (Sully and Taylor, 1987).

General corrosion	Localized corrosion	Metallurgically influenced corrosion	Mechanically assisted degradation	Environmentally induced cracking
Corrosive attack dominated by uniform thinning. i.e., Atmospheric corrosion, Galvanic corrosion, Stray-current corrosion, General biological corrosion, Molten salt corrosion, Corrosion in liquid metals, High-temperature corrosion	High rates of metal penetration at specific sites. i.e., Crevice corrosion, Filiform corrosion, Pitting corrosion, Localized biological corrosion	Affected by alloy chemistry and heat treatment. i.e., Intergranular corrosion, Dealloying corrosion	Corrosion with a mechanical component. i.e., Erosion corrosion, Fretting corrosion, Cavitation and water drop impingement, Corrosion fatigue	Cracking is produced by corrosion, in the presence of stress. i.e., Stress corrosion cracking, Hydrogen damage, Liquid metal embrittlement, Solid metal-induced embrittlement

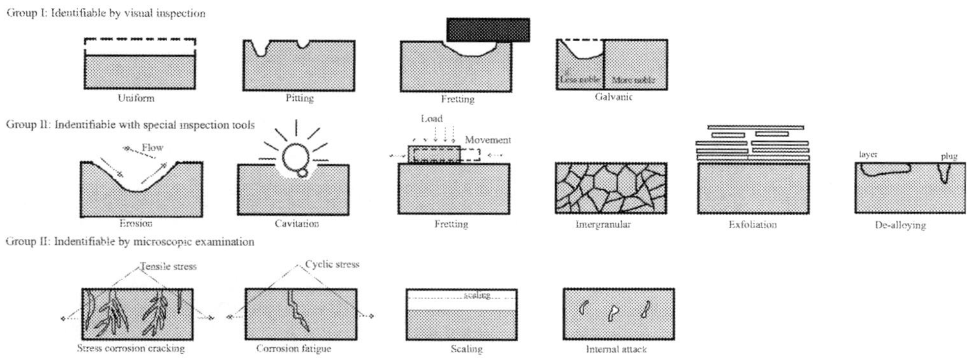

FIGURE 2–1 The most common types of corrosion attacks are grouped for easy identification.

The forms of corrosion listed above are discussed in detail in the sections below:

2.1.1 Uniform or general corrosion

2.1.1.1 Occurrence

The loss of metal in a consistent pattern throughout the entire surface is known as uniform corrosion. However, because it is relatively easy to forecast, it is not considered the most dangerous type of corrosion. Corrosion that spreads equally throughout the whole surface area, or a considerable percentage of it, is known as uniform corrosion. This type of corrosion is usually characterized by a general thinning until it fails. It is a factor that contributes to the tonnage loss (Roberge, 2000).

On the other hand, uniform corrosion is simple to analyze and predict, making catastrophic failures rare unless the corroding material is buried. Internal pipeline corrosion, for example, concealed component corrosion and buried or submerged structures, are ideal illustrations of how even the simplest corrosion process must be monitored. It is just unsuitable in many situations due to its appearance. The failure of protective coating systems on structures is a common cause of this form of corrosion. The image in Fig. 2–2A and B shows homogeneous rusting on an iron gas cylinder and roadside signboard.

Corrosion-resistant metals and stainless steel can discolor or oxidize in corrosive environments. On the other hand, surface corrosion may indicate a breakdown in the protective coating system and should be checked carefully for signs of a more serious attack. If uniform corrosion is allowed to continue, the surface will become rough, and more harmful types of corrosion will develop.

It is not, however, typically regarded as a significant kind of corrosion for the following reasons:

1. Simple tests can be used to predict the rate of thickness decrease.

(A) (B)

FIGURE 2–2 Uniform corrosion of (A) iron gas cylinder (B) roadside signboard.

2. The various protective measures typically reduce the rate of corrosion to a manageable level. The practical methods include coatings, cathodic protection, and maybe a change in the environment or material.

2.2 Examples of uniform corrosion

Fig. 2–3 showed the various examples of uniform corrosion.

2.3 Mechanism of uniform corrosion

It is essential to comprehend how iron and steel rust in water because they are essential building materials. For the sake of simplicity, consider one anode and one cathode in a piece of iron immersed in aerated water, as indicated in Fig. 2–4.

The anodic reaction would be

$$Fe \rightarrow Fe^{2+} + 2e^{-} \quad (2-1)$$

The loss of electrons leaves positively charged ions on the anode, which migrates from the anode to the cathode via the water and carry the positive current. The electrons released at the anode migrate from the anode to the cathode via the metallic circuit. These electrons

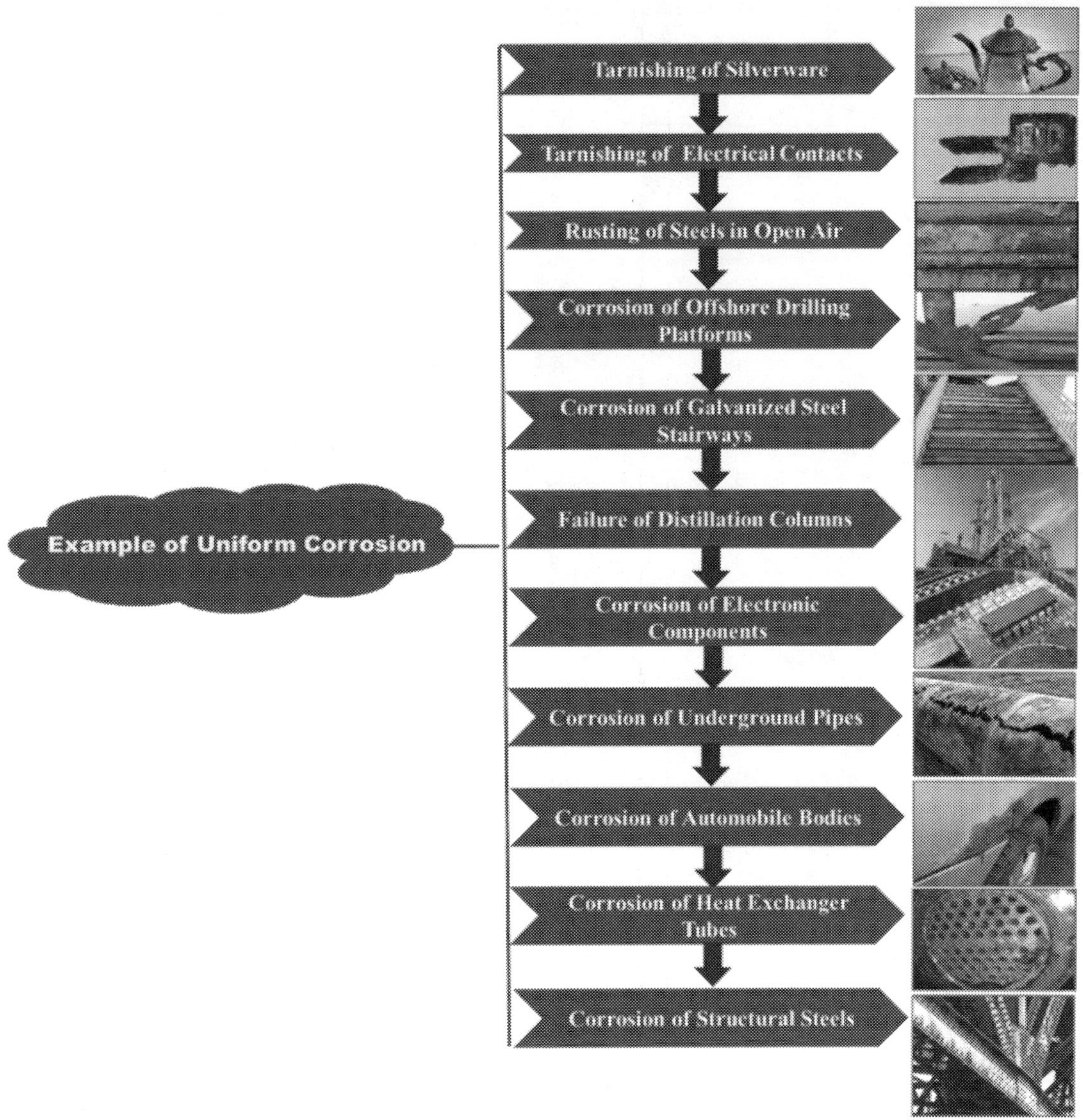

FIGURE 2–3 Various examples of uniform corrosion.

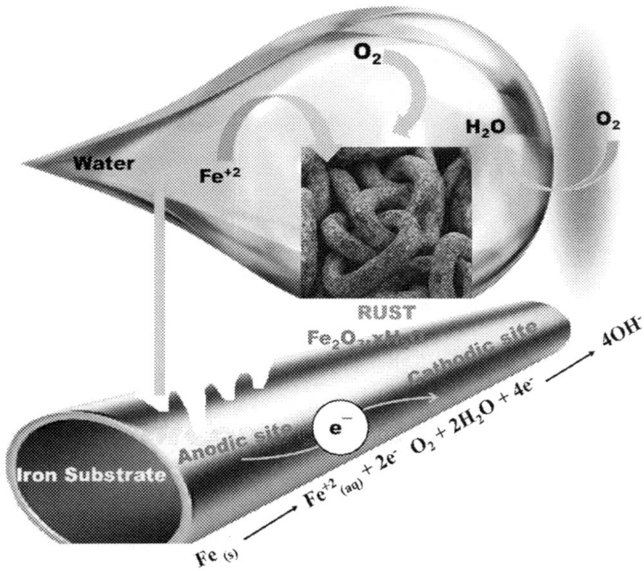

FIGURE 2–4 Mechanism of uniform corrosion.

are used in the reduction of oxygen that is present in water which is in contact with the cathode. Therefore, one of the following reactions can take place at the cathode. The cathodic reactions would be

1. In acidic environments, that is, $pH < 7$, the cathodic process is mainly the reduction of hydrogen ions:

$$2H^+ + 2e^- \rightarrow H_2\uparrow \tag{2-2}$$

2. In an neutral or alkaline environment, that is, $pH = 7$ or $pH > 7$, reduction of dissolved oxygen is the predominant cathodic process that causes uniform corrosion:

$$O_2 + 2H_2O + 4e^- \rightarrow 4OH^- \tag{2-3}$$

The negatively charged OH^- ions react with the positively charged Fe^{2+} ions and form Fe $(OH)_2$.

$$4Fe^{2+} + 2OH^- \rightarrow 4Fe(OH)_2\downarrow \tag{2-4}$$

With more access to oxygen in the air, $Fe(OH)_2$ oxidizes to $Fe(OH)_3$

$$4Fe(OH)_2 + O_2 + 2H_2O \rightarrow 2(FeOH)_3\downarrow \tag{2-5}$$

Ferrous hydroxide is converted to hydrated ferric oxide by oxygen

$$4Fe(OH)_3 + O_2 \rightarrow 2Fe_2O_33H_2O\downarrow + 2H_2O \quad (2-6)$$

The overall outcome is $2Fe_2O_3 \cdot 3H_2O$, which is popularly referred to as "rust." In a metallic circuit, electrons move from the anode (drop center) to the cathode (periphery). The ferrous ions on the surface of iron are soluble, but those in solution are oxidized by oxygen into insoluble hydrated ferric oxides known as rust. Rust develops away from the corroding area. If the ferrous ion is rapidly oxidized to ferric oxide, the corrosion rate is quite high. The corrosion reaction accelerates faster when $Fe(OH)_3$ occurs distant from a metal surface because equilibrium must be maintained by delivering more ferrous ions (Fe^{+2}) from the surface. However, if $Fe(OH)_3$ is generated on the metal's surface relatively quickly, corrosion is prevented (a passive film).

In addition, zinc can uniformly corrode forming a White Rust according to the following reactions:

$$2Zn \rightarrow 2Zn^{+2} + 4e^- \quad (2-7)$$

$$O_2 + 2H_2O + 4e^- \rightarrow 4OH^- \quad (2-8)$$

$$2Zn + 2H_2O + O_2 \rightarrow 2Zn^{+2} + 4OH^- \quad (2-9)$$

$$2Zn^{+2} + 4OH^- \rightarrow 2Zn(OH)_2\downarrow \quad (2-10)$$

$$2Zn(OH)_2 + 2CO_2 + O_2 + H_2O \rightarrow (Zn)_4CO_3 \cdot (OH)_6\downarrow \quad (2-11)$$

$(Zn)_4CO_3{\cdot}(OH)_6orZnCO_3{\cdot}Zn(OH)_2$ is known as white rust or wet storage stain (porous).

Atmospheric corrosion of aluminum is due to a passive oxide film formation instead of a porous layer. The gray/black-color film may form as follows

$$2Al \rightarrow 2Al^{+3} + 3e^- \quad (2-12)$$

$$3/2O_2 + 3H_2O + 6e^- \rightarrow 6OH^- \quad (2-13)$$

$$2Al + 3/2O_2 + 3H_2O \rightarrow 2Al^{+3} + 6OH^- \rightarrow Al_2O_3{\cdot}3H_2O\downarrow \quad (2-14)$$

2.4 Factors affecting uniform corrosion

1. Effect of pollutants: even in a dry atmosphere with no moisture, corrosion can occur if residues of sulfur components, H_2S, or other pollutants are present in the air.
2. Effect of humidity: because this is the equilibrium level for a saturated NaCl solution, and NaCl is typically found on surfaces, corrosion can occur in the atmosphere when the

humidity level is around 70%. A thin layer of moisture forms on the surface of a metal in the presence of such humidity. Moisture is an electrolyte, which allows electricity to flow through it. Dampness has an impact on structures that are exposed to the elements. Corrosion speeds up dramatically when the relative humidity rises above 80%.

3. Water layer: corrosion happens when visible water layers build on a metal surface. A wet atmosphere is created by the splashing of waves, rain, and dew droplets. Like in a corrosion cell, the water coating on the metal surface works as an electrolyte and provides a path for current transmission.
4. Dew formation: the rate of corrosion accelerates if the dew becomes acidic owing to the presence of SO_2. Automobiles exposed to the elements may corrode due to the production of acidic dew.
5. Corrosion products: the rate of corrosion increases as the dew becomes acidic due to the presence of SO_2. Due to the production of acidic dew, automobiles exposed to the elements may corrode.

2.5 Prevention methods

Because corrosion occurs uniformly throughout the whole metal component's surface, it may be efficiently controlled using thicker materials, cathodic protection, metallic coatings such as plating, galvanizing, or anodizing, or paints and corrosion inhibitors or modifying the environment (Shreir, Jarman, & Burstein, 1994). Continuous rusting can give a surface color and a pleasing appearance in some situations. The patina developed by naturally tarnishing copper roofs and the rust colors formed by weathering steels is two classics in this regard.

2.6 Failure case histories

2.6.1 Case 1: failure of a distillation column wall

The rate of corrosion accelerates if the dew becomes acidic owing to the presence of SO_2. Automobiles exposed to the elements may corrode due to the production of acidic dew.

1. Environment: the relative humidity level is at 90%. The conditions are condensed moisture and oxygen. A pressure of 0.1 MPa was present. The insulation became moist after being exposed to the air.
2. Micro- and macroexaminations: a macrocheck using magnification revealed that the exterior surface was extensively damaged. After the surface had been lowered to the penetration level, it was perforated. It was considered that microscopy was unnecessary.
3. Remarks: the distillation tower's bottom working temperature is around 60°C, while the highest operating temperature is 0°C. Moisture penetrated below the insulating surface due to water leakage in the insulation material and ice development. The corrosion rate was significantly enhanced because of the wetness of the insulation. If the temperature range had been 60°C–80°C, the situation could have been much worse.

4. Remedy: The entire steel column should be coated with an epoxy-phenolic coating. Because the polarity of zinc can change at temperatures above 60°C, creating zinc cathodic and iron anodic, cathodically protective pigments should not be used in the coating system, especially in the hotter portions. Zinc is resistant to iron corrosion.

2.6.2 Case 2: Severe corrosion of carbon steel pipe: A carbon steel pipe carrying brackish water was found to be severely corroded after 6 months of service

1. Equipment: 6-inch diameter carbon steel pipe.
2. Environment: Brackish water containing 900 g/L chlorides and 1500 T.D.S. (Total Dissolved Solids).
3. Conditions: Water is a major species. T.D.S., oxygen, minor species temperatures range from 25°C–35°C (ground temperature in summer) 10 PSIG pressure. The time it takes to fail is 6 months.
4. Observations and remarks: Corrosion was discovered both internally and outside. High chloride levels and dissolved oxygen-induced interior corrosion. The damage was hastened by the hot weather. Corrosivity and low conductivity dirt harmed the external surface.
5. Remedy: Use PVC pipes. This history is written from the perspective of a town in the Arabian Gulf. The problem was handled by using PVC pipes. Most carbon steel pipes have been replaced with PVC pipes. This can be used with nonpotable water.

Useful websites

1. https://www.sciencedirect.com/topics/engineering/uniform-corrosion
2. https://www.corrosionclinic.com/types_of_corrosion/uniform_corrosion.htm

Short questions

Q.1. What is the important basis for the classification of corrosion?
Q.2. Which types of corrosion are uniform?
Q.3. Write some examples of uniform corrosion?
Q.4. What happens when uniform corrosion occurs?
Q.5. Why uniform corrosion is assumed to be the most common form of corrosion?
Q.6. What are the prevention methods of uniform corrosion?

References

Dillon, C. P. (1982). *Forms of corrosion: recognition and prevention.*

Fontana, M. G., Greene, N. D., & Klerer, J. (1968). Corrosion engineering. *Journal of the Electrochemical Society, 115*(5), 142C. Available from https://doi.org/10.1149/1.2411256.

Roberge, P. R. (2000). *Handbook of corrosion engineering.*

Shreir, L. L., Jarman, R. A., & Burstein, G. T. (1994) *Corrosion control.*

Sully, J. R. & Taylor, D. W. (1987). Electrochemical methods of corrosion testing. *Metals handbook.* 13.

3

Galvanic corrosion

3.1 Occurrence

Due to the increasing demands for equipment and structures, there is a wide range of materials that are needed to meet these needs. Unfortunately, many of these materials can lead to the development of galvanic corrosion. American society for testing and materials (ASTM) published the first special technical publication containing extensive galvanic information in 1976. The first reference to the galvanic effect dates to 1799 by Giovanni Fabbroni, in Journal de Physique (Fabbroni, 1799) where he wrote, "I have observed that the alloys used for soldering roof copper plates of the Observatory of Florence have rapidly transformed, altered to a white oxide, only at the junctions." Davy proposed avoiding corrosion attack using a galvanic coupling a few years later in 1805, and cathodic protection was developed.

Galvanic corrosion referred to as "dissimilar metal corrosion" takes place when two dissimilar materials are combined in an electrolyte that is corrosive. Galvanic corrosion is defined by ASTM International Committee G01 on Metal Corrosion and NACE International as "accelerated corrosion of a metal because of an electrical contact with a more noble metal or nonmetallic conductor in a corrosive electrolyte." Galvanic corrosion is defined by the International Organization for Standardization as "Corrosion due to the action of a corrosion cell (a short-circuited combination of different electrodes connected in series with an ionic conductor)" (Hack, 2016). The various organizations such as the Electrochemical Society, the American Water Works Association, the American Chemical Society, and other organizations dealing with this topic have their own definitions of galvanic corrosion. The one common definition is "galvanic corrosion is a type of electrical interaction that occurs between two or more metals or nonmetallic conductors." Various factors, such as electrical contact, dissimilar metals, and conductive contact, can lead to the development of galvanic corrosion. These factors can also accelerate the development of other forms of metal corrosion, such as pitting and crevice.

3.2 Causes

Galvanic corrosion is caused by three basic factors:

1. Difference in electrical potential of the two metals in contact with one another.
2. There is an electrical path between the two metals.
3. A fluid such as atmospheric humidity or salt fog exists that can break down the metal.

The four elements anode, the cathode, the electrolyte, and the return current path are required to form galvanic corrosion. If one of the component is removed the flow will be

Handbook of Corrosion Engineering. DOI: https://doi.org/10.1016/B978-0-323-95185-2.00003-4

interrupted, and galvanic corrosion will not occur. The current flow may reverse if the anode or cathode is replaced with a different metal, resulting in a switch to the corroded electrode. In the galvanic series of metals, metals and alloys are listed in decreasing order of electrical activity. Metals near the top of the table have fewer noble metals and a higher tendency to lose electrons than those at the bottom.

3.2.1 Galvanic series

The energy emitted as metals corrodes is linked to their electrode potential. The term corrosion potential "E_{corr}" refers to the electrode potential of various metals and alloys. The galvanic series (in saltwater) is an ordered list of the corrosion potentials of different pure metals and alloys.

The corrosion potential of dissimilar metals in different environments can be measured by observing their current direction by the galvanic action in seawater. An arrangement of metals in a series Fig. 3–1 (SSPC-VIS, 2002) is commonly used to model the expected direction of different metals' galvanic effects in different environments. For instance, in a galvanic pair comprising any two metals in a galvanic series, the corrosion of a metal higher in the list is accepted to be accelerated, while the lower metal in the list will experience the reduction in corrosion rate. Noble or cathodic metals have higher positive corrosion potentials, while active or anodic metals have higher negative corrosion potentials. The more apart the various types of metal and alloys are in the series, the faster the anodic metal's corrosion accelerates. When aluminum is connected to copper or silver, galvanic corrosion occurs. Different types of stainless steels are found at the anodic or active end of the series. On the other hand, some can be found at the noble or passive end. Due to their dual character, some types of stainless steels form a barrier against the effects of some aggressive solution such as nitric acid. In acids like HCl or HF, or other nonoxidizing acids, these films are destroyed, and the steels corrode quickly. Therefore, it is recommended that before choosing stainless steel for usage in a specific environment, it is crucial to consider the potential impact of the conditions on their behavior. It is important to determine if it will be in an active or passive phase. If two metals in the galvanic series that are close to one other, such as copper and brass, are linked, galvanic corrosion may not occur. Corrosion is minimized when metals are close together.

Potential values can be affected by various factors such as temperature, velocity, and aeration. These can result in values shifting from one solution to another. Without taking into account the potential values of the various metals in an environment, researchers are not able to predict the direction of galvanic effects. For instance, at ambient temperatures, zinc exhibits a negative or anodic effect on steel as illustrated in the galvanic series in Fig. 3–1. However, as the temperature rises, the potential difference reduces until it is either zero or reversed at 60°C (Capp, 1914; D714, 2002).

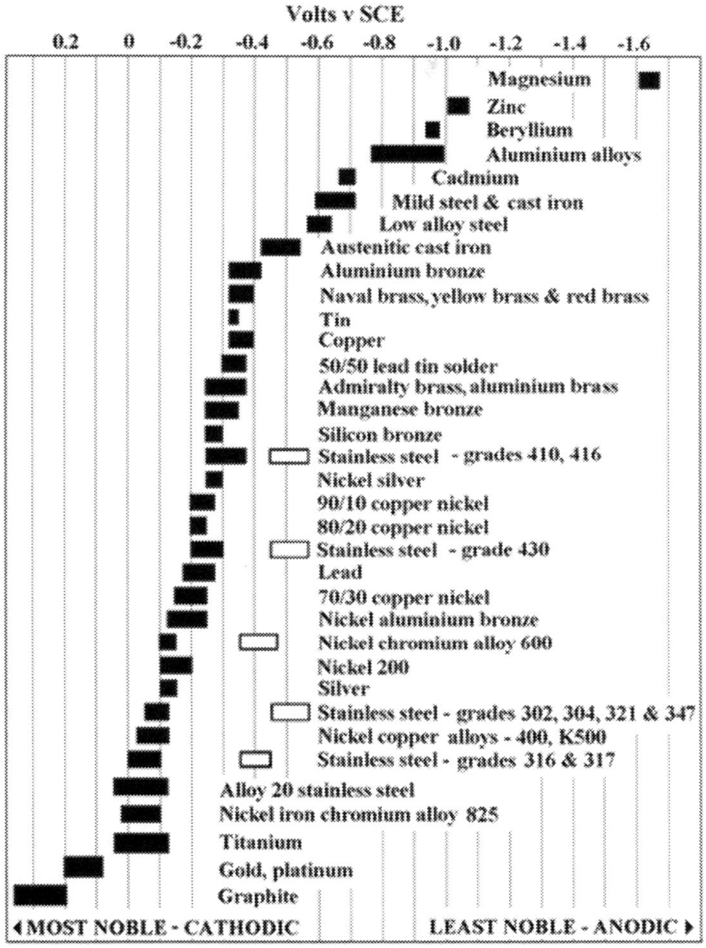

FIGURE 3–1 The galvanic series of some commercial metals and alloys in seawater. *Redrawn from Tugsataydin, CC BY-SA 4.0, via Wikimedia Commons https://commons.wikimedia.org/wiki/File:Galvanic_series_noble_metals.jpg.*

3.2.2 Limitations of galvanic series

The galvanic series cannot be utilized for predicting the corrosion rate of certain materials in solutions that are significantly different from saltwater. As a result, expanding it to include other solutions, such as hydrochloric acid, should be approached with caution. Also, it does not provide information on the rate of corrosion.

The galvanic series in seawater is, to date, the most comprehensive set of corrosion potentials established for any corrosive environment. However, galvanic series exist for metals in flowing hot water (Butler & Ison, 1966), in phthalate buffers (Kaesche, 1985), in solutions of $NaHCO_3$ or Cl^--containing $NaHCO_3$ solutions (Butler, Francis, & McKie, 1969).

3.3 Factors affecting galvanic corrosion

In addition to the potential difference between the two metals, many other factors such as temperature, pressure, and chemical composition may also play a role in galvanic corrosion (Fig. 3–2).

3.3.1 Electrode potential

The potential difference between two metals must be considered to provide the driving power of a large galvanic current. The quantity of galvanic corrosion that is observed in the couple's electrodes also depends on the reaction kinetics of the two metals. However, if the reaction kinetics of the two metals change, the potential of the two metals can be affected. A change in the anodic or cathodic reaction kinetics can result in a change in the potential of both metals. Alterations in pH, temperature, and extreme aeration can all lead to potential changes. Due to the presence of a biofilm, which can grow in seawater, the potential of most electropositive alloys can be affected by temperature. This biofilm, which takes a few days to a few weeks to form, boosted the metals' potential by stimulating the cathodic reduction of dissolved oxygen. On the other hand, adding a small quantity of Cl or raising the temperature can prevent the formation of biofilms. As the dissolved O_2 content in deaerated saltwater decreases, the potential of certain high-alloy steels, such as super duplex and 6% Mo austenitic, becomes electronegative. In certain metals, temperature variations can cause their

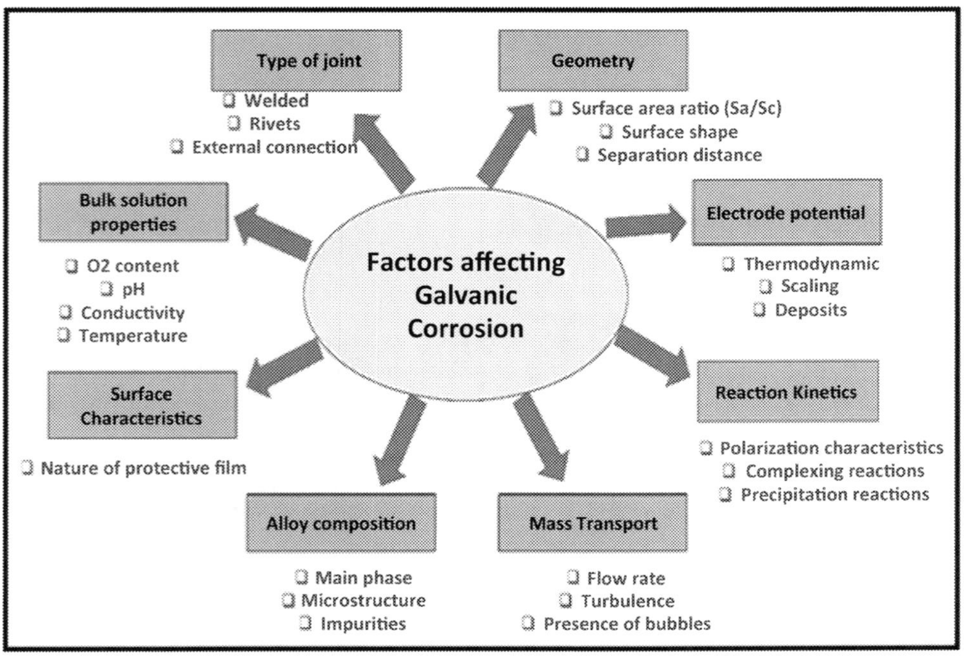

FIGURE 3–2 Factors affecting galvanic corrosion.

potential to change. For instance, iron becomes anodic to zinc when it reaches temperatures of over 60°C resulting in galvanized steel corrosion in hot water systems.

3.3.2 Electrode efficiency

The rate of anodic or cathodic reactions is determined by the kinetics of metal reactions. The cathodic reaction is more essential than the anodic response in galvanic corrosion. When the galvanic current produced by the two linked metals is followed by a shift in the potential of the anodic member to a more electropositive value and the potential of the cathodic member to a more electronegative value, polarization occurs. Because the sacrificial anode's small volume can efficiently port a large cathodic surface in neutral electrolytes, the cathode is always polarized substantially more than the anode. The degree of polarization of any metal as a cathode affects its ability to generate anode corrosion.

3.3.3 Area ratio

The severity of galvanic corrosion is influenced by the ratio of exposed anode and cathode portions in a galvanic couple. More O_2 reduction happens when the cathode is larger than the anode, resulting in a stronger galvanic current. In stagnant or moderately flowing circumstances, the rate of diffusion of dissolved O_2 to the cathode is often a factor of the galvanic corrosion current. The severity of galvanic corrosion depends on the surface area of the cathode. For instance, an increase in the iron cathode area magnifies the zinc galvanic corrosion as seen in Fig. 3–3. On the other hand, zinc galvanic corrosion shows no significant changes in its severity after increasing the zinc anode area.

The corrosion current remains constant at a constant cathode area. When the anode area decreases, the current density increases. Extremely small anodic zones may form at discontinuities in cathodic coatings, such as cracks or pores, when exposed to a highly conductive electrolyte. A thickness with enough integrity for protection when submerged in strongly conducting electrolytes must be specified in this situation.

3.3.4 Environments

3.3.4.1 Electrolyte

Corrosion intensity and distribution are influenced by electrolyte and electrical conductivity. Corrosion is localized to an area near the intersection of two metals in an electrolyte with low conductivity. As a result, while overall corrosion levels may be low, corrosion is likely to be extremely focused and intense. Corrosion will be more prevalent in high-conductivity electrolytes than in low-conductivity electrolytes. If conductivity is high, such as in salt water, the anodic metal's galvanic corrosion is evenly distributes across the surface. On the other hand, when the conductivity goes down, the corrosion can concentrate in a restricted area around the joint, as seen in Fig. 3–4A poorly conducting electrolyte usually has less total galvanic corrosion than a highly conducting one. Noble element ions in solution can cause galvanic corrosion of a less noble metal immersed in the solution because the noble element can precipitate

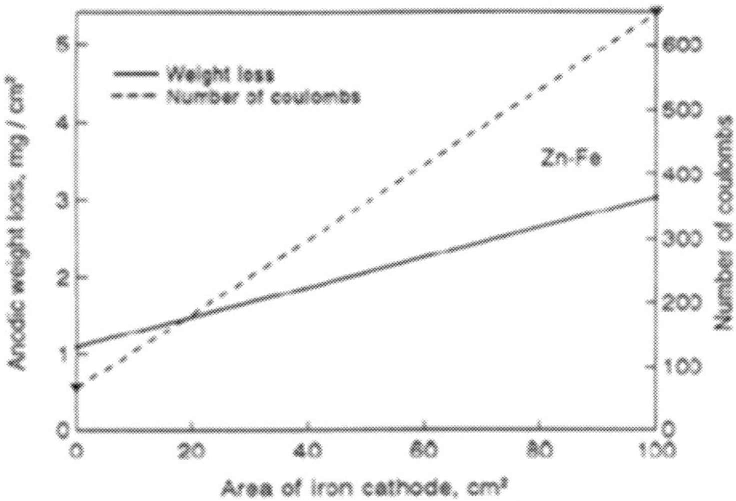

FIGURE 3–3 Effect of the area of mild steel cathode on weight loss of Zn anode (area of 100 cm^2) and on a number of coulombs flowing between Zn–steel couple over a 96-h period in 1 N NaCl solution at 25°C. *Republished with permission from Pryor, M.J. & Keir, D.S. (1957). Journal of the Electrochemical Society, 104, 269.*

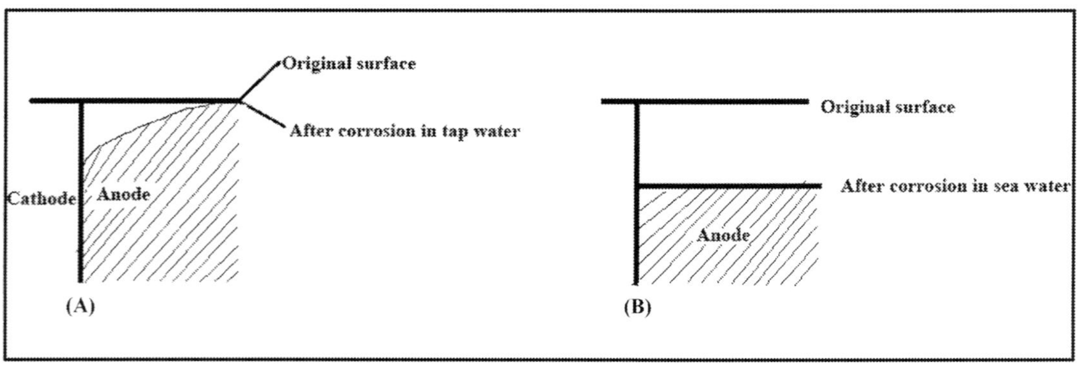

FIGURE 3–4 Effect of solution conductivity on distribution of galvanic corrosion: (A) low conductivity and (B) high conductivity.

and form tiny galvanic cells (Kenworthy, 1943). Electrochemical processes can cause the electrolyte's composition to vary dramatically if there is only a little amount of it. After a period of galvanic action for a zinc–steel couple, Massinon and Dauchelle (1989) discovered an increase in pH in a contained electrolyte. Because the mass transit of the reactants and reaction products is limited when the distance between the anode and cathode is modest relative to the dimension of the electrodes, Pryor, Keir, and Electrochem (1958) found that galvanic corrosion is minimal.

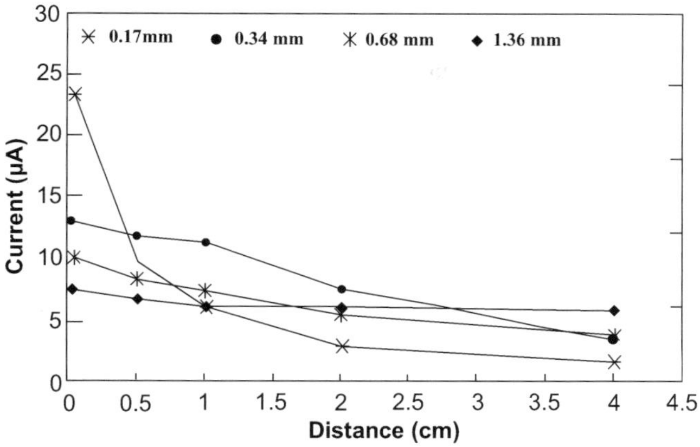

FIGURE 3–5 Galvanic current as function of distance between zinc and steel in 0.001 M Na_2SO_4 solutions of different electrolyte thickness. *Republished with permission from Massinon, D. & Dauchelle D. (1989). Proceedings of the international conference on zinc and zinc alloy coated steel sheet. GALVATECH'89 585–595 recent progress towards the understanding of underfilm corrosion of coaled steels used in the automotive industry.*

This process is like local cell corrosion and is sensitive to the changes in the stability of the metal ions caused by the presence of certain electrolyte elements such as silicate, hydrogen, and bicarbonate. It can trigger the formation of insoluble or sticky corrosion products. Crevice corrosion can occur when insoluble corrosion products from the anodic metal deposit on the cathodic metal of a pair.

The rate-controlling procedure for a given cell dimension is also determined by variations in electrolyte thickness. When the anode and cathode are near each other, the galvanic current is larger (Southwell, Bultman, & Alexander, 1976). The change in the relative values of galvanic currents for small and large distances is due to the rate-limiting process moving from oxygen diffusion (close distance) to ohmic conduction (large distance) (Fig. 3–5) because the oxygen diffusion rate under thin layer electrolytes changes with electrolyte thickness.

3.3.4.2 Aeration and flow rate

The cathodic reaction in most practical situations of galvanic corrosion is the reduction of dissolved oxygen in aqueous solutions. The rate at which oxygen diffuses from the bulk electrolyte to the cathodic surface affects galvanic corrosion. The oxide layer that forms on the cathodic surface of a flowing, aerated electrolyte is likely to thicken, preventing galvanic corrosion of the connected metal. In neutral electrolytes, complete deaeration can typically prevent galvanic corrosion. If an alternate reaction to the reduction of dissolved O_2, such as metal ion reduction and deposition, is present under such anaerobic conditions, cathodic depolarization and corrosion may ensue.

3.3.5 Natural waters

There are two types of natural waters: freshwater and saltwater. The latter has a higher conductivity due to its high salt content. Water as corrosive environments is homogeneous in comparison to other environments (e.g., seawater is almost constant with respect to time and geographic location). Due to its appreciable conductivity and uniform distribution, the action of galvanic substances in saltwater is long-range and widely dispersed on the full surface area of a metallic structure. Because of the reduced conductivity, the galvanic impact in freshwater is often significantly lower than in seawater (Zhang & Valeriote, 1993).

3.4 Examples of galvanic corrosion

Galvanic corrosion affected the Statue of Liberty in New York, United States. Corrosion had taken place between the wrought iron support framework and the statue's copper exterior. The iron support structure rusted as a result. The combat cruiser "USS Independence" was another victim of galvanic corrosion. A steel water jet propulsion system was fitted to the ship's aluminum hull. In this corrosion, the aluminum hull worked as an anode against the steel.

3.5 Prevention of galvanic corrosion

If any of the three preceding requirements for dissimilar metal corrosion are avoided, bimetallic or galvanic corrosion can be avoided (Southwell et al., 1976).

1. Electrochemically dissimilar metals with different anodic indexes.
2. Dissimilar metals must be in electrical contact.
3. Metals must be exposed to the electrolyte for metal ions to shift from anodic to cathodic.

However, it is extremely difficult to avoid these situations in product design in the actual world. As a result, during product design, the following aspects are considered to reduce the influence of galvanic corrosion.

3.5.1 By incorporating insulation in between dissimilar metal

Because galvanic corrosion requires electrical contact between dissimilar metals, insulation between dissimilar metals aids in the prevention of bimetallic corrosion (Fig. 3–6).

3.5.2 Absence of electrolyte

Dissimilar metal corrosion can be avoided by avoiding contact between dissimilar metals and electrolyte/water or any other conductive fluid. This prevents the passage of metal ions from anodic to cathodic metal.

3.5.3 By selecting the materials closer in galvanic series

Materials should be chosen for part design based on their anodic index value to avoid galvanic corrosion. Because the greater the gap in anodic index values, the more quickly corrosion occurs. As a result, materials with the smallest anodic index variances are selected.

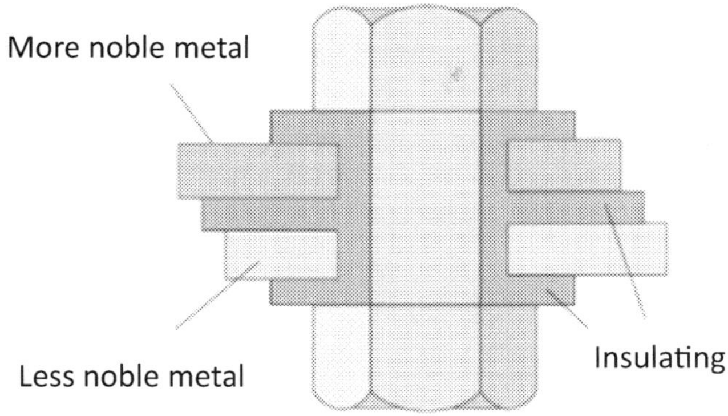

FIGURE 3–6 Insulating flange.

3.5.4 By controlling the size of cathode and anode in dissimilar metals

The smaller the surface area of the anode compared to the cathode, the higher the concentration of electron flow from anode to cathode. As a result, corrosion will spread quickly. As a result, by limiting the contact of small anodes with larger cathodes, dissimilar material corrosion can be prevented.

3.5.5 Metallic and nonmetallic coating

Metal substrates are protected by two types of noble metal coatings. These are noble metal coatings and sacrificial metal coatings. The first category, sacrificial coatings, is employed to prevent cathodic protection from damaging the substrate. The employment of a zinc outer layer over a steel base is the most prevalent example (Fig. 3–7A). In most circumstances, when zinc is coupled to steel, it has a stronger negative electrode potential than steel, therefore zinc becomes the anode once more. When a pinhole or crack appears in the outermost zinc covering, sacrificial corrosion of zinc protects the underlying steel, as can be seen schematically in Fig. 3–7B.

A coating of nickel on steel, as seen in Fig. 3–8A, illustrates the second form of metallic coating, the noble metal coating. The positive electrode potential of nickel makes it a better choice than steel for aquatic applications. When nickel and steel are combined, steel becomes the anode. The nickel coating's protective function is due to its increased corrosion resistance in corrosive environments. When defects occur in the noble metal coating Fig. 3–8B, the steel substrate is corroded, and the attack is concentrated at the base of the defect in the steel substrate. This is not a desirable circumstance because the attack could be extensive and localized. This contrasts with sacrificial coatings, which safeguard the substrate when a breach occurs in the coating and disperse the attack to some extent along with the coating.

These coatings, on the other hand, are not as effective as sacrificial coatings at preventing dissimilar metal corrosion. Painting can be employed as a preventative measure if it is applied to the cathodic areas, which are made of a more noble metal, or to the entire

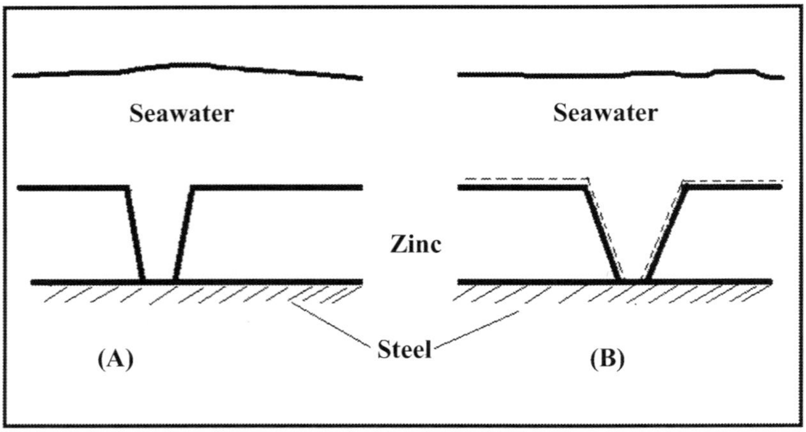

FIGURE. 3–7 A sacrificial protective coating: (A) when a crack or flaw appears in the coating and flows down to the substrate and (B) the zinc coating will galvanically corrode to protect the underlying substrate.

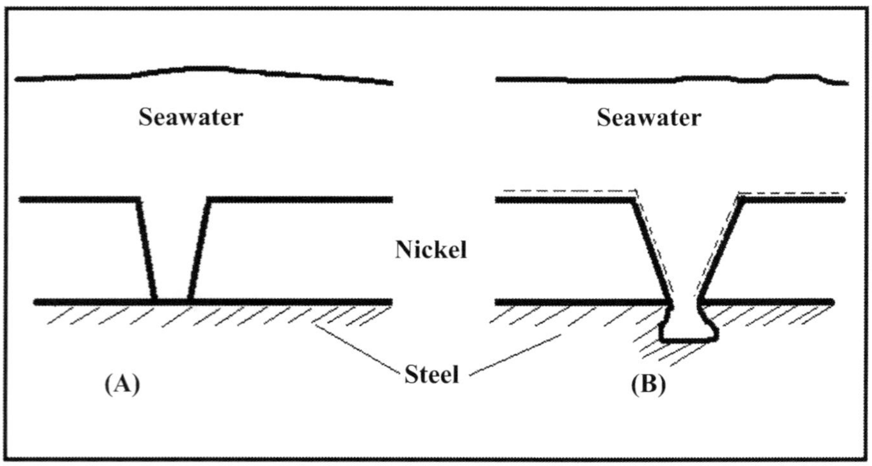

FIGURE 3–8 A noble metal protective coating: (A) when a break or an imperfection develops in the nickel coating and extends down to the substrate and (B) the underlying steel substrate will galvanically corrode.

coupling, which includes both anodic and cathodic portions. Painting only the less noble metal (as is commonly done to preserve the corroding zone) is extremely harmful since the surface area ratio is considered unfavorable in the presence of flaws (Fig. 3–9).

3.5.6 Corrosion inhibitors

In order to control galvanic corrosion, either anodic inhibitors are used to prevent anode corrosion or cathodic inhibitors are used to lower the driving force of the anode. An anodic inhibitor will shift the coupled potential in a cathodic direction, while a cathodic inhibitor will shift in an anodic direction. Steel, copper alloys, and aluminum comprise the components of automobile

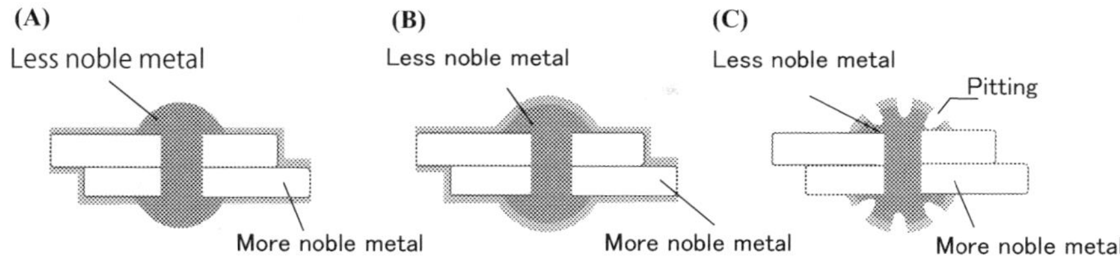

FIGURE 3–9 Example of corrosion prevention by protective coatings: (A) correct; (B) correct; (C) incorrect.

cooling systems, all of which are directly in contact with an electrolyte, typically antifreeze or coolant. This is because the manufacturers of antifreeze/coolants have been able to develop effective corrosion inhibitors that can be used on vehicles every few years.

3.5.7 Countercurrent devices

In the domain of galvanic corrosion control, countercurrent devices are relatively recent. In a galvanic pair, they generate an ohmic potential (IR) drop in the electrolyte that is opposite to the IR drop generated between the anode and cathode. This essentially prohibits current from flowing through the electrolyte, possibly preventing galvanic corrosion. A bioelectrode, for example, has been the subject of study and patents to date. The galvanic couple geometry that can be regulated with this technology must be well managed, and the most difficult aspect of designing such a system is controlling the counter current.

3.6 Advantages of galvanic corrosion

When employed in engineering applications, galvanic reactions between two different metals can provide various advantages. Anodic metallic coatings protect the base material from galvanic corrosion by acting as a sacrificial barrier. To safeguard the base material, the metallic coating makes a compromise. Zinc coating on steel, for example, reduces galvanic corrosion. Even if the sacrificial coating is destroyed, the base material is still protected. Corrosion resistance is determined by the coating material and thickness. In comparison to an 8-μm zinc coating, 12-μm zinc plating will provide higher corrosion resistance.

3.7 Case histories

3.7.1 Case I: buried type 304L stainless steel drain line failure

3.7.1.1 Description of the problem

One of the subterranean drain lines carrying acidic wastewater from a chemical plant broke after a few months of installation (Buried Type 304L Stainless Steel Drain Line by Galvanic Attack, 2019). Each line was made of 304L stainless steel and had a wall thickness of 5 mm.

The various sections of the pipe were joined using shielded metal arc welding. The results of the tests were then analyzed by water back-pressure testing.

3.7.1.2 Identification

A pressure drop was detected in the backpressure test. This issue revealed a failure site.

3.7.1.3 Conclusion

It was concluded that the failure was caused by galvanic corrosion due to the high corrosivity of the soil. This issue had a voltage drop from 1.3 to 1.7 V near the point of failure.

3.7.1.4 Remedy

1. Pipelines should be asphalt coated and enclosed in a concrete trough with a concrete cover.
2. It was suggested that a series of magnesium anodes should be installed at intervals along the line to provide cathodic protection.

3.7.2 Case II: failure of aluminum alloy spacers

3.7.2.1 Description of the problem

The issue was traced to the galvanized steel hot water pipes of a housing society (Failure of Aluminum Alloy Spacers, 2019). The water temperature inside the pipe was about 80°C. The water turned red and stained the hands of the residents.

3.7.2.2 Identification

The metallographic analysis revealed that the pipe's internal surface had been corroded. The analysis also showed that the iron was leaching from the pipe.

3.7.2.3 Conclusion

The amount of iron that was deposited in the water increased overnight, and in the morning, when the tap was opened, a large amount of iron was found in the water and making the water unsuitable for use. Unfortunately, the zinc did not prevent the corrosion of the steel pipe after it became anode. The reason why zinc became the anode was due to the reverse of polarity, which occurs when hot water flows through it.

3.7.2.4 Remedy

1. The temperature of the water in the pipe was regulated to a minimum of 60°C.
2. It was additionally treated with sodium hexametaphosphate prior to entering the house's plumbing system.
3. Copper is typically used for hot water pipelines and tanks to avoid this problem.

Useful Websites

1. https://www.corrosionpedia.com/an-introduction-to-the-galvanic-series-galvanic-compatibility-and-corrosion/2/1403.
2. https://corrosion-doctors.org/Definitions/galvanic-series.htm.

Short questions

1. What are the four main factors affecting galvanic corrosion?
2. What happens during galvanic corrosion?
3. What is the effect of area ratio and distance on the magnitude of galvanic corrosion?
4. State three main differences between the emf series and galvanic series.
5. Explain why:
   a. When steel rivets are placed into a copper plate, the rivets corrode completely after a period of time?
   b. Galvanic corrosion is most severe at the point where the two metals connect?
   c. A new pipe joined to an old pipe corrodes?
   d. Welded joints are better than riveted joints?
   e. A metallic couple with an unfavorable area ratio (small anode/large cathode) corrodes faster than a couple with a large anode and small cathode?
   f. As the distance between the two metals widens, the magnitude of galvanic corrosion reduces?
   g. In the presence of NaCl and humidity, the galvanic attack is accelerated?
   h. When dissimilar metals are to be coated, the nobler must be coated?

References

Butler, G. & Ison, H. C. K. (1966). *Corrosion and its prevention in waters.*

Butler, G., Francis, P. E., & McKie, A. S. (1969). *Corrosion Science, 9.*

Buried Type 304L Stainless Steel Drain Line by Galvanic Attack, (2019). 2019 978-1 10.31399/asm.fach.modes.c0046535 ISBN electronic.

Capp, J. A. (1914). A rational test for metallic protective coatings. *Proceedings American Society for Testing of Materials, 14*, 474–481.

D714 (2002). Standard test method for evaluating degree of blistering of paints. *American Society for Testing of Materials.*

Fabbroni, G. (1799). Sur l'action chimique des diff érents métaux entr'eux, à la temperature commune de l'atmosphère; et sur l'explication de quelques phenomènes galvanique. *Journal de Physique, 49.*

Failure of aluminum alloy spacers by galvanic attack, ASM Failure Analysis Case Histories: Failure Modes and Mechanism (2019). ISBN electronic. 978-1

Hack, H. P. (2016). *Galvanic corrosion.* Elsevier BV. Available from 10.1016/b978-0-12-803581-8.01594-0.

Kenworthy, L. (1943). *Journal of the Institute of Metals, 69.*

Kaesche H. (1985). Metallic corrosion. *National Association of Corrosion Engineers.*

Massinon, D. & Dauchelle D. (1989). GALVATECH'89 585–595 recent progress towards the understanding of underfilm corrosion of coaled steels used in the automotive industry. In: Proceedings of the international conference on zinc and zinc alloy coated steel sheet.

Pryor, M. J., & Keir, D. S. (1958). *Journal of the Electrochemical Society, 105.*

SSPC-VIS (2002). The society for protective coatings standard method of evaluating degree of rusting on painted steel surfaces 2.

Southwell, C. R., Bultman, J. D., & Alexander, A. L. (1976). *Materials Performance, 15.*

Zhang, X. G., & Valeriote, E. M. (1993). *Corrosion Science, 34.*

4 Crevice corrosion

4.1 Occurrence

Crevice corrosion is a type of localized corrosion produced by the presence of cracks, gaps, screens, or deposits on a metal surface. It is also known as interstitial corrosion or corrosion under deposit. Corrosion of this type is generally overlooked and underestimated. Crevice corrosion is considered particularly dangerous, as it has the potential to cause premature component failure. Metals that are relatively resistant to corrosion, such as titanium, stainless steel, and aluminum, are commonly affected. Small gaps or crevices produced by various components such as bolts, threads, and joint heads are frequently targeted by this type of attack. It can also happen when there is a lack of oxygen in a passivating environment. Some examples of crevice corrosion are shown in Fig. 4–1.

Crevice corrosion is a condition that occurs when steel is subjected to seawater or a chloride-containing solution in various industrial processes such as chemical, petrochemical, pharmaceutical, and food processing, as well as biomedical, nuclear, and civil engineering. This can cause equipment to shut down or contaminate a liquid. It can also lead to the formation of stress corrosion cracking.

4.2 Causes

Crevice corrosion can develop in geometrical constraints such as (1) metal-to-metal or nonmetal-to-metal components with microscopic spaces between them; (2) metals with fractures, cavities, or other difficulties; (3) dirt, mud, barnacles, biofouling organisms, and other deposits can cause crevice corrosion.

The main cause of crevice corrosion is the development of tiny fissures or gaps, commonly a few micrometres wide, in which a very restricted solution chemistry is predominant. This chemistry differs significantly from that of the bulk solution, and it may result in a metal attack within the creviced area. Cracks resulting from other localized corrosion processes such as corrosion fatigue and stress corrosion cracking, in addition to naturally produced fissures such as behind gaskets or nuts, may also develop occluded cells and exhibit similar electrocatalytic behavior. Crevice corrosion occurs when electrolyte flow is significantly restricted. This is typically followed by a reduction process, the most common of which is the reduction of oxygen on the outer surface of a passive metal containing a fissure, such as stainless steel. Waterline corrosion, differential aeration corrosion, under-deposit corrosion, and pitting corrosion are all like crevice corrosion. Although oxygen reduction is the most performed reaction due to crevice corrosion, there are also other reduction processes that

Handbook of Corrosion Engineering. DOI: https://doi.org/10.1016/B978-0-323-95185-2.00004-6

FIGURE 4–1 Examples of crevice corrosion.

can be triggered by certain factors. These include water reduction, hydrogen ion reduction, chlorine reduction, etc.

4.3 Theories of crevice corrosion

4.3.1 Critical crevice solution theory

The high concentration of chloride and pH ions in the solution has caused the passive coating to become rupture and begin developing active crevice corrosion. The active stage of crevice corrosion started, with high rates of corrosion occurring within the crevice. The anodic sites are linked to a nearby cathodic site, which could also accelerate the anodic processes. This could result in a faster rate of anions entering the crack and a drop in pH causing structural failure.

4.3.2 IR drop theory

This concept was proposed by Pickering (1986). He noted that an anodic potential would be activated if the difference between the interior and mouth of a crevice is large enough. This led to crevice corrosion occurring abruptly. The size and composition of the solution are the factors that influence the reaction kinetics of an IR drop. In addition, the concentration gradients and geometry of the solution can also affect its size. The effect of the IR drop on reaction kinetics is depicted in Fig. 4–2 which showed no effect of IR drop on the cathodic polarization curve (Kennell, Evitts, & Heppner, 2008). This curve only cuts the anodic polarisation curve within the passive zone, resulting in a very low corrosion current. It is shown that the cathodic polarisation curve with IR effects cuts the anodic polarisation curve below the critical potential in the active region, leading to noticeably higher corrosion current magnitudes.

4.3.3 Modeling studies for crevice corrosion

The existence of mathematical models can help us understand the various steps involved in the formation and prediction of crevice corrosion. For instance, the critical crevice solution

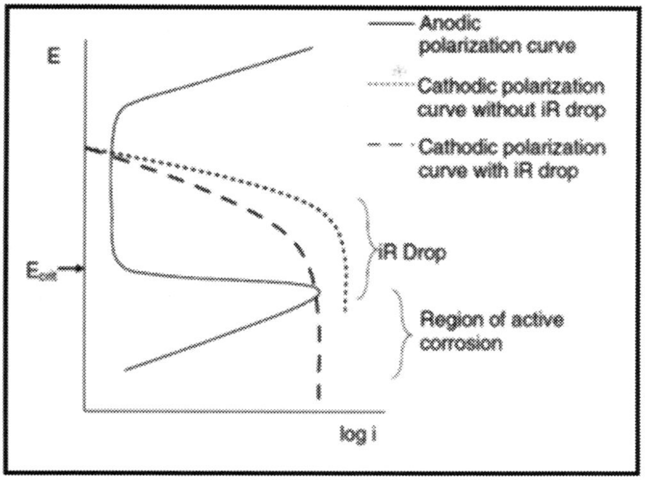

FIGURE 4–2 Idealized kinetic corrosion diagram for a passive metal. *Republished from Kennell G.F, Evitts, R.W., Heppner, K.L. (2008). A critical crevice solution and IR drop crevice corrosion model. Corrosion Science. 50 (6) 1716–1725. https://doi.org/10.1016/j.corsci.2008.02.020.*

theory (CCST) model developed by Oldfield and Sutton (1978) predicted the onset of the incubation period for the corrosion process for stainless steel. The authors of the study assumed that the mass transmission was limited to the bulk fluid and that the crevice did not have concentration gradients. They assumed that the mass transmission did not occur within the crevice. Galvele (1976, 1981) and Gravano and Galvele (1984) then developed three computational models of pitting corrosion, the last of which considered diffusion, migration, and electroneutrality. According to scientists, the pH within the pit was significantly lower than the surrounding environment, with the greatest dip occurring near the pit mouth. In a moderately concentrated solution, Walton, Cragnolino, and Kalandros (1996) employed an IR drop theory model to account for diffusion and electromigration. The simulation, which was run on stainless steel, assumed that the current was passive and unaffected by the crevice's potential. The results for AISI 304 stainless steel were quite like experimental findings of Alavi and Cottis (1987). Watson and Postlethwaite (1990, 1991) solved Poisson's equation for electroneutrality and established a CCST model for describing the incubation period of a crevice with few empirical restrictions. Heppner, Evitts, and Postlethwaite (2002a, b) developed this idea further to create a generic model for multiple metals and electrolytes. The model precisely described the chemistry within the crevice.

4.4 Mechanism

One of the most widely recognized mathematical models is that published by Oldfield and Sutton (1978) for stainless steel in neutral media. According to this, mathematical model mechanism of crevice corrosion can be split into two sections: initiation and propagation (Fig. 4–3). According to Oldfield and Sutton, the initiation phase is separated further into

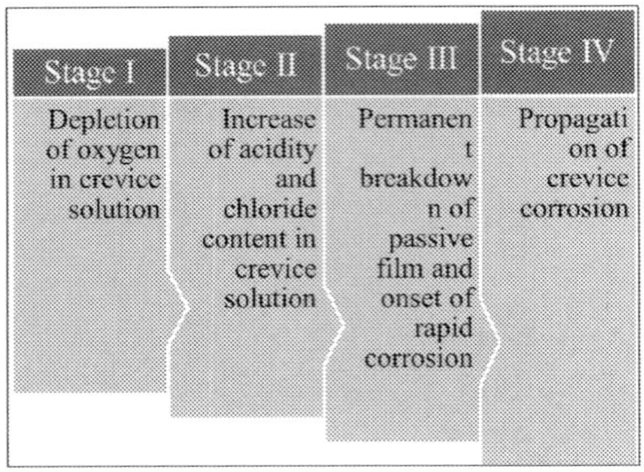

FIGURE 4–3 Steps involved in the mechanism of crevice corrosion.

three phases: deoxygenation, hydrolysis-acidification, and activation (formation of a critical crevice solution).

4.4.1 Deoxygenation of crevice

The quantities of soluble oxygen inside and outside the fissure are initially identical, and anodic/cathodic reactions occur on all metal surfaces equally. Cathodic processes are unable to continue because oxygen transport in the crevice is hindered, resulting in oxygen depletion within the crevice. The uneven distribution of oxygen between the inside (crevice) and external bulk portion forms a differential aeration cell. Oldfield and Sutton uncovered several crucial parameters influencing the course of this deoxygenation phase in their modeling treatment, including passive current density, oxygen concentration, and crevice shape. Because of the oxygen depletion within the crevice, oxygen reduction is mainly limited to the stainless-steel exterior surfaces, while the anodic breakdown of the alloy elements occurs within the boundaries of the crevice.

4.4.2 pH decrease/acidification in the crevice

The reduction in oxygen in the fissure causes the metal's surface to become overpopulated with positive ions. The presence of positive ions on the surface of the metal causes the microenvironmental conditions within the crevice to become anodic. This could result in the migration of chloride ions into the area to balance the anodic charge. The chloride ions then attack the surface and interact with the positive ions to form acids, which are very harmful to the metal. This process is commonly referred to as autocatalytic. The dissolution triggers the formation of new acids, which can then interact with the remaining chloride ions to generate more corrosion. The hydrolysis of the metal ions generated while dissolving will lower the pH of the crevice solution. Passivation is prevented by reducing the pH to extremely low levels. The oxygen reduction helps in promoting passivity by increasing the alkalinity outside the crevice. This step is influenced by the alloy's bulk solution components such as pH and chloride concentration, the crevice depth (longer depths allow for larger concentration gradients), and the alloy's chromium content (Peterson, Lennox, & Groover, 1970).

4.4.3 Passive layer breakdown

The third stage of the deoxygenation–acidification mechanism happens when a sufficiently aggressive crevice environment, that is, low pH, high chloride concentration is achieved, causing the protective passive coating to permanently collapse for rapid corrosion to begin. The concept of an initiation or incubation time, which is generated by the creation of such crucial crevice solution chemistry, is one of the main implications of this mechanism. The pH must be extremely low, and the chloride concentration must be extremely high, to cause a film breakdown in stainless steels with relatively high quantities of nickel and chromium. At this point, the initial phase comes to an end.

4.4.4 Propagation of the crevice corrosion

After the passive layer has been removed, the metal substrate is exposed to the acidic solution within the crevice. The removal of the passive layer has allowed the corrosion process to continue without interruption.

4.5 Factors affecting crevice corrosion

Crevice corrosion is influenced by several factors, as shown in Fig. 4–4. The following are a few of them:

4.5.1 The type of crevice

Whether the crevice is between metal and metal, metal and nonmetal, or has marine growth on the metal surface, such as barnacles or other marine biofouling organisms, is determined by the crevice type. It is critical to determine whether the components affecting crevices are man-made or natural before choosing effective preventative strategies.

4.5.2 The geometry of the crevice

The geometry of a crevice, which comprises the gap's size, depth, and surface roughness, determines its size. One of the most important components is the crevice gap. Corrosion is more aggressive when the gap is large enough to allow the solution to soak in but small enough to keep the solution stagnant in the fissure. A gap that is either too large or too little will not provide the best conditions for corrosion to start. Fig. 4–5 illustrates this. As the material grows more resistant, the critical gap size, below which crevice corrosion is possible under the actual situation, reduces (Bardal, 2004).

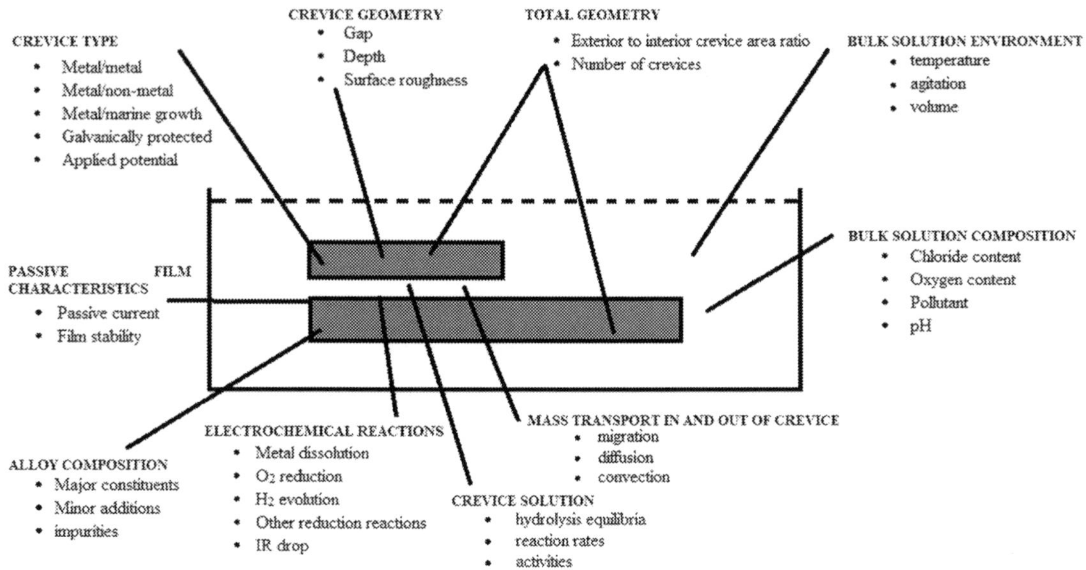

FIGURE 4–4 Factors affecting crevice corrosion. *Redrawn from Oldfield J.W, Sutton, W.H. (1978). Crevice corrosion of stainless steels: I. A mathematical model. British Corrosion Journal. 13 (1) 13–22. https://doi.org/10.1179/000705978798358671.*

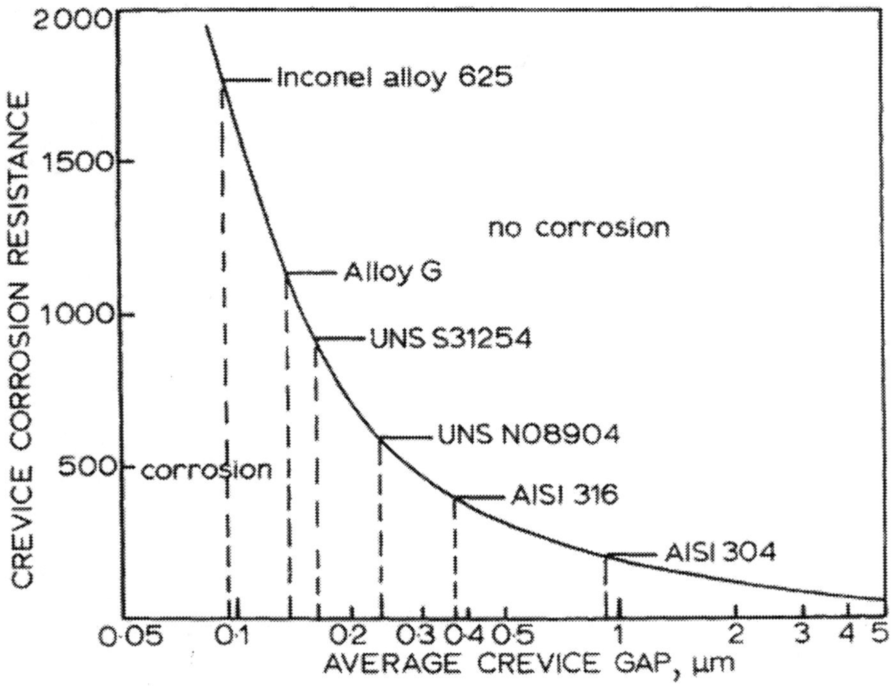

FIGURE 4–5 Prediction of whether or not crevice corrosion will occur in crevices 5 mm deep in seawater at ambient temperature as a function of crevice gap. *Redrawn from Corrosion and Protection, n.d.*

4.5.3 The composition of the metal

It is critical to figure out if the alloy will corrode in crevices. For example, studies using SS 304, carpenter alloy, Incoloy (Alloy 825), Hastelloy (Alloy G), and Inconel (625) indicated that the latter two alloys were exceptionally resistant to crevice corrosion in seawater at both ambient and elevated temperatures. Cr, Mo, iron, and nickel can be added to steels to make them more resistant to crevice corrosion. Crevice corrosion is uncommon with 6% molybdenum. If chromium causes the acidity in the crevice to rise owing to hydrolysis, it can be dangerous. A higher chromium concentration improves the stability of the passive film, lowers the pH below the critical threshold for fissure development, and reduces the passive current density. As a result, the rate of chromium ions entering the crevice decreases, as does hydrolysis, which causes acidic conditions in the crevice.

4.5.4 The environment

pH, halide ions, temperature, and oxygen are all factors to consider. Outside the crevice, the more aggressive the liquid, the more likely the crevice will be attacked. As a result, crevice attacks may occur in a saltwater pool but not in a freshwater tank. Crevices near the sea

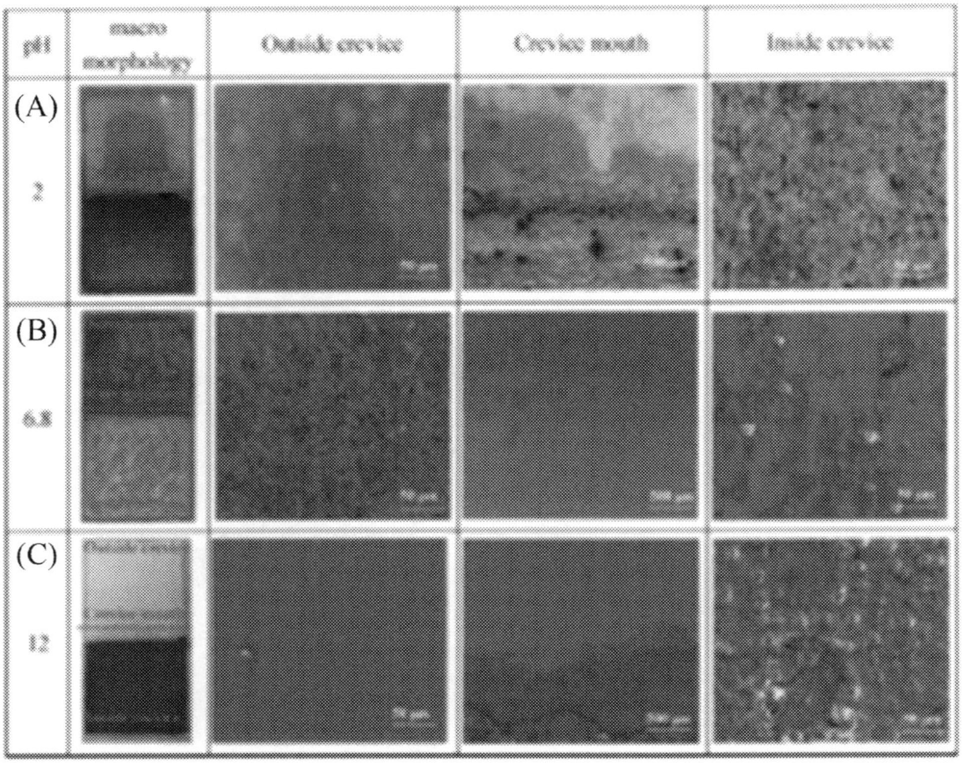

FIGURE 4–6 The micro- and micromorphology of crevice specimen after corrosion for 72 h in the 0.34 mol/L NaCl with different pH solution: (A) pH = 2, (B) pH = 6.8, (C) pH = 12. *Redrawn from Mu J, Li, Y. Z., Wang X. (2021). Crevice corrosion behavior of X70 steel in NaCl solution with different pH. Corrosion Science. 182.*

cause more issues in the atmosphere than in rural areas. Crevice attacks will be more severe if the liquid outside the crevice is highly oxidizing, such as bleach, hydrogen peroxide, or ozone. The macro- and micromorphology of an X70 steel crevice specimen after 72 hours in a 0.34 mol/L NaCl solution with varied pH is shown in Fig. 4–6 (Mu, Li, & Wang, 2021). The specimen inside the crack suffers from minor corrosion in the solution with pH = 6.8, while the solution with pH = 2 and pH = 12 exhibits significant corrosion.

The effect of chlorine accumulation on crevice corrosion is also a key element to consider. The rate of corrosion increases as more Cl^- enters the crevice, and the concentration of chlorine in the shelter rises (2004).

4.5.5 Temperature

The critical crevice temperature (CCT) is a measure that can be used to evaluate a material's resistance against crevice corrosion. It is commonly done in line with the ASTM standards G48-03. It can also be used to rank a material's resistance to pitting corrosion. The CCT is generally lower than the critical pitting temperature (CPT), which can trigger faster corrosion. When the CCT gets surpassed, it can result in faster corrosion like pitting corrosion. The thumb rule is that if the temperature increases by 10°C, the rate of corrosion doubles. This is because the increasing temperature causes the oxygen solubility to decrease. When the temperature of seawater rises to 70°C, the rate of crevice corrosion propagation of stainless-steel alloys such as types 304 and 316 is reduced (Anderson, 1978). This phenomenon is caused by the decrease in oxygen solubility and changes in the physical properties of the steel, that is, the formation of the passive layer.

4.5.6 Bulk solution composition

The aggressive behavior of the electrolyte can cause a film to break off on a metal surface. All chloride-based solutions are quite aggressive and contribute to crevice corrosion. Seawater and brackish water are extremely aggressive, causing the steel to corrode in crevices.

4.5.7 Mass transfer in and out of crevice

The three mass transfer phases are migration, diffusion, and convection. Migration and diffusion carry the majority of the current. The concentration of chloride can greatly affect the start of crevice corrosion. The concentration of chloride that is concentrated in a small crevice can also increase the aggressiveness of the solution's electrolyte. The solution at a stage becomes highly acidic and the pH is lowered.

4.5.8 Conductivity

It controls the size of the caves that form around the gap; the higher the conductivity, the larger the caverns become because the corrosion process can employ a larger cathodic surface. Aside from conductivity, the corrosion process includes transport phenomena that promote ion penetration inside the interstice, which can lead to self-stimulation and metal dissolution. Unlike pitting, where the size of the pit controls the corrosion process, the simple presence of the gap is a required and sufficient requirement for establishing the "occluded cell's" self-stimulating mechanism.

4.5.9 Agitation

The exact design and size of the gap can also affect the role of the solution. For instance, if the gap is designed to allow oxygen passage, the agitation can improve the crevice corrosion resistance. If, on the other hand, solution agitation enhances the oxygen cathode reaction, this factor has a detrimental impact on crevice corrosion resistance.

4.5.10 Oxygen

The nature of the passive film on the metal surface has a considerable influence on the development of crevice corrosion. Crevice corrosion is prevented if the passive film is very stable. The development of a passive layer on a metal surface necessitates the presence of oxygen. As a result, it has a big influence on when crevice corrosion starts. As oxygen is consumed within the crevice, a possible difference between creviced and boldly exposed regions of the crevice that are exposed to oxygen arises. As a result, a concentration disparity is created, causing oxygen to accelerate. However, as the temperature rises, the oxygen's solubility drops, which can slow down the crevice formation. The link between temperature and oxygen is unclear. In thermal insulation, water and oxygen can get into the crevices of the plant.

4.5.11 Stagnancy of the solution in the crevice

The faster the influx of chloride ions and the rate of corrosion attack, the more stagnant the solution in the crevice becomes.

4.5.12 Crevice critical gap size

The crevice critical gap size (CCGS) is an important parameter in crevice corrosion, defined as the smallest size that permits the aggressive environment to penetrate the interstice while preventing oxygen diffusion. The critical gap size varies depending on the metal composition and ranges from 0.1 m to 0.1 mm. With a crevice depth of 5 mm, the critical crevice gap size for stainless steel can be calculated using the following equation based on data from Oldfield and Sutton (1978):

$$\text{CGCS} = (17/\text{PREN})^2 \tag{4.1}$$

where CCGS is in mm and Pitting Resistance Equivalent Number (PREN) is the PREN, which is calculated from stainless steel composition. This is an empirical concept that is defined as

$$\text{PREN} = \%\ \text{Cr} + 3.3\%\ \text{Mo} + 30\%\ \text{N} \tag{4.2}$$

As a result of the PREN values, a rating system based on alloying elements known to have a substantial effect on pitting/crevice corrosion was developed. The greater the PREN value, the more resistant the alloy is against pitting and crevice corrosion. When the gap size is smaller than the crucial one, crevice corrosion occurs. The higher the likelihood of crevice formation, the narrower and deeper the gap.

4.6 Evaluation of a material's resistance to crevice corrosion

The CCT is evaluated by following the ASTM Standard G48-03: Standard Test Methods for Pitting and Crevice Corrosion of Stainless Steels and Alloys by Use of FeCl3 to evaluate and rank a material's resistance to crevice corrosion. The CCT is the minimal temperature (°C) to

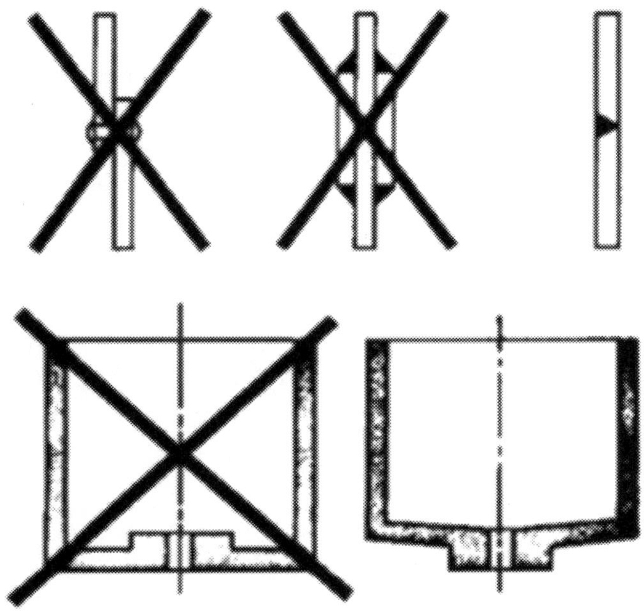

FIGURE 4–7 Examples of bad and good design. *Redrawn from Prevention of Corrosion, n.d.*

initiate a crevice attack and CCT is usually lower than the CPT. For a particular temperature and chloride concentration, computer software such as CRA-Compass can be used to analyze an alloy's crevice corrosion resistance.

4.7 Prevention of crevice corrosion

Crevice corrosion can be avoided by using good design, manufacturing, and operating techniques.

4.7.1 Design

To minimize the occurrence of crevices, maintain as much space as feasible between junction locations, avoid flanged or bolted connections with "designed-in fissures." To avoid crevices forming between the two materials, ensure that nonmetallic connections or gaskets are properly fitted and adhered. Joints are best served by full-penetration butt welds. Gaps between pipes and fittings should be avoided. Bolted connections and other fasteners should be used as little as possible. A couple of examples are given in Fig. 4–7 (Prevention of Corrosion, n.d.). Instead of overlap joints, use butt welds. Continuous welds can also be used to seal overlap junctions. When welding, employ caution when using backing gas and double-check that the penetration is complete. After welding stainless steel, remove the less protective and resistant surface layers by pickling, brushing, blasting, or grinding. Design in a way that allows for proper drainage and avoids deposits from forming. Make it simple to check and clean.

4.7.2 Fabrication

With a smooth weld bead, make sure to maintain the full root penetration of all your weld connections. When welding, avoid cracks and undercuts and avoid using sticky labels or markers. Also, make sure to keep it clean and smooth. It is also important to avoid staining and grease or oil streaks. Crevice corrosion expertise is tested by ASSDA Accredited Fabricators.

4.7.3 Operation

Crevices can be caused by various factors such as scale or sediment. Having a regular cleaning and maintenance program can help prevent these issues. Having a proper circulation system can also help prevent the formation of bacterium film. Doing so can also help minimize the risk of materials getting collected in dead legs or low-flow zones and therefore can also help to prevent the formation of fissures.

4.7.4 The chemical composition

The composition of an alloy can also affect its crevice corrosion resistance. To improve the corrosion resistance of steel, various alloying metals such as Ni, Cu, and Mo are added to steel. High molybdenum content, that is, 2%–4% can also help improve the resistance of various steels to crevice corrosion (Hibner & Fende, 1999; Practical Guidelines for Fabrication of Stainless Steel, 1999). Mo primarily increases the corrosion resistance and temperature strength. This improvement is due to the shifting of metal potential toward a more negative side, that is, nobler side. Stainless steels with a high molybdenum content, such as the 6% molybdenum austenitic and super duplex grades, are expected to have the best resistance against crevice corrosion attacks.

4.7.5 Deposition-prevention measures

1. During and between service periods, inspection, and cleaning.
2. In process plants, separation of solid material from flowing medium.
3. Filling up underground pipelines and infrastructure with gravel.

4.8 Failure case histories

4.8.1 Case I: penetration of AISI 316 steel pipe walls of thickness 3–4 mm occurred after 2–3 months of installation

Equipment

150–400 mm diameter AISI 316 steel pipe.

Environment

Seawater.

Conditions

The flow velocity £1.2 m/s and the water temperature in the range 7°C–30°C.

Observation and remarks

The most common cause of these failures was the failure of the weld. They were mainly responsible for the appearance of small gray-black bulges on the pipe wall. The diameters of the holes varied from 1 to 5 mm. The issue was caused by the improper usage of gas. The low flow velocity during the welding process allowed the biofilm to form and cause the bulges to appear. The lack of backing gas causes the weld zone to oxidize, which lowers the Cr concentration in the material and its corrosion resistance. This is revealed by the color pattern on the pipe surface at the welds (dark brown–blue regions).

Remedy

After calculating ohmic potential dips, sacrificial anodes made of zinc alloy were installed to discover optimal sites for the anodes. The plant functioned adequately if the anodes were replaced after 12–14 months. Using higher-alloyed stainless steel as an alternative to cathodic protection is an option.

4.8.2 Case II: development of leaks at the rolled joint of the pipe and pipe bottom in a heat exchanger

Equipment

25/30 mm diameter pipe of 35–29 stainless steel.

Environment

Oil.

Conditions

The pipe was heated externally from 90°C to 170°C by superheated steam of 8–10 atm.

Observation and remarks

The surface of the pipe was pitted around the rolled joint, which was close to the steam chamber. Crevice corrosion was the cause of the attack. Due to the lack of tightness in the joint, steam condensate entered the gap. The lack of oxygen in the fissure caused it to form a differential aeration cell.

Remedy

To prevent further damage to the plant, an alternative metal such as Inconel 625 or Alloy 20 Cr-25 Ni-4.5 Mo-1.5 Cu could be used. And minimize the gap by a change in the design of the rolled joint.

Useful websites

1. https://www.materials.sandvik/en/materials-center/corrosion/wet-corrosion/crevice-corrosion/
2. https://www.corrosionclinic.com/types_of_corrosion/crevice_corrosion.htm
3. http://www.corrosion-doctors.org/Forms-crevice/Crevice.htm
4. https://www.corrosionpedia.com/whats-the-inside-scoop-on-crevice-corrosion/2/6588
5. https://www.nitty-gritty.it/en/crevice-corrosion/
6. https://www.ampp.org/resources/what-is-corrosion/forms-of-corrosion#crevice
7. https://feoinc.com/learning-center/recognizing-and-preventing-crevice-corrosion

Short questions

1. Why does crevice corrosion occur?
2. What are the three important metallurgical factors which affect crevice corrosion? Explain the reasons.
3. Explain the mechanism of crevice corrosion.
4. How is oxygen depleted in the crevice?
5. How the crevice corrosion can be prevented?
6. What design feature will reduce crevice corrosion?
7. What is the incubation period associated with a crevice attack?
8. Name a metal most susceptible to crevice corrosion.

References

Alavi, A., & Cottis, R. A. (1987). The determination of pH, potential and chloride concentration in corroding crevices on 304 stainless steel and 7475 aluminium alloy. *Corrosion Science, 27*(5), 443–451. Available from https://doi.org/10.1016/0010-938X(87)90088-6.

Anderson D.B., (1978). Statistical aspects of crevice corrosion. Paper TL-26.

Bardal, E. (2004). Corrosion and protection. Springer Verlag London Ltd.

Galvele, J. R. (1976). Transport processes and the mechanism of pitting of metals. *Journal of the Electrochemical Society, 123*(4), 464–474. Available from https://doi.org/10.1149/1.2132857.

Galvele, J. R. (1981). Transport processes in passivity breakdown-II. Full hydrolysis of the metal ions. *Corrosion Science, 21*(8), 551–579. Available from https://doi.org/10.1016/0010-938X(81)90009-3.

Gravano, S. M., & Galvele, J. R. (1984). Transport processes in passivity breakdown-III. Full hydrolysis plus ion migration plus buffers. *Corrosion Science, 24*(6), 517–534. Available from https://doi.org/10.1016/0010-938X(84)90034-9.

Heppner, K. L., Evitts, R. W., & Postlethwaite, J. (2002a). Prediction of the crevice corrosion incubation period of passive metals at elevated temperatures: Part I - Mathematical model. *Canadian Journal of Chemical Engineering, 80*(5), 849–856. Available from https://doi.org/10.1002/cjce.5450800508, http://onlinelibrary.wiley.com/journal/10.1002/(ISSN)1939-019X.

Heppner, K. L., Evitts, R. W., & Postlethwaite, J. (2002b). Prediction of the crevice corrosion incubation period of passive metals at elevated temperatures: Part II - Model verification and simulation. *Canadian Journal of Chemical Engineering, 80*(5), 857–864. Available from https://doi.org/10.1002/cjce.5450800509, http://onlinelibrary.wiley.com/journal/10.1002/(ISSN)1939-019X.

Hibner E. L. & Fende D. S. (1999). Highly resistant 6% molybdenum austenitic stainless steel for desalination and chemical processes applications. In *Proceedings of the IDA congress on desalination and water re-use. I.*

Kennell, G. F., Evitts, R. W., & Heppner, K. L. (2008). A critical crevice solution and IR drop crevice corrosion model. *Corrosion Science, 50*(6), 1716–1725. Available from https://doi.org/10.1016/j.corsci.2008.02.020.

Mu, J., Li, Y. Z., & Wang, X. (2021). Crevice corrosion behaviour of X70 steel in NaCl solution with different pH. *Corrosion Science, 182.*

Oldfield, J. W., & Sutton, W. H. (1978). Crevice corrosion of stainless steels: I. A mathematical model. *British Corrosion Journal, 13*(1), 13–22. Available from https://doi.org/10.1179/000705978798358671.

Peterson, M. H., Lennox, T. J., & Groover, R. E. (1970). *Material Protection, 9.*

Pickering, H. W. (1986). On the roles of corrosion products in local cell processes. *Corrosion, 42*(3), 125–140. Available from https://doi.org/10.5006/1.3584892.

Practical Guidelines for Fabrication of Stainless Steel. International Molybdenum Association. (1999).

Walton, J. C., Cragnolino, G., & Kalandros, S. K. (1996). A numerical model of crevice corrosion for passive and active metals. *Corrosion Science, 38*(1), 1–18. Available from https://doi.org/10.1016/0010-938X(96)00107-2, https://www.journals.elsevier.com/corrosion-science.

Watson, M., & Postlethwaite, J. (1990). Numerical simulation of crevice corrosion of stainless steels and nickel alloys in chloride solutions. *Corrosion, 46*(7), 522–530. Available from https://doi.org/10.5006/1.3585142.

Watson, M. K., & Postlethwaite, J. (1991). Numerical simulation of crevice corrosion: The effect of the crevice gap profile. *Corrosion Science, 32*(11), 1253–1262. Available from https://doi.org/10.1016/0010-938X(91)90136-D.

5

Pitting corrosion

5.1 Conditions, characteristics, and occurrence

One of the most dangerous forms of corrosion is pitting. This occurs when cavities are formed in a structure that is subjected to stress corrosion (Zhu, Yan, Qiao, & Volinsky, 2013). This condition can lead to the failure of steel and iron hydraulic structures. It can also cause the formation of cavities in a structure, which can lead to the failure of the water pipe. Even though the structural integrity of the structure is intact, pitting can still cause it to fail due to its effects on the remaining sound materials. The term pitting is often used to refer a localized corrosion attack, but it should also be used more properly for the type of corrosion that occurs in chloride-contaminated environments (Pradhan, Bhuyan, & Mandal, 2019). Examples of pitting corrosion are given in Fig. 5–1.

In addition, the image of the pitting area taken from the scanning electron microscope can be seen in Fig. 5–2.

When the protective layer on a material's surface fails due to the interaction of two different metals or concentration cells, a pit forms. This contact creates a potential difference, which allows an electric current to travel from the metallic anode to a nearby cathode through the water or over moist steel. The cathode can be brass or copper, mill scale, or any

FIGURE 5–1 Examples of pitting corrosion.

Handbook of Corrosion Engineering. DOI: https://doi.org/10.1016/B978-0-323-95185-2.00005-8

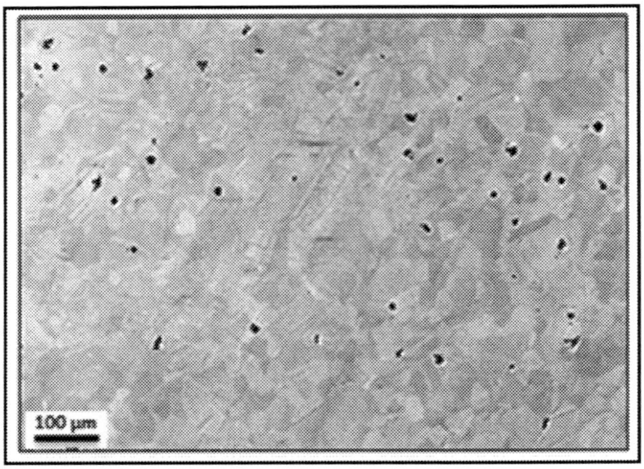

FIGURE 5–2 Surface of a pitted specimen (scanning electron microscopy, 304-SA austenitic stainless steel exposed to a 3.5 wt.% NaCl solution.

other piece of a metal surface that is cathodic to the more active metal sections. The damage is spread out and usually small when the anodic zone is greater than the cathodic area. The metal loss is concentrated and can be significant when the anodic area is tiny. It is possible to predict breakdowns in the continuity of the protective material when significant areas of the surface are typically covered in mill scale, applied coatings, or other types of deposits (Papavinasam). Unscrupulous elements in the chemical or physical structure of the metal can cause pitting on bare metal surfaces. Also, different soil conditions can cause pitting on steel.

The most common causes of pitting corrosion are cracks in the protective coating, scuffs and minor chips, and nonuniform stress. Other factors such as a defective metal substrate, turbulent fluid flow, and a chemical attack can also cause the damage.

Among the metals that are prone to pitting corrosion are copper, aluminum, chromium, stainless steel, passive iron, mercury, cobalt, and various other metals alloys.

5.2 Types of pitting corrosion

According to the international standard ASTM G46, the term pit is usually employed to describe an area that is shallow, deep, or narrow (https://www.ampp.org/resources/impact/corrosion-basics/group-1/pitting-corrosion). It can also be described as an elliptical, transverse, sub-skin, vertical, or horizontal, as depicted in Fig. 5–3.

The shape of the pit can vary depending on the type of material that is affected and the direction of its grain.

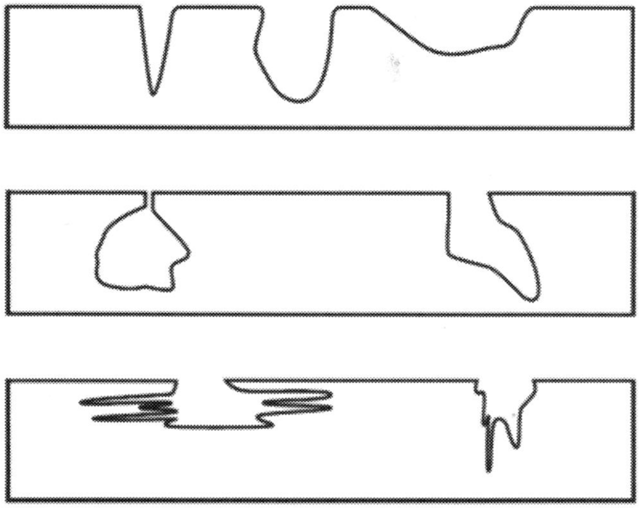

FIGURE 5–3 Typical forms of pitting attack.

5.2.1 Trough pitting corrosion

Hemispherical, cup-shaped, or irregular shaped trough pits are common. These troughs are usually formed when the protective film of a metal is compromised, causing its flat walls to crack and expose its crystal structure. Due to their shape, these troughs can quickly alter a material's thickness such as a roof sheet, truss, or gutter component (Akpanyung & Loto, 2019).

5.2.2 Sideway pitting corrosion

Sideway pits develop in horizontal grain attack, undercutting, and subsurface shapes and are covered with a semi-permeable corrosion product. Sideway pitting corrosion can swiftly corrode the metal. It is hard to spot the damage caused by this type of corrosion since the metal's surface appears to be intact.

5.3 Methods to examine pitting corrosion

1. *A visual examination* is the first step in determining the exact size and distribution of the pits. Using a microscope, count the number of pits over a defined surface area such as 20 cm^2 on the metal surface to get an estimate of the size and distribution of the pits. The quantity of pits is not the most crucial element in determining the size and distribution of the pit. However, the depth of the pit is the most dangerous aspect of pitting. One narrow deep pit can be more dangerous than many shallow pits. A metallographic cross-section is also used to measure the depth of the pit.

2. *Sonics testing* is a type of test that uses ultrasonic pulses to generate and reflect waves on the metal surface. This test involves the transmission of sound energy through an oil or water couplant onto a metal surface. The resulting waves reflect and convert the sound energy into the electrical signals, which can be used to identify the locations of various flaws and pits in the metal.
3. *Electromagnetic testing* is a type of test that uses electromagnetic fields to detect defects in the structure of materials such as steel and iron. The goal of this type of test is to detect defects in the structure of materials that are electrically conducting. When these materials have defects, their magnetic field will vary from the reference material without defects.
4. *Electrochemical testing*, such as cyclic polarization and potentiostatic tests, is also an option to measure pitting in various metals. These tests can be performed in short bursts and provide instant results.
5. *Another method* is to perform an immersion test. This involves immersing a metal sample in a solution for a couple of days. After that, it can be removed, and its corrosion rates can be analyzed. One can observe the pits and pit depth under a microscope and can make the necessary calculations.

There are various standards that specify the conditions under which a steel sample should be subjected to an electrochemical test and the most popular ones are ASTM G61 Cyclic Polarization Test (https://www.astm.org/g0061-86r18.html), ASTM G150 (https://www.astm.org/g0150-18.html), and Electrochemical pitting temperature test.

5.4 Factors affecting pitting corrosion

pH and chloride concentration: The formation and propagation of a pit are influenced by the chemical environment around it. Anions such as chlorides and bromides can contribute to the formation and propagation of aggressive pits (Virtanen, Kobayashi, & Boehni, 2001). In oil field applications, the chloride ion is commonly used. It can bind to cations and enhance their solubility in water. This can cause the passivated layer to break down (Enerhaug, Grong, & Steinsmo, 2001). Due to its small size, the chloride ions can easily go to anodic sites. The reduction in the pH level around the pit due to the cations' hydrolysis can cause it to increase. The pitting resistance and pitting potential of a material can increase with the increasing pH. This occurs when the concentration of chloride drops.

Flow velocity: An aggressive environment is present within a pit when pitting is initiated, and this is accompanied by an increase in the pH level at the adjacent free surface. This allows for the formation of new pits around the previous one. If the liquid does not move, large pits will form. Liquid flow offers a better chance of washing away the pit's aggressive environment while simultaneously improving oxygen transmission to the active area, allowing the pit to be passivated before it grows to a major size. Pit initiation is also easier in the vicinity of an active pit because the alkaline layer is washed away to a higher extent. Pitting corrosion is less severe because of the greater flow velocity because smaller but more numerous pits emerge (Papavinasam, 2013).

The gravity force: Horizontal top surfaces are frequently deeply pitted, while underneath surfaces are either firm or unaffected. In terms of pitting, vertical surfaces are in the middle. The reason for this is that the pits' hostile environment has a higher density than clean water.

Cu^{2+} and Fe^{3+} ions: Because they lead to enhanced potential, they prefer pit initiation and expedite pit expansion. Because Cu is precipitated on the material surface and produces effective cathodes, the Cu ions have a dual effect.

Metallurgical properties: Impurities and inclusions, such as the AlFe secondary phase, play a crucial role in the localization of pits on aluminum.

1. The oxide film on top of and around the inclusion is weak, thin, and stressed,
2. These inclusions function as efficient cathodes. This form of pitting normally only lasts until the pits become so large that the inclusions fall out.

The insulating ability of the oxide: The surface is inactive as a cathode if the oxide insulates well. This is true to some extent for aluminum in seawater, which is the direct cause of pit growth being rather sluggish in this scenario.

Surface roughness: The general trend is for smooth surfaces to have a few large pits and rough surfaces to have a lot of smaller pits.

Temperature: Diffusion and convection are aided by the temperature. As the temperature rises, the pitting potential decreases, and the risk of pitting corrosion rises (Brigham & Tazar, 1973).

Alloy composition: The structure and composition of an alloy also affect its pitting tendency. The presence of certain alloying elements can also help to reduce the pitting susceptibility of an alloy. For instance, the presence of chromium can enhance the passivation of a particular type of steel. In addition, the presence of certain elements such as Mo, nitrogen, and tungsten can also improve the pitting resistance of stainless steel.

5.5 Theories of pitting corrosion

It may be broadly classified into classical and nonclassical theories.

5.5.1 Classical theory

According to this theory, the process involves three phases (Pradhan et al., 2019):

1. *A passive film formation on the metal surface*: Bilayers are usually composed of a compact interior layer and an outer layer made up of anions and/or cations. Passivity can still be observed in the film even though the outer layer is unstable. In the absence of an outer precipitated phase, the interior layer exhibits passivity.
2. *The pits formation at localized regions on the metal surface where films or surface layers break down*: There are a lot of discussions right now about how pitting corrosion starts. One approach (a priori) emphasizes the metal surface's inherent microscopic defects,

such as inclusions, grain boundaries, scratches, and so on, whereas the other acknowledges that a nonuniformity on the metal surface and its development to visible dimensions occurs after a passive metal is exposed to corrosive medium (a posteriori). The prior method assumes heterogeneities on the passive metal, whereas the posterior method assumes induced heterogeneity at the passive metal-corrosive fluid interface. The importance of stochastic or fluctuation processes as the first phase of pitting has been emphasized in relation to the posteriori assumption. The results of studies conducted on the initial pit initiation revealed that the dispersion of the material was higher than that of other types of corrosion. Both stochastic and deterministic techniques have their advantages, and they are not mutually exclusive.

3. *Pit propagation*: The formation of stable and continually growing pits is the final step in pitting corrosion. It occurs when the conditions and processes that lead to this happen. This process is like those that can develop in crevice corrosion.

5.5.2 Nonclassical model

The nonclassical pitting corrosion theory can explain localized pitting corrosion in carbon steel. According to this hypothesis, the primary passive layer associated with corrosion-resistant alloys (CRA), classical theory of pitting corrosion does not exist, and surface layers are created because of carbon steel corrosion in each environment. These surface layers can be thought of as the precipitated outer layers in the standard pitting corrosion model.

1. The metal will undergo uniform corrosion until the surface layer is formed.
2. As surface layer(s) form, the corrosion rate decreases.
3. The protection of metal beneath the surface layer occurs if the surface layer is completely covered. This method can reduce the rate of corrosion to around 4 mpy (0.1 mm/y).
4. If the surface layers are destroyed or removed, accelerated corrosion will occur. This process is referred to as pit initiation. Once the additional layers are formed, the rate of corrosion decreases and reaches its lowest point in Step 3. This process may be known as repassivation or reformation of the surface layer.
5. The repassivation and pit initiation processes will continue until the pit reaches its critical depth. At this point (pit propagation phase), the repassivation phase will stop, and the pit will continue to grow. However, the rate of growth will likely outpace that of Step 3.

The various factors that affect the formation of the layers are also considered to determine their various characteristics. These include the flow rates; composition; pressure of the gas, oil, and brine; microbial activity; and operation time. The morphology and types of carbon steel are also considered to determine their susceptibility to pitting corrosion. If the surface layer is compact, adherent, and uniform, the carbon steel is protected from pitting corrosion. However, nonuniform, loosely adherent, and noncompact layers can increase the susceptibility of the steel to pitting corrosion.

5.6 Models for predicting pitting corrosion rates

Aside from the likelihood of pitting corrosion, field operators are also concerned with the rate at which it occurs. Most of the models used for predicting pitting corrosion in the oil and gas industry are based on nonclassical theories, whereas those developed for CRAs are based on classical theory. The kinetic models used for predicting the presence of corrosion can also be categorized into different categories such as corrosion science, electrochemical science, and corrosion engineering concepts depending on their data and procedures.

5.7 Corrosion science model

There are several corrosion science models available, but the de Waard and Milliams model is the most commonly employed to assess internal corrosion (de Waard & Lotz, 1993; de Waard & Milliams, 1975; de Waard, Lotz, & Dugstad, 1995). This method was first introduced in 1975 and has been updated multiple times since then. The de Waard and Milliam model and numerous of its derivatives are helpful in determining the general corrosion rate (Step 1 of the nonclassical theory) and comprehending the different types of surface layers that occur (in Step 2). According to their model, the rate of corrosion increases with CO_2 partial pressure and temperature until it reaches a maximum value at 60°C–70°C, after which it falls to 90°C. Due to short-term laboratory tests and nonevaluation of corrosion rates after surface layer development, corrosion science models do not provide rates in the presence of surface layers, frequency of pit initiation and repassivation, and their kinetics, as well as pit propagation rates.

5.7.1 Electrochemical models

Most of the models used to predict pitting corrosion are based on the classical pitting corrosion resistance theory. Some of them are described in this chapter.

A point-defect model was developed by Macdonald and MacDonald (1990) to study the steady-state properties of films that are formed on metals and alloys in aqueous conditions. It predicts the varying log and steady-state currents of the barrier film. The model is used to calculate the steady-state thickness (Lss) [Eq. (5–1)] of various surface layers and surface layer breakdown and pit nucleation using the equations.

$$\mathrm{Lss} = \frac{1}{\epsilon}\left[1 - \alpha - \frac{\alpha\alpha_{\mathrm{d}}(\delta - \chi)}{2}V - \frac{1}{2K}\left[n - \frac{2\beta\gamma}{2.303}\right]\ln C_{\mathrm{H+}} + \frac{1}{2K}\ln\left[\frac{(4KD_0C_0^L)}{\chi^k K_{\mathrm{s0}}}\right]\right. \tag{5–1}$$

where

$$\gamma = F/RT \tag{5–2}$$

$$K = \epsilon\gamma \tag{5–3}$$

where ε, electric field strength; Lss, steady-state surface layer thickness; a, constant depending on electric potential drop across film-solution interface; a_d, transfer coefficient; d, constant depending on passive film nature; c, stoichiometry of film; b, constant depending on potential drop across the film-solution interface; g, activity coefficient; K, Macdonald constant; V, potential; n, kinetic order of primary film dissolution; C_H^+, hydrogen ion concentration; C_o, concentration of chloride ions; D_o, diffusion coefficient; L, thickness of passive film; and K_s^o, reaction rate constant.

Macdonald and MacDonald (1986) presented the deterministic extreme value statistics approach to characterize surface layer breakdown and pit nucleation. The breakdown potential can be computed using an equation according to the model.

$$V_c = \frac{4.606RT}{c^{aF}} \log\left(\frac{J_m}{(J^0 u^{-C/2}}\right) - \frac{2.303RT}{aF} \log(a_x) \tag{5-4}$$

where T, temperature; V_c represents breakdown potential; F, Faraday constant; R, gas constant; a, depends on the electrical potential drop across the film-solution interface; J_m, vacancy flux in metal phase; J_o and u, Macdonald constants, respectively; c, stoichiometry of the film; a_x, thermodynamic activity of the aggressive anion (e.g., Cl^-) in the solution.

5.7.2 Corrosion engineering model

Engineering models are used to predict the internal pitting corrosion in pipelines used in the oil and gas industry using a nonclassical concept. These models are based on a four-phase approach.

1. The low-corrosion stage occurs when the surface of a pipeline has been covered with hydrocarbons.
2. The formation of surface layers on the pipeline's steel surface is caused by corrosion reactions once the water has covered the surface.
3. Initiation of pits at localized regions on the steel surface when a surface-layer breakdown occurs, and
4. Pit propagation and eventual penetration of the pipe wall.

5.8 Prevention of pitting corrosion

1. Since pitting is an electrochemical process, it can be stopped by cathodic protection.
2. In some cases, pitting can be prevented by treating the environment with an inhibitor to prevent the formation of environmental differences. This method can also remove the driving force that is responsible for the pitting process.
3. In other cases, pitting can be prevented using a sacrificial coating or a protective coating. For instance, zinc-rich paint can prevent pitting on aluminum or steel due to its sacrificial nature.

4. Choosing a more corrosion-resistant material is another technique to avoid pitting corrosion. An empirical ranking scheme for austenitic and duplex stainless steel corrosion pitting susceptibility can be helpful. The pitting resistance equivalent number calculated by Eq. (4–2), given in Chapter 4 forecasts these alloys' pitting resistance as a function of their chromium, molybdenum, tungsten, and nitrogen content.

5.9 Failure case histories

5.9.1 Case I

1. *Equipment*—thickness thinning of underground API 5L-X42 transmission pipeline (shown in Fig. 5–4) (Zaferani, 2015).
2. *Environment*—flow line.
3. *Conditions-fluid*—sweet gas pressure: 1200 psi.
4. *Observation and remarks*—in this pipeline nonmetallic inclusions act as an initiation point for pitting. Further liquid accumulation in the bottom of the pipeline intensified the growth.
5. *Remedy*—separation of liquids from the flow line using a separator.

5.9.2 Case II

1. *Equipment*—Gas leakage from API 5L- X42 transition pipeline (Fig. 5–5).
2. *Environment*—flow line.
3. *Conditions*—fluid: sweet gas, pressure: 1100 psig, time in service: 7 years.
4. *Observation and remarks*—the main reason for this phenomenon was because of lacking cathodic protection. The thickness thinning causes a failure in the static test in 900 psig. As well metallic and nonmetallic inclusion played a decisive role in the start of a corrosion pit.
5. *Remedy adequate inspection*—appropriate cathodic protection according to suitable standards.

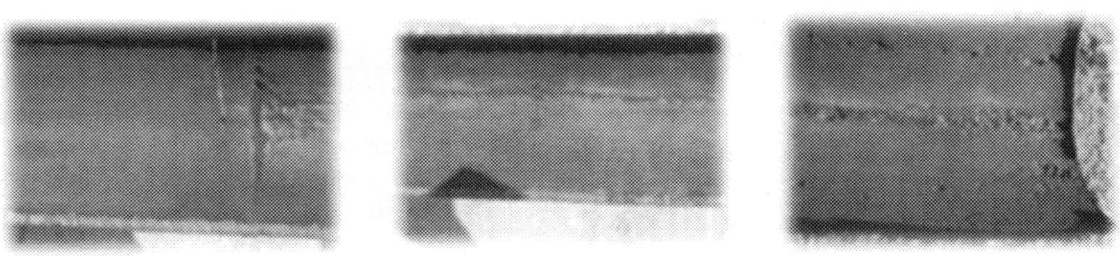

FIGURE 5–4 Underground API 5L-X42 transmission pipeline. *From Zaferani, S. H. (2015). Failure analysis of corrosion case histories. http://chrome-extension://efaidnbmnnnibpcajpcglclefindmkaj/https://digital.library.adelaide.edu.au/dspace/bitstream/2440/123817/3/hdl_123817.pdf.*

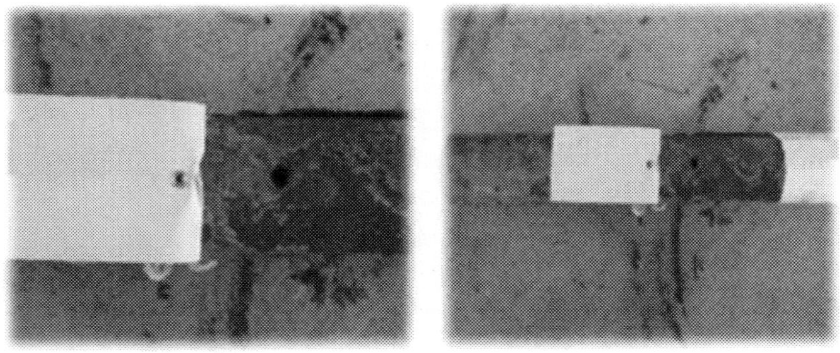

FIGURE 5–5 Intense pit on API 5L- X42 transition pipeline. *From Zaferani, S. H. (2015). Failure analysis of corrosion case histories. http://chrome-extension://efaidnbmnnnibpcajpcglclefindmkaj/https://digital.library.adelaide.edu.au/dspace/bitstream/2440/123817/3/hdl_123817.pdf.*

Short questions

1. What factors affect pitting corrosion? And where does it occur?
2. Is pitting caused by the environment?
3. How a pit once initiated can be repassivated?
4. What are the classical and nonclassical corrosion theories?

References

Akpanyung, K. V. 1, & Loto, R. T. (2019). Pitting corrosion evaluation: A review. *Journal of Physics: Conference Series, 1378,* 022088.

Brigham, R. J., & Tazar, E. W. (1973). *Corrosion, 29,* 33.

de Waard, C. & Lotz, U. (1993). Prediction of CO2 corrosion of carbon steel. In *NACE corrosion 93,* Paper #69, Houston, Texas.

de Waard, C., Lotz, U., & Dugstad, A. (1995). Influence of liquid flow velocity on CO2 corrosion: a semi-empirical model. In *NACE corrosion 95,* Paper #128, Houston, Texas.

de Waard, C., & Milliams, D. E. (1975). Carbonic acid corrosion of steel. *Corrosion, 31*(5), 177–181, NACE.

Enerhaug, J., Grong, Ø., & Steinsmo, U. M. (2001). Factors affecting initiation of pitting corrosion in super martensitic stainless-steel weldments. *Science and Technology of Welding and Joining, 6,* 330–338.

Macdonald, D. D., & MacDonald, M. U. (1986). Distribution functions for the breakdown of passive films. *Electrochimica Acta, 31,* 1079.

Macdonald, D. D., & MacDonald, M. U. (1990). Theory of steady-state passive films. *Journal of the Electrochemical Society, 137,* 2395.

Papavinasam, S. (2013). *Corrosion control in the oil and gas industry.* Gulf Professional Publication, ISBN 978-0-1239-7022-0 (Imprint of Elsevier).

Papavinasam S., *Pitting corrosion, trends in oil and gas corrosion research and technologies production and transmission,* Ed. A.M. El-Sherik, ISBN978-0-08-101105-8.

Pradhan, S. K., Bhuyan, P., & Mandal, S. (2019). Influence of the individual microstructural features on pitting corrosion in type 304 austenitic stainless steel. *Corrosion Science, 158*, 108091.

Virtanen, S., Kobayashi, Y., & Boehni, H. (2001). *Electrochemical Society Proceedings, 99-42*, 665.

Zaferani S.H. (2015). Failure analysis of corrosion case histories, OMICS Group eBooks.

Zhu, L. K., Yan, Y., Qiao, L. J., & Volinsky, A. A. (2013). Stainless steel pitting and early-stage stress corrosion cracking under ultra-low elastic load. *Corrosion Science, 77*, 360–368.

6

Intergranular corrosion

6.1 General conditions, occurrence, and causes

It is also known as "interdendritic corrosion" or "intercrystalline corrosion." For alloys and in some environments, intergranular corrosion (IGC) is preferential corrosion at the grain boundaries of the material (Roberge, 2000). Most commonly, alloys undergo IGC when they have been subjected to certain heat treatments or processes. Some of the most common types of steels that are prone to this type of corrosion are highly alloyed austenitic chromium-nickel and ferritic chromium steels (Cihal, 1984). Passivation is then used to protect these materials from the effect of corrosion. The Fe-Cr alloy must have a minimum chromium content of roughly 10.5% by weight for passivation to occur and remain stable; below this level, passivity is impossible. Within the temperature range of 550°C–750°C, IGC can occur due to the precipitation of chromium carbides (Cr23C6) or intermetallic phases. The phenomenon is known as "sensitization," and it refers to the alloy's sensitivity to grain boundary attack. Fig. 6–1 depicts an intergranular attack of 0.06% C austenitic stainless steel sensitized at 600°C for several hours (Pedeferri (Deceased), 2018).

The causes of sensitization can be varied. Some of these include the slow cooling of the steel's annealing temperature, which can affect the components' structural properties. In addition, stress relieving can occur when the steel is treated with sensitized austenitic materials. Another common cause is welding operations. The condition known as weld deterioration of AISI 304 and 316 (UNS S31600) stainless steel components caused by the heat-affected zone of the weld has been an issue in various industrial applications (Steigerwald, 1978). The presence of Cr-rich carbides at the grain boundaries can cause sensitization in certain austenitic materials. This phenomenon can also occur when Cr depletion takes place near the areas where the deposits are located (Jones, 1996).

The slow diffusion rate of chromium has caused the free chromium content to fall to a level that no longer allows for passivation due to the presence of carbide precipitations, specifically below roughly 13%. These precipitations can occur at various grain boundaries of steel, but they only occur at the edges of a weld seam during heat treatment. The duration and temperature of the depletion of chromium and carbide can also affect the resistance of steel to grain separation. Low-carbon steels with a relatively carbon concentration of less than 0.03% (so-called L grades) are known to be resistant to this process. Because of this, they should be treated or welded to prevent their grain separation (Shreir et al., 1994). However, chromium carbides may form in grain borders, resulting in chromium depletion in zones adjacent to grain boundaries due to the slow chromium diffusion rate. The corrosion resistance of the chromium-depleted zones decreases compared to the rest of the matrix. The depleted areas may be triggered in a corrosive environment, and corrosion will occur in

Handbook of Corrosion Engineering. DOI: https://doi.org/10.1016/B978-0-323-95185-2.00006-X

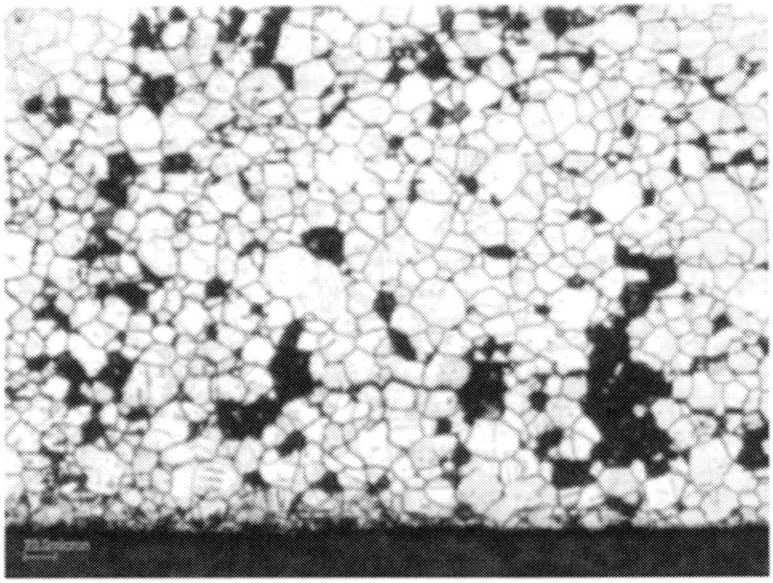

FIGURE 6–1 Intergranular corrosion of an austenitic stainless steel AISI 304 with 0.06% of C, sensitized at 600°C for several hours. *From Pedeferri (Deceased), P. (2018). Intergranular and selective corrosion. In Engineering materials (pp. 297–311). Springer Science and Business Media B.V. https://doi.org/10.1007/978-3-319-97625-9_15.*

extremely small places between the grains. IGC, often known as weld degradation, is a particularly serious problem in stainless steel welding. Also, sensitized is stainless steel that has been heat-treated to produce grain boundary precipitates and adjacent chromium-deficient zones. To prevent the formation of chromium carbide precipitations, stainless steels can be stabilized by adding various types of titanium, nickel, and tantalum. These materials, which are known to form titanium carbide, niobium carbide, and tantalum carbide are preferred over chromium carbide due to their reduced carbon content. In addition, by quenching the water with a high temperature above 1000°C, the steel can prevent the formation of chromium carbide in the grains and then prevent its precipitation (Ashby & Cebon, 2007).

6.2 Factors affecting intergranular corrosion

Composition, processing, and environmental conditions all have an impact on IGC. In terms of metallurgy, electrochemistry, and chemistry, these components contribute to the differential in thermodynamic and kinetic characteristics between the matrix and the sensitized grain boundary. The following factors have been observed to affect IGC start and propagation in Al-Mg alloys:

6.2.1 Solution chemistry

The bulk chemical makeup of the electrolytes has a role in determining whether IGC develops. Aggressive acidic test settings were created so that a metallurgical state, such as the one that leads to IGC susceptibility, may be clarified in the accelerated test by direct evidence of grain boundary attack. The electrochemical response of the matrix and the sensitive grain boundary differs in 72 test settings, indicating IGC. It is safe to assume that aggressive acids will attack nearly all the particles exposed. As aggressive acidic test conditions for Al-Mg alloys, solutions of HNO_3 (70 vol.%) and H_3PO_4 (10 vol.%) have been developed and standardized (Vetrano et al., 1997). The phase is robust in the sense that it is the most vulnerable phase in many situations. Aside from that, anions such as chloride ions (Cl^-) have been shown to infiltrate and damage the oxide passive layer, as well as contribute to the beginning of localized corrosion events (McCafferty, 2003). Cl^- ions are also known to obstruct passive film reformation and play a role in attacks on component particles (Aballe et al., 2001).

6.2.2 Solution pH

Corrosion at a very alkaline pH includes a more homogeneous matrix breakdown, whereas corrosion at a slightly acidic pH is limited to grain boundary attack *Birbilis*.

6.2.3 Applied potential

In a wide variety of aluminum alloys, applied voltage affects localized corrosion morphology (Huang & Frankel, 2007; Zhang & Frankel, 2002; Buchheit et al., 1990; Muller & Galvele, 1977). In IGC-susceptible materials, the concept of a potential window for IGC is frequently used to characterize the zone of IGC (Little et al., 2007; Urushino & Sugimoto, 1979). The IGC zone refers to the potential range of breakdown potentials between the anodic and Cu-depleted zones around an Al-Cu intermetallic. In AA5083, the Cu-depleted zone is more important than the S-phase. If the phase and the amount of segregation are not enough, then the sensitive alloy will have a lower breakdown potential (Jain et al., 2012).

6.3 Mechanism

IGC at a grain boundary is thought to go through multiple stages, like any other type of localized corrosion (Bumiller, 2011). IGC can begin on exposed metal surfaces if the protective passive coating has deteriorated or the near-surface deformed layer (NSDL) has dissolved. Rapid cathodic dissolution or comparatively slower passive dissolving can be used to remove the 1 m NSDL (Seong et al., 2015). Primary cathodic sites are intermetallic complexes or component particles that include relatively noble elements (Chen et al., 1996). The cathodic reaction produces hydroxide ions, which raises the local pH in the intermetallic particles, leading to chemical breakdown of the oxide layer and anodic dissolution of the surrounding metal (Guillaumin & Mankowski 1998). The intermetallic particles are surrounded by matrix trenching and cavity development, or phase pitting, because of anodic dissolution. The dissolution of the phase releases metal cations, which react with the water in the pits, generating local acidity and high chloride content. The reduction in the breakdown potential of the phase leads to the accelerated dissolution rate of the phase, which results in the formation of corrosion-causing substances within the grain boundaries (Jain et al., 2012). This process can also lead to the spread of corrosion-causing substances across the surfaces of pits. Aside from the surface IGC, the local electrolyte chemistry of the pits and the subsurface IGC sites also influence the propagation and penetration of this toxic substance. The presence of cations and anions in the subsurface IGC sites and their migration of them from the fissure tip can affect the propagation and penetration rate of this toxic substance. Depending on the depth of the subsurface IGC penetration, the rate at which this substance can be generated may vary (Lim et al., 2015). The steady-state anodic dissolution of a fissure supports stable IGC propagation through the presence of critical chemistry. However, if the chemistry at the fissure tip is not conducive to maintain a steady rate of dissolution, or if the voltage or electrolyte combination does not maintain a certain level of penetration, then the IGC may be restrained.

6.4 Prevention

IGC can be prevented in two ways: by lowering carbon content (below 0.03%) and by adding stabilizing materials that form stable carbides. The most common commercial grade is AISI 304L or 316L (L stands for low carbon). Low carbon grades have the disadvantage of having reduced mechanical strength. An element that can be used to form dispersed carbides in the matrix can be added to prevent chromium carbides from precipitation. These elements, which are titanium and niobium, are 10 and 5 times more powerful than carbon content. AISI 321, titanium stabilized and AISI 349, niobium stabilized as variants of classic AISI 304 and AISI 316Ti and AISI 316Nb as variants of classic AISI 316 are examples of stabilized stainless steels.

6.5 Intergranular corrosion testing

Although IGC tests are commonly used to identify carbide compounds at grain boundaries, they should not be used for detecting intermetallic phases. Because the responses of different test solutions to varying carbon contents can vary, they should not be utilized for this purpose. When the carbon percentage in corrosion-resistant alloys reached 0.08%, these tests were developed to improve their weldability. Most CRAs now have a 0.03% carbon maximum and are commonly 0.01% to 0.02% carbon due to AOD (argon oxygen decarburization) melting. This eliminates the possibility of carbide formation during welding, and modern CRAs should easily pass one of these tests. However, because of the growing sourcing of CRAs from the far East, the carbon content may be more than stated on the certificate. IGC testing is becoming more significant because of this.

1. *A262 method B and the G28 method A* tests are the same, except for their duration. ASTM A262 was designed for austenitic stainless steels, specifically the 300 series alloys. ASTM G28 lists alloys and specifies the test exposure for each. It is mostly 120 hours, but some alloys such as C-276 (UNS N10276) have a test exposure of 24 hours. These tests are like ISO 3651 method C and SEP 1877 method 2, except for the shorter duration and the lower acid content.

2. *ASTM A262 method C (Huey test)* is the same as that specified in ISO 3651-1. Technique E of ASTM A262, often known as the Strauss test, is identical to ISO 3651-2 method A except for the duration. Because 15 hours is a challenging test duration in practice and is merely a minimum, running the test for 20 hours overnight is equivalent.
3. *ASTM A262 method F* is a more aggressive version of method E, as is ISO 3651-2 method B and SEP 1877 method 1. None of these are identical and cross-substitution of results is not possible.
4. *ASTM G28 method B* is a more aggressive test than method A and is intended for more corrosion-resistant alloys, as listed in the document.
5. *SEP 1877 method 3* is a special environment only suitable for the Ni-Mo alloys.

6.6 Failure case histories

6.6.1 Case I

1. *Equipment*—thickness thinning of underground API 5L-X42 transmission pipeline (Zaferani, 2015).
2. *Environment*—flow line.
3. *Conditions*—fluid: sweet gas; pressure: 1200 psi.
4. *Observation and remarks*—in this pipeline non-metallic inclusions act as an initiation point for pitting. Further liquid accumulation in the bottom of the pipeline intensified the growth.
5. *Remedy*—use of separator to remove liquids from the flow line.

6.6.2 Case II

1. *Equipment*—gas leakage from API 5L- X42 transition pipeline (Zaferani, 2015).
2. *Environment*—flow line.
3. *Conditions*—fluid: sweet gas, pressure: 1100 psig, time in service: 7 years.
4. *Observation and remarks*—the main reason for this phenomenon was because of lacking cathodic protection. The thickness thinning causes a failure in the static test in 900 psig. As well metallic and nonmetallic inclusion played a decisive role in the start of a corrosion pit.
5. *Remedy adequate inspection*—appropriate cathodic protection according to suitable standards.

Short questions

1. What is the cause of intergranular corrosion?
2. Discuss the effect of temperature, time, and carbon content on the sensitization of austenitic stainless steel.

3. Why do grain boundaries become sensitive to the corrosion attack in case of precipitation of chromium carbides?
4. How do you test for intergranular corrosion?
5. What causes intergranular corrosion in stainless steel? Discuss the two approaches used to prevent intergranular corrosion of stainless steels.

References

Aballe, A., Bethencourt, M., Botana, F. J., Cano, M. J., & Marcos, M. (2001). Localized alkaline corrosion of alloy AA5083 in neutral 3.5% NaCl solution. *Corrosion Science, 43*(9), 1657–1674. Available from https://doi.org/10.1016/S0010-938X(00)00166-9.

Ashby, M., Shercliff, H., & Cebon, D. (2007). *Materials: engineering, science, processing and design.* Butterworth-Heinemann.

Buchheit, R. G., Moran, J. P., & Stoner, G. E. (1990). Localized corrosion behavior of alloy 2090. The role of microstructural heterogeneity. *Corrosion, 46*(8), 610–617. Available from https://doi.org/10.5006/1.3585156.

Bumiller, E. (2011). *Intergranular corrosion in AA5XXX aluminum alloys with discontinuous precipitation at the grain boundaries.*

Chen, G. S., Gao, M., & Wei, R. P. (1996). Microconstituent-induced pitting corrosion in aluminum alloy 2024-T3. *Corrosion, 52*(1), 8–15.

Cihal, V. (1984). Intergranular corrosion of steels and alloys. *Materials Science Monographs, 18.*

Guillaumin, V., & Mankowski, G. (1998). Localized corrosion of 2024 T351 aluminium alloy in chloride media. *Corrosion Science, 41*, 421–438.

Huang, T. S., & Frankel, G. S. (2007). Kinetics of sharp intergranular corrosion fissures in AA7178. *Corrosion Science, 49*(2), 858–876. Available from https://doi.org/10.1016/j.corsci.2006.04.015.

Jain, S, et al. (2012). *Corros. Sci., 59*, 136–147.

Jain, S., Lim, M. L. C., Hudson, J. L., & Scully, J. R. (2012). Spreading of intergranular corrosion on the surface of sensitized Al-4.4Mg alloys: A general finding. *Corrosion Science, 59*, 136–147. Available from https://doi.org/10.1016/j.corsci.2012.02.018.

Jones, D. A. (1996). *Principles and prevention of corrosion.* N.J.

Lim, Mary Lyn C., Kelly, Robert G., & Scully, John R. (2015). Overview of intergranular corrosion mechanisms, phenomenological observations, and modeling of AA5083. *CORROSION, 72*, 198–220.

Little, D. A., Connolly, B. J., & Scully, J. R. (2007). An electrochemical framework to explain the intergranular stress corrosion behavior in two Al-Cu-Mg-Ag alloys as a function of aging. *Corrosion Science, 49*(2), 347–372. Available from https://doi.org/10.1016/j.corsci.2006.04.024.

McCafferty, E. (2003). Sequence of steps in the pitting of aluminum by chloride ions. *Corrosion Science, 45*(7), 1421–1438. Available from http://researchgate.net/publication/223723422_Sequence_of_Steps_in_the_Pitting_of_Aluminum_by_Chloride_Ions.

Muller, I. L., & Galvele, J. R. (1977). Pitting potential of high purity binary aluminium alloys-I. AlCu alloys. Pitting and intergranular corrosion. *Corrosion Science, 17*(3), 179–193. Available from https://doi.org/10.1016/0010-938X(77)90044-0.

Pedeferri P. (Deceased) (2018). *Intergranular and selective corrosion engineering materials. springer science and business media* B.V., Italy. 297–311. Available from: http://springer.com/series/4288.10.1007/978-3-319-97625-9_15.

Roberge, P. R. (2000). *Handbook of corrosion engineering.* McGraw-Hill.

Seong, J., Yang, F., Scheltens, F., Frankel, G. S., & Sridhar, N. (2015). Influence of the altered surface layer on the corrosion of AA5083. *Journal of the Electrochemical Society, 162,* C209.

Shreir, L. L., Jarman, R. A., & Burstein, G. T. (1994). *Corrosion.* Butterworth-Heinemann.

Steigerwald, R. F. (1978). Intergranular corrosion of stainless alloys, ASTM special publication N656. American Society for Testing of Materials, Philadelphia.

Urushino, K., & Sugimoto, K. (1979). Stress-corrosion cracking of aged AlCuMg alloys in NaCl solution. *Corrosion Science, 19*(4), 225–236. Available from https://doi.org/10.1016/0010-938X(79)90008-8.

Vetrano, J. S., Williford, R. E., Bruemmer, S. M., Jones, R. H. (1997). *12 1997/12 TMS annual meeting 77–85 minerals, metals & materials Soc (TMS) United States Influence of microstructure and thermal history on the corrosion susceptibility of AA5083*

Zaferani, S. H. (2015). *Failure analysis of corrosion case histories, 2015.* OMICS Group eBooks, Foster City.

Zhang, W., & Frankel, G. S. (2002). Journal of the Electrochemical Society, 149.

7

Selective leaching or selective corrosion

7.1 General conditions, occurrence, and causes

The term selective corrosion refers to the process of removing one element from a solid material by means of selective dissolution. This process can be done in different phases of an alloy or in a single-phase or multiphase material. Depending on the type of corrosion, it can be either a plug type or uniform. The result of selective corrosion is the formation of a nanoscale structure on the surface of the material. One of the most common examples of this process is the dezincification of brass, which involves selectively removing zinc from a copper-zinc alloy (Callister & Rethwisch, 2013). Aluminum, iron, cobalt, chromium, and other elements are eliminated in a similar way in different alloy systems. In an alloy, the elements that are more resistant to the environment are left behind. Many alloys and conditions have been shown to exhibit selective corrosion. Table 7–1 contains a representative list.

Some of the common dealloying examples such as decarburization, decobaltification, denickelification, dezincification, and graphitic corrosion are discussed below:

Table 7–1 Environments causing dealloying for various alloys.

Alloy	Environment	Element removed
Brasses	Many water, especially under stagnant conditions	Zinc (dezincification)
Gray cast iron	Soils, many water	Iron (graphitic corrosion)
Tin bronzes	Hot brine or steam	Tin (destannification)
Aluminum bronzes	Hydrofluoric acid, acid chloride solutions	Aluminum (dealuminification)
Copper-nickels	High heat flux and low water velocity (in refinery condenser tubes)	Nickel (denickelification)
Gold alloys	Sulfide solutions, human saliva	Copper, silver
High-nickel alloys	Molten salts	Chromium, iron, molybdenum, tungsten
Medium, high-carbon steels	Oxidizing atmospheres, hydrogen at high temperatures	Carbon (decarburization)
Monels (Ni-Cu)	Hydrofluoric and other acids	Copper in some acids, nickel in others
Iron-chromiumalloys	High-temperature oxidizing atmospheres	Chromium which forms a protective film
Nickel-molybdenum alloys	Oxygen at high temperature	Molybdenum

Handbook of Corrosion Engineering. DOI: https://doi.org/10.1016/B978-0-323-95185-2.00007-1

7.1.1 Decarburization

When metal gets heated to around 700°C, the carbon in it gets mixed with hydrogen or oxygen gases, causing a reaction. This reaction can lead to the loss of carbon from the surface layer of the alloy. This occurs due to an interaction between the surface of the metal and various chemical compounds. Low-carbon, medium-carbon, and high-carbon steels are the three types of steel. The strength and hardness of steel diminish as the carbon concentration decreases, according to a general rule. High temperatures and a considerable amount of oxygen in the environment are important factors. The diffusion rate of carbon is so high at higher temperatures that longer annealing durations result in decarburization. The carbon atoms migrate from the outer layer to the surface, where they are oxidized to CO or CO_2, which are volatile gases, allowing the carbon to escape. Increased ferrite fractions and a lower number of carbides or lamellar pearlite in spheroidized annealed hypereutectoid steels are decarburization characteristics. The depth of partial and complete decarburization is added together to form decarburization. Complete decarburization refers to an area where carbon has been removed almost entirely. Partial decarburization can be defined as a reduction in carbon content that is not total decarburization (Edenhofer, Joritz, Rink, & Voges, 2015). The depth of the decarburization process is determined by the change in the microstructure of the metal. If the border layers are too soft, the process can be reversed. However, in vacuum or atmospheric environments, the process is not performed because carbon cannot burn due to a lack of oxygen.

One method to check for decarburization is to conduct a hardness test on the material. Because the metal will be softer than in its original form, this will provide a rough estimate of how much carbon diffused from the metal's surface. This is due to the fact that the lower the carbon concentration in the material, the less hard the material will be. The depth of decarburization along a steel's surface is determined by various tests such as ASTM E1077. These tests are used to determine if the steel's surface has undergone complete or partial decarburization. Based on the type of decarburization, the surface has been referred to as Type 1 (complete decarburization), Type 2 ($>$ 50%), or Type 3 ($<$50%). Decarburization can be avoided by heat-treating the component in an inert environment, covering it in stainless steel foil, or coating it with an isolating paint.

7.1.2 Decobaltification

It is the leaching of cobalt from cobalt-base alloys like Stellite or from cemented carbides. In the manufacturing industry, where cobalt is commonly used, it can be leached by various amino alcohols and amine-based additives found in water-miscible machining fluids. This can cause problems in the areas where cobalt is used, such as in the production of carbide tools.

7.1.3 Denickelification

This is the process of selective nickel leaching from nickel-containing alloys. The majority were found in copper-nickel alloys that had been exposed to seawater or other aquatic environments for a lengthy period.

7.1.4 Dezincification

This is the process of selective zinc leaching from zinc-containing alloys. After extended service in water containing dissolved oxygen, the majority are found in copper-zinc alloys containing less than 85% copper. As the zinc is eliminated in this case, the brass takes on a crimson coppery hue. It also becomes porous and brittle without causing any changes to the part's overall dimensions. The reduction of zinc in a copper structure leads to a weakening of the copper structure. This can be done on a small scale or large scale.

Dezincification occurs in two forms as can be seen in Fig. 7−1:

1. *Uniform or layer dezincification.* The front of the selective attack passes equally through the material in this category. High Zn concentration and acidic liquids prefer this form (Fig. 7−1A).
2. *Localized or plug-type dezincification.* In this category, the attack's location and form are chosen at random (Fig. 7−1B). This is more common in alloys with a lower Zn content, as well as in neutral, alkaline, and slightly acidic solutions. A stagnant solution, porous surface layers, and oxygen in the corrosive medium enhance dezincification.

There were two competing theories for describing the mechanism of dezincification for many years (Weisser, 1975). In one, zinc corrodes faster than copper and is thus removed from the alloy. On the other hand, the copper and zinc ions in the solution corrode and are removed from the alloy, but the copper ions plate back onto the surface. Following the initial suggestion of dezincification, the metal's surface should become porous, but nothing else should change. The second strategy, however more complex, is necessary to explain circumstances where copper crystals form on the surface after dezincification (Walker, 1977). Another mechanism that is commonly used to explain the formation of copper crystals on the surface following the dezincification process is by removing the zinc from the brass. Although the attraction of the zinc does not dissolve the copper, it allows it to move around freely. As the copper on the surface is drawn by negative ions in the solution, this rearrangement is feasible. Although the attraction is insufficient to dissolve the copper, it does reduce the copper's bond to the surface, allowing it to move around more freely (Erlebacher, Newman, & Sieradzki, 2012). Dezincification is most common in moderate environments, such as slightly acidic or alkaline solutions. When brass is exposed to fluids containing chloride ions, such as seawater, dezincification can occur (Moss, 1969). Removing oxygen from

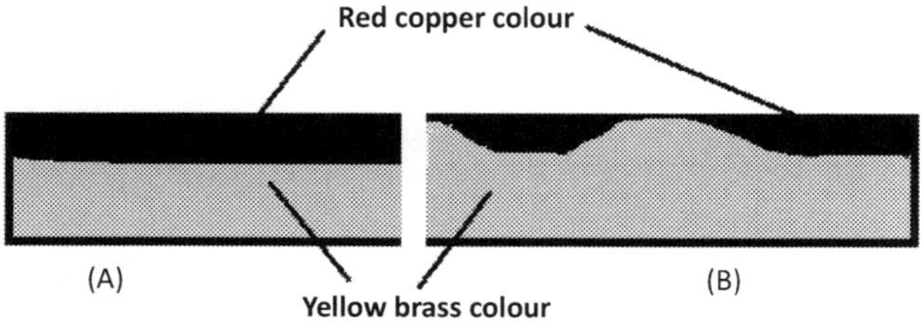

FIGURE 7–1 (A) Uniform (layer) dezincification and (B) localized (plug–type) dezincification of brass.

the solution can help prevent the dezincification process. Other factors such as the selection of an appropriate alloy and the cathodic protection can also help minimize the effects of this process.

7.1.5 Graphitic corrosion

Gray cast iron deterioration is a type of corrosion that involves the selective removal of the iron while preserving the intact carbon. This process is referred to as graphitic corrosion. It is not the same as graphitization, which is a type of corrosion caused by the disintegration of iron carbide.

7.1.6 Dealuminification

It refers to the process of aluminum leaching in aluminum-bronze alloys. The dealuminification process is believed to be electrochemical in nature. The microstructure of the material has both copper-rich α and aluminum-rich $\gamma 2$ phases when dealuminification occurs in aluminum-bronze alloy components. The aluminum-rich $\gamma 2$ phases will preferentially erode when these alloys are exposed to solutions containing chlorine ions or an oxidizing acid, such as hydrofluoric acid, resulting in substantial aluminum losses. Dealuminization has a negative impact on component's ultimate tensile strength, which is a result of concern. Dealuminification has thus been a problem for nuclear power plants that use aluminum-bronze alloy pipelines and piping components to transmit brackish water (Foster, Zanis, & Crisci, 1967). A variety of analytical tools can be used to determine the presence of dealuminification in a sample. These include chemical analysis, electrochemical spectroscopy, and electrical resistance and etching.

7.2 Mechanism and prevention

The dealloying process can be performed in tandem with two primary mechanisms. One of these is the diffusion mechanism, which allows one of the alloy's species to leave the bulk material and re-enter the surface. The other is the dissolution mechanism, which allows one of the alloy's constituents to dissolve within the material. Preventive measures to dealloying are:

1. Limiting the proportions of certain elements in a material helps prevent its various properties from being affected by corrosion. For instance, in brasses, 1% tin can be added to stop the leaching of zinc. This gives superior resistance to dealloying when compared to alpha brass. In cast irons, ductile iron has a higher resistance to graphite corrosion than gray cast iron.
2. One of the most effective ways to prevent dealloying is to control the environment. However, this method can be very challenging since it involves controlling the various elements in the material.
3. The design of a component can also affect its ability to resist dealloying. For instance, in areas where high stress is placed on the structure, reducing the turbulence can help prevent corrosion.

4. The use of a sacrificial anode to safeguard the alloy is a final step of corrosion protection when dealloying. This can be accomplished by choosing a material with higher electrochemical reactivity to corrode, therefore safeguarding the main component. This anode can then be replaced on a regular basis.
5. Implementing impressed current cathodic protection to induce a regulated current into the alloy to counterbalance the flow of electrons detected in the corrosion cell is another method of corrosion prevention.

Short questions

1. What is decarburization and why is it a user concern? And how do you prevent steel decarburization?
2. At what temperature does decarburization occur?
3. What is the most common form of selective leaching?
4. What type of pipe material can be affected by graphitic corrosion?
5. What is the main reason for dezincification and how can we stop it?
6. Why does brass dealloy at 20% zinc?

References

Callister, W. D., & Rethwisch, D. G. (2013). *Materials science and engineering: An introduction* (9th ed.). Wiley.

Edenhofer, B., Joritz, D., Rink, M., & Voges, K. (2015). *Carburizing of steelsthermochemical surface engineering of steels: Improving materials performance* (pp. 485–553). Germany: Elsevier Inc. Available from http://www.sciencedirect.com/science/book/9780857095923; https://doi.org/10.1533/9780857096524.3.485.

Erlebacher, J., Newman, C., & Sieradzki, K. (2012). Fundamental physics and chemistry of nanoporosity evolution during dealloying. *RSC Nanoscience and Nanotechnology*. 11–29. Available from: http://www.rsc.org/shop/books/series/18.asp?seriesid = 18; https://doi.org/10.1039/9781849735285-00011.

Foster, M. L., Zanis, C. A., & Crisci, J. R. (1967). *Dealuminization of cast aluminum bronzes*.

Moss, A. K. (1969). Corrosion of copper and copper alloys. *Australasian Corrosion Engineering, 13*(5), 5–11.

Walker, G. D. (1977). An SEM and microanalytical study of In service dezincification of brass. *Corrosion, 33* (7), 262–264. Available from https://doi.org/10.5006/0010-9312-33.7.262.

Weisser, T. S. (1975). The de-alloying of copper alloys. *Metals, 20*, 207–214. Available from https://doi.org/10.1179/sic.1975.s1.035; https://www.tandfonline.com/doi/abs/10.1179/sic.1975.s1.035.

8

Erosion and abrasion corrosion

8.1 General conditions, occurrence, and causes

Flow corrosion is a type of metal damage that can occur when a flowing liquid gets exposed to a metal (Roberge & Syrett, 2004). It can be caused by various factors such as flow velocity and mass transport. There are four different types of flow-induced corrosion. These include mass transport-controlled, phase transport-controlled, erosion–corrosion, and cavitation–corrosion. The rapid flow of a liquid on metal can cause erosion–corrosion. This type of damage occurs when metal gets exposed to a combination of physical and mechanical forces that can affect its surface (Heitz, 1996). Erosion–corrosion is a type of metal damage that is caused by the combined effects of corrosion and erosion. The other types of damage caused by flow corrosion include the erosion of protective films on a surface due to the presence of a liquid or a solid particle in a turbulent environment. If the flow of a liquid over a surface causes mass transfer and increased turbulence, then this type of corrosion is referred to as flow induced. The continuous damage to the protective film can be triggered by the mechanical action of the environment, which can lead to the oxidation of bare areas. This process can also be triggered by the actions of the surrounding passive areas.

Erosion corrosion can occur due to damage to the protective coating. In the case of hard metals, erosion–corrosion can be triggered by intensified general corrosion. Air bubbles or solid particles in liquids are known to be destructive in the process of erosion–corrosion. This can be distinguished from other types of flow patterns by the appearance of smooth pits, waves, or gullies (Fig. 8–1). According to ASTM G 40-77 "erosion is a progressive loss of original material from a solid surface due to mechanical interaction between that surface and a fluid, a multicomponent fluid, or impinging liquid or solid particles."

Most commonly, erosion–corrosion occurs in various components such as pumps, control valves, and impellers. It can also affect the inlet ends of heat-exchanger tubes and agitated tanks. In systems with sudden changes in flow patterns, such as those involving heat exchangers, the locations where water enters and exits the tubes can be prone to erosion–corrosion. Unlike erosion, which is the process of removing a surface's mechanical abrasion by the suspension of particles, erosion–corrosion is a chemical reaction. It occurs in a corrosive environment.

8.2 Types and mechanism

Erosion corrosion can be divided into two types as described below:

1. *Impingement corrosion*, sometimes, in systems with a multiphase or two-phase flow, the flow is forced to change its direction. This phenomenon can be caused by various factors

Handbook of Corrosion Engineering. DOI: https://doi.org/10.1016/B978-0-323-95185-2.00008-3

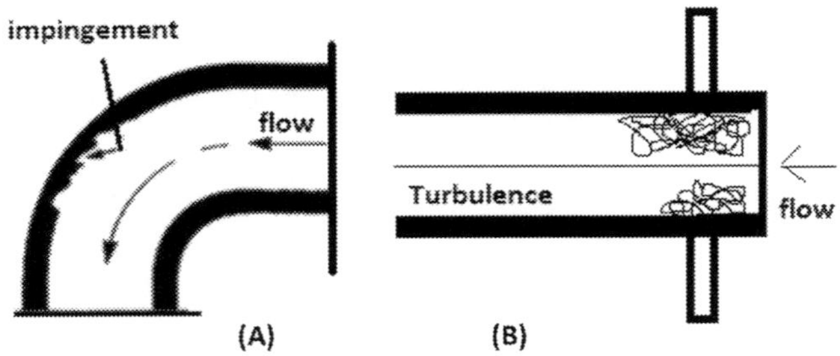

FIGURE 8–1 (A) Impingement, and (B) turbulence corrosion.

such as gas bubbles or liquid drops. In Fig. 8–1A, the impact of these substances on a surface can lead to pits with a directional pattern.

2. *Turbulence corrosion* occurs in areas with particularly strong turbulence such as the inlet end of heat exchanger tubes (Fig. 8–1B).

8.3 Factors affecting erosion–corrosion

The following factors affect erosion–corrosion:

Turbulence: When the liquid moves from a larger area to a smaller diameter pipe, it can cause turbulence. This phenomenon can lead to erosion–corrosion. It can also cause the liquid to exert more pressure on the metal surface, which can lead to more contact between the environment and the metal. This phenomenon is typically limited to a few inches at the tubing inlet.

Metal hardness: The degree to which a metal's hardness affects its ability to endure various types of corrosion, such as abrasion, turbulence, and cavitation, is known to be a factor that contributes to its overall resistance. Although it does not directly affect a pure process, its effects can be influenced by certain factors, such as the composition of the materials. For instance, if a metal's structure is homogeneous, its resistance to erosion can increase. On the other hand, if a structure is inhomogeneous, its resistance to erosion can decrease. Cast iron is known to be better against abrasion and corrosion as they have high tensile strength which make them incredible strong and durable. Despite its poor workability, this material is commonly used in the chemical industry.

Velocity: The low velocity of the metal surface can affect the erosion–corrosion rate. At high velocity, the steel may be more prone to experiencing corrosion. This can be caused by the presence of various elements such as CO_2, H_2S, and O_2 on the metal surface. It can also improve the effectiveness of corrosion inhibitors by treating the surface with higher levels of chemicals. One of the most effective ways to prevent the deposition of dirt and silt on the metal surface is by preventing the suspension of solid particles. This can prevent the surface protection from getting damaged at high velocity.

Impact of solid particles: Solid particles that are trapped in a material surface can damage metal in addition to the liquid flow damage.

Liquid droplets in gas flow: Due to repeated contact with liquid droplets, solids can get eroded and corroded. It produces damaging and impulsive contact pressure. Low-pressure turbines and aircraft damage from rain are issues. The localized film damage caused by air bubbles in a copper alloy heat exchanger is what causes corrosion and erosion. The same harm may also be done by nitrogen bubbles.

Galvanic effect: If the two metals are dissimilar, can affect corrosion. If metals come in contact but under static conditions, there is no effect. But under moving circumstances, this effect is more obvious.

Nature of metal and alloy: Erosion corrosion may be impacted by chemical composition, corrosion resistance, and hardness. We can learn about a metal's corrosion resistance from its composition. By the action of fluid or other reagents encouraging erosion–corrosion, soft metals are easily eroded. The best factor to control wear is hardness.

8.4 Prevention

The most important measures to avoid or reduce the extent of erosion and abrasion corrosion are:

1. *Appropriate design*. Numerous examples of good design of flow systems and other equipment are given by Pludek (1977). To keep the flow velocity below the critical level, pipe size, and flow cross-sections must be carefully considered. It should be remembered that changing nearby geometrical components could result in localized turbulence and, eventually, localized attack. As a result, such components must be kept to a minimum, and any change in cross-section or flow direction ought to be sufficiently seamless. It is best to prevent turbulence, especially in the system's most crucial areas, by, for example, placing bends sufficiently away from turbulence-causing components. Regions, where corrosion is impossible to prevent, should, if possible, be positioned so that the assault does not affect anything. For example, outlet pipe ends could be extended into the tank.
2. *Change of environment by* (1) filtration or precipitation of solid particles (2) precipitation of water droplets from air, gases, or steam (3) and removal of air or other gases from liquid flow. The pipe diameter should be large (low velocity) or resistant/replaceable sections should be employed. To prevent gas in liquid systems, it is also crucial to (1) lower the temperature and (2) maintain high pressure. Inhibitor usage and air removal may be efficient but not always cost-effective techniques.
3. *Application of corrosion-resistant coatings*, for example, metallic coatings by welding, metallic, metallic–ceramic, and ceramic coatings by thermal spraying, concrete, or rubber/polymer coatings.
4. *Cathodic protection*. The cathodic reaction happens very effectively at high flow velocities, which presents a challenge because a very high current density is typically required. Heat exchanger tubes' inlet portions have historically been covered by sacrificial anodes made of steel or zinc plates. Plugs made of sacrificial anodes can be used in pump housings and other components.

8.5 Evaluation of erosion–corrosion

To assess erosion–corrosion, numerous specific tests have been applied. The shape and exposure conditions can have a significant impact on the attack from erosion–corrosion and/or velocity-accelerated corrosion. The test outcomes, as well as the test/service conditions, must therefore always be thoroughly scrutinized. Spinning cylinder and disk apparatus are the most popular approaches because they are reasonably simple to set up and provide circumstances that can be easily assessed. They do not always provide conditions that are reflective of those in actual service. In recent years, actual pipe flow cells and jet impingement have been extensively used to model turbulent flow and multiphase settings more precisely. These tests should be carried out to generate precisely calculated wall shear stress conditions that correspond to those in the expected service. The damage or removal of normally protective corrosion products and inhibitor films is most directly related to the wall shear stress, which is a measurement of the mechanical action produced on the surface of the material by the flowing media.

8.6 Standard test methods

ASTM G-32 method of vibratory cavitation erosion testing
ASTM G-73 practice for liquid impingement erosion testing
ASTM G-75 test method for slurry abrasivity by miller number
ASTM G-76 practice for conducting erosion tests by solid particle impingement using the gas jet
NACE TM0170 method of conducting controlled velocity laboratory corrosion tests
NACE TM0286 cooling water test units incorporating heat transfer surfaces

Short questions

1. What is the difference between erosion and erosion-corrosion?
2. How can material selection and design prevent erosion corrosion?
3. Why are copper alloys less resistant than stainless steels to erosion-corrosion?
4. Discuss the effect of the presence of oxidizing or passivating species on erosion corrosion.

References

Heitz, E. (1996). Mechanistically based prevention strategies of flow-induced corrosion. *Electrochimica Acta, 41*(4), 503–509.

Pludek, V. R. (1977). *Design and corrosion control.* London: The Macmillan Press.

Roberge, P. R., & Syrett, P. B. C. (2004). *Erosion–corrosion, corrosion testing made easy* (Vol. 8). Houston, TX: NACE International.

9 Stress corrosion cracking

9.1 Occurrence

Corrosion cracking is a phenomenon that occurs when a corrosive substance and tensile stress are present simultaneously. Stress corrosion cracking (SCC), corrosion fatigue (CF), and hydrogen-induced cracking (HIC) are the three types of corrosion cracking (Fig. 9–1).

The SCC and CF differ depending on whether tensile stress is constant (SCC) or varies cyclically (CF). HIC is another type of corrosion cracking. When cracking develops as a result of atomic hydrogen being formed on a metal surface for multiple reasons and then diffusing into the metal, the process is known as hydrogen embrittlement (HE). In addition to cracks, atomic hydrogen in a metal can induce various sorts of damage such as blisters, bulges, and small interior fissures (blistering and stepwise cracking).

This chapter concentrates on SCC. SCC is defined as the crack failure of alloys by subjecting them to an aggressive environment in the presence of static tensile stress (nominally static and tensile). For metals, "reactive" does not cover gaseous hydrogen or cathodic polarization, but it does include aqueous and nonaqueous electrolytes, as well as reactive atmospheres. Similar effects have been observed in inorganic glasses and ceramics, particularly in water, and are assumed to be implicated in key geological processes such as earthquakes and mid-ocean volcano. Slow monotonic straining or low-amplitude cycling (ripple loading) are not excluded by static stress which accelerates SCC in many metallic systems, for example,

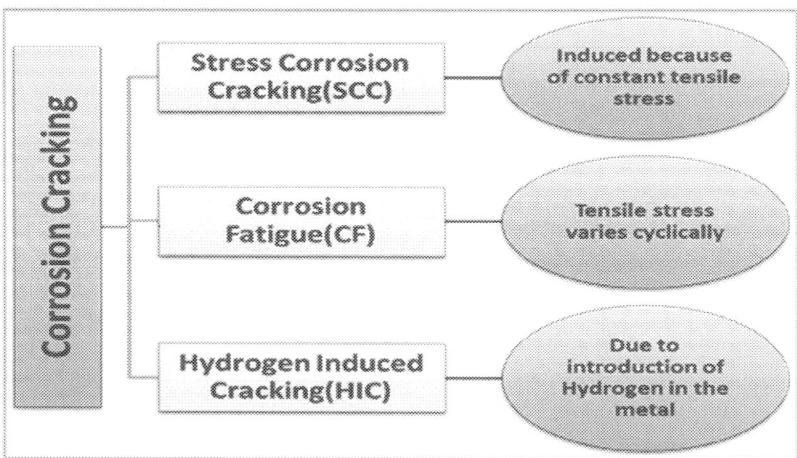

FIGURE 9–1 Schematic representation of types of corrosion cracking.

Handbook of Corrosion Engineering. DOI: https://doi.org/10.1016/B978-0-323-95185-2.00009-5

by promoting oxide film rupture at the crack tip. In practice, stress occurs from applied loads or residual stress caused by welding or inhomogeneous plastic deformation. In contrast to inorganic glasses and ceramics, which are brittle materials and crack via reactivity of the corrodent with highly stressed bonds at an atomically sharp crack tip, plasticity plays a crucial role in SCC of metals.

The relevance of a combined action of corrosion and stress is reflected in the term; stress and corrosive environment applied separately will not result in SCC. The stress needed for a mechanical failure in the absence of corrosion is much lower. The threshold stress is the lowest stress below which SCC will develop, and it may be as low as 10% of the yield stress in certain systems. SCC is not caused by corrosion alone in the absence of tension. SCC develops under stressful conditions. The metal or alloy is usually corrosion-free across most of its surface, but tiny cracks penetrate the surface at stress areas. The cracking might be intergranular or transgranular, depending on the alloy system and corrodent combination (Schweitzer, 2006). Stress levels, temperature, and the concentration of the corrodent all impact the rate of propagation. This attack occurs in certain types of media. Caustic embrittlement of steels in riveted boilers of steam-driven locomotives and season cracking of brasses in brass cartridge casings owing to ammonia in the atmosphere are two famous instances of SCC. Depending on the type of corrosive environment, stress cracking of various alloys may develop. Stainless steel cracks in a chloride environment (Natarajan, Srinivasan, & Shivakumar, 2014). The composition of the solution, the structure and composition of the metal/alloy, stress, and temperature are all influential in determining SCC (Pedeferri, 2018). Tensile fracture and inter- or transgranular cracking have been found in SCC failures (Fig. 9–2). When the stresses are higher, it takes a crack less to develop.

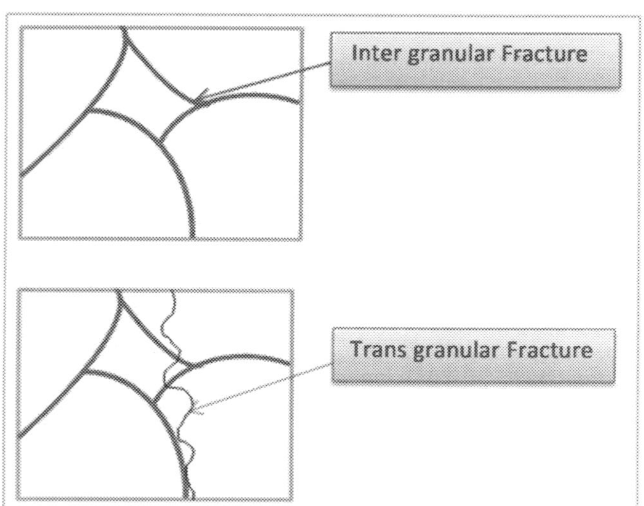

FIGURE 9–2 SSC cracking morphology for intergranular and transgranular fracture.

9.2 Initiation of stress corrosion cracking

Many pre-existing conditions are required for SCC initiation on metal: an unstable passivity in a corrosion condition, surface flaws, and local tensile stress. Cracks usually start because of surface flaws. Pits, scratches, and inclusions are examples of defects. Some flaws, particularly in stainless steel, might cause a Cr depletion zone, altering the local passivity. Corrosion activity is frequently greater on these defects than in other places. The proper chemistry/electrochemistry is created on the surface to enhance the process for local passivity breakdown. To induce SCC, a local mechanical loading threshold, such as a threshold stress or a critical concentration factor (KISCC), must be satisfied (Lou, 2010). Many methods have been proposed to explain the fracture initiation process in SCC. Due to the high local stress intensity, cracks may form at scratches and dents on the metal surface.

9.3 Propagation of stress corrosion cracking

The combined effects of crack tip local stress/strain, chemistry/electrochemistry, and microstructure influence the crack's future development after SCC initiation. Continuous passivation and breakdown may be a primary driving force in the propagation of brittle failure. A plastic zone forms at the fracture tip because of local stress, which aids lattice straining. The direction of crack growth is usually determined by microstructure and defects. As discussed in the sections below, there are many hypothesized SCC processes, each with its own sequence of events and reasons of local embrittlement (Lou, 2010). Table 9–1 lists some of the conditions that might induce SCC of different metals.

Table 9–1 Certain environments that cause stress corrosion cracking of materials.

Material	Environment
Aluminum alloys	NaCl, H_2O_2 solutions, air, water vapor, seawater
Gold alloys	$FeCl_3$ solutions, acetic acid solutions
Inconel	Caustic soda solutions
Copper alloys	Ammonia vapors and solutions, amines, water, water vapor
Lead	Lead acetate solution
Monel	Fused caustic soda, hydrofluoric acid, hydrofluorosilicic acid
Nickel	Fused caustic soda
Magnesium alloys	Seawater, NaCl-K_2CrO_4 solutions, distilled water, rural and coastal atmospheres
Stainless steels	Seawater, acid chloride solutions (e.g., $MgCl_2$ and $BaCl_2$), NaCl-H_2O_2 solutions H_2S, condensing steam from chloride waters, NaOH-H_2O solutions
Ordinary steels	Seawater, NaOH solutions, NaOH-Na_2SiO_2 solutions, calcium, ammonium, and sodium nitrate solutions, mixed acids (H_2SO_4-HNO_3), acidic H_2S solutions, HCN solutions, molten Na-Pb alloys
Titanium alloys	Red fuming nitric acid, seawater, N_2O_4, methanol HCl, naphthenic acid, polythionic acid, NH_3

9.4 Mechanism of stress corrosion cracking

The distinguishing characteristics of the process, such as an alloy's susceptibility to a specific range of environments, cracking characteristics, the role of stress, and the effects of numerous ecological and mechanical factors on the cracking process, are illustrated by a corrosion cracking mechanism. SCC fundamentally requires a harsh environment, so mechanisms based on localized metal breakdown have been proposed. However, a mechanism based on cleavage has been suggested because SCC is distinguished by brittle fracture. These two fundamental processes may be combined in a variety of ways. According to certain hypotheses, fracture initiation and propagation are both regulated by dissolution, the so-called continuous electrochemical processes.

The periodic electrochemical–mechanical mechanism states: The beginning of fractures or the deformation of the area ahead of the crack tip is controlled by dissolution, whereas the propagation is controlled by a mechanical brittle fracture event. There is a third category in which the involvement of dissolution in the embrittlement mechanism is minimized. Fundamentally, the bulk of SCC phenomena may be described in two ways: dissolution based and cleavage based (Lou, 2010).

Fig. 9–3 is a flow chart illustrating many forms of SCC mechanisms.

9.4.1 Dissolution-based mechanism

9.4.1.1 Pre-existing active path mechanism

This is the first suggested mechanism. As noted, before, the segregation at grain boundary or precipitation alters the microchemistry of the grain border significantly in comparison to the

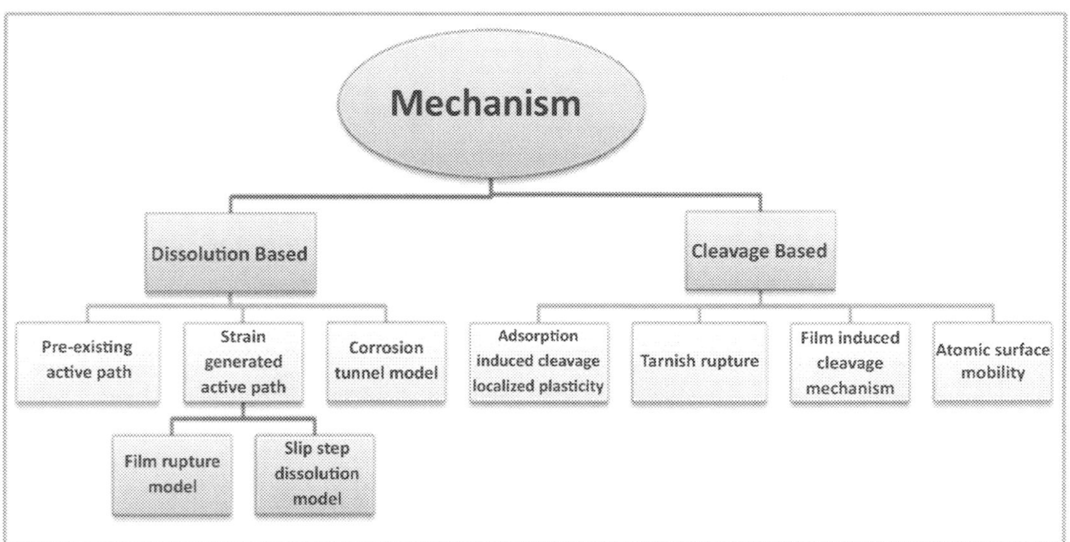

FIGURE 9–3 Various forms of SCC mechanism.

bulk alloy that may operate as a cathode in the local cell or, serving as an effective anode, may lead to dissolution to be confined to the nearby matrix. The purpose of stress in this case is to maintain the fracture exposed, allowing the harsh media to reach the crack tip. These terms have been used to describe the intergranular stress corrosion cracking (IGSCC) of sensitized stainless steel, aluminum alloys, and carbon steel in a variety of corrosive surroundings such as ultrapure water, chloride solution, and hydroxide, nitrate, or carbonate–bicarbonate medium respectively. However, a proclivity for intergranular corrosion does not always imply the presence of IGSCC.

9.4.1.2 Strain-generated active path mechanism

Two distinct models have been proposed to explain SCC in alloys that do not have a pre-existing route: the film-rupture model and the slip-step dissolution model. Referring to the mechanism of film rupture stress deepens the fracture, and localized plastic deformation at the crack tip punctures the passivating coating, exposing the metal's surface, which soon dissolves, leading to crack propagation. The crack tip remains exposed if the rate of film rupture at the fracture tip exceeds the rate of passivation. The crack tip, on the other hand, repassivates completely and is occasionally perforated by the formation of slip steps, while the bare slip step dissolves until the next repassivation. Most scientists currently feel that film rupture is essential to start cracking because most stress-corroding systems are related to film formation and the SCC begins in a borderline passive state. As a result of deformation, transgranular crack initiation occurs owing to slip steps. Nevertheless, disagreement continues over the spread of cracks by ongoing disintegration.

9.4.1.2.1 Film rupture model

Film rupture is the most often referenced SCC process based on dissolution. It entails the breakdown of protective layers along grain boundaries or low-index crystallographic planes by slip bands crossing the crack tip, followed by their dissolution until repassivation occurs (Fig. 9–4). If the rate of strain of crack tip is large relative to the repassivation rate, crack tips may stay basically film-free. Slip procedures improve the displacement of opening of crack tip, enabling solvent and solvated ions to flow in and out from crack tips more quickly, preventing dissolution from being inhibited by salt films or other corrosion products.

9.4.1.2.2 Slip-step dissolution model

According to dissolution-based SCC processes, atoms must be removed largely parallel to the applied force at fracture locations, with minor dissolution beyond the crack tip. Dissolution is promoted because of precipitation or segregation occurring at the grain boundaries as well as due to the simultaneous stress operating at the crack tips. The dissolution is restricted beyond the crack tips as a result of a number of factors (1) dissolution of microstructural/microchemical characteristics of anodic materials, (2) due to less stresses/strains near the fracture tip, or (3) comparatively less harsh conditions behind crack tips than those at crack location (Logan, 1952).

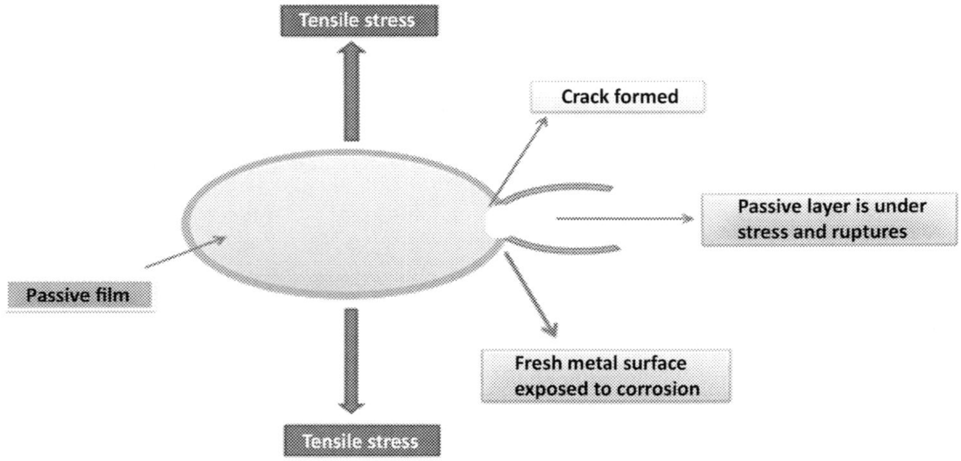

FIGURE 9–4 Diagrammatic representation of film rupture model.

9.4.1.3 Corrosion tunnel model

The combination of dissolution and mechanical fracture has resulted in a corrosion tunnel model. It is assumed that at the developing slip steps, a fine range of tiny corrosion tunnels develops. The remaining ligaments are stressed, causing ductile deformation and fracture, and the tunnels increase in diameter and length. Such a hypothesis, on the other hand, should yield grooved fracture surfaces with proof of microvoid convergence at the damaged ligaments, which has not been seen empirically. Therefore this model has been refined (Silcock & Swann, 1979) to propose that applying a tensile stress causes the morphology of corrosion destruction to shift from tunnels to thin, flat slots.

9.4.2 Cleavage-based mechanism

Cleavage-type cracking has been seen in SCC, and many processes have been proposed to explain it. These processes include tarnish-rupture, hydrogen-assisted cracking, adsorption-induced cleavage, film-induced cleavage, and atomic surface mobility.

9.4.2.1 Adsorption-induced cleavage localized plasticity

The stress-sorption cracking process, also known as adsorption-induced cleavage, is grounded on the assumption that ambient molecules get adsorbed on the metal and reduce the interatomic bond strength as well as the stress necessary for cleavage rupture, comparable to the decohesion mechanism of HIC. Adsorption is thought to be reliant on potential, which explains why cathodic polarization below a certain potential prevents SCC. The preferential adsorption theory may also be used to explain the uniqueness of the species for generating SCC in each alloy. This concept, on the other hand, does not describe how a fracture retains an atomically pointed tip in a usually ductile substance nor does it clarify why crack propagation is interrupted (Stress Corrosion Cracking: Theory & Practice, 2011).

9.4.2.2 Tarnish rupture

The tarnish-rupture process was first postulated to elucidate why single crystals of brass in ammoniacal solutions developed discontinuous transgranular cracks (Edeleanu & Forty, 1960). As per this hypothesis, a fragile surface coating or tarnish builds on the metal, which cracks when the tension is applied. The bare metal exposed because of cracking interacts quickly with the corrosive environment, the layer thickens, and the tarnish loop of growth and crack continues. The model was then adjusted (McEvily & Bond, 1965) for IGSCC, with the tarnish penetrating along the grain boundary before reaching the fracture tip. Experiment findings are still inadequate to validate or deny this process, while some researchers (Pinchback, Clough, & Heldt, 1976) have questioned grain boundary penetration's validity in all environments.

9.4.2.3 Film-induced cleavage mechanism

As a result of the tarnish-rupture model (Pugh, 1985), the film-induced cleavage mechanism was proposed. In this scenario, dissolution results in the creation of a surface layer, which then propagates a fracture further into the underlying metal. The crack progresses for a short distance via cleavage from an initial arrest location. The fracture is then gradually sharpened by plastic deformation until the propagation stage occurs again. The layer might be a dealloyed film or an oxide, with nanoporosity being the critical property that causes a severe fracture to form in the substrate. Although experimental data suggesting a link between SCC and dealloying have been published for Cu-Al and Cu-Zn alloys (Sieradzki & Newman, 1987) as well as stainless steel (Newman, Corderman, & Sieradzki, 1989), several critics have raised concerns about this process. The fracture progress distance, as determined from arrest marks, is generally a few micrometers, and it has been questioned if a cleavage fracture can penetrate so far from an initial layer that is just 30 nm thick.

9.4.2.4 Atomic surface mobility

Numerous forms of cracking generated by environmental factors, notably HE, might be explained by the atomic surface mobility process (Galvele, 1987). SCC and liquid-metal embrittlement both develop because of surface vacancies being captured at the crack tip and atoms diffusing away from the fracture tip in a countercurrent mode. The environment's role is to influence that diffusivity, which is directly related to the presence of surface components that melt at low temperatures. The coefficient of surface self-diffusion determines the speed of crack formation. Furthermore, it has been discovered that carbon steel fractures when high melting points are practised, despite the fact that it does not fracture when nitrates are present because they combine with iron to form low-melting compounds (Newman & Procter, 1990).

9.5 Protection of SCC in different metals and alloys

The main rule for avoiding SCC is to avoid using a vulnerable alloy under load in an SCC-producing environment. In the case of magnesium alloys, research has revealed that they are vulnerable in distilled water, with a threshold stress roughly half that of the yield stress. The goal of this research is to get scientific knowledge that will enable the creation of Mg alloys that are more resistant to SCC. Meanwhile, the authors would like to provide some broad principles. To prevent SCC, it is common practice to begin with a design that minimizes stress concentrations at the outset, during use, or by increasing sensitivity to various types of deterioration in service, such as localized corrosion, galvanic corrosion, (corrosion) fatigue, and erosion–corrosion. In dry atmospheres, SCC is unlikely to be a concern if the relative humidity is less than 90% and there are no cracks. At lower relative humidity levels, capillary condensation may result in the development of a liquid in fissures. Filling cracks with corrosion-resistant putty is a good idea. It is also worth noting that fissures are a feature of diecast alloy's microstructure. The following are the many techniques of SCC protection:

1. *Choice of material:* The first step in preventing SCC is to be aware of its tendency throughout the design and construction phases. SCC difficulties can be prevented by selecting a material that is not sensitive to SCC in the service environment and correctly processing and manufacturing it. Unfortunately, it is not always so easy.
2. *Critical stress intensity factor (threshold stress):* The metal may be protected if the stress level is maintained below the crucial threshold where SCC does not occur. Stress-relief annealing, which is extensively used for carbon steels, can remove residual stresses. Because they have a relatively high threshold stress for most environments, it is relatively easy to remove residual stresses to a tolerable level. Austenitic stainless steels, on the other hand, have a very low chloride SCC threshold stress. This suggests that stress relief as a method of managing SCC for this system is rarely successful due to the high annealing temperature necessary to prevent other problems such as sensitization and sigma phase embrittlement.
3. *Avoiding residual stress (tensile):* SCC is caused by the tensile component. In the case of brass, for example, there occurs seasoning cracking in the ammonical solution. The residual stress may be reduced if residual stress-reducing operations are performed.
4. *Removal of the SCC-causing element:* SCC may be reduced to a large degree by removing corrosive species from the environment that produce it, such as in the case of stainless steel pipes in boilers where SCC can be managed by removing Cl^- and O_2. For example, degasification, demineralization, and distillation. If the species that cause cracking are a necessary component of the environment, environmental management methods include adding inhibitors, altering the metal's electrode potential, or coatings that isolate the metal from the environment.
5. *Inhibitors:* To manage SCC, effective inhibitors may be used.
6. *Cathodic protection:* Cathodic protection can be employed to avoid SCC if the mechanism involved is anodic dissolution because negative current is given which does not allow the dissolution to happen. Cathodic protection should be utilized properly if the material is prone to HIC.
7. *Protective coating:* Material can also be protected against SCC by protective coatings. SCC can be prevented by using metallic coatings that insulate the metal from the environment.

However, the risk of the coating being penetrated in service due to improper application or mechanical damage must be considered. Zinc is a preferred coating for carbon steel. Zinc's usual corrosion potential is low, and if any of the underlying steel is exposed, it will be cathodically protected. The low electrode potential, on the other hand, will stimulate hydrogen evolution, which could lead to HE. As previously mentioned, HE can occur because of hydrogen evolution during the initial electroplating procedure. As a result, zinc plating on steels must be done with caution. Paints and other polymeric coatings protect the underlying metal primarily through their high electrical resistance, which limits current flow from the anode to the cathode.

9.6 Failure case histories

9.6.1 Case I

1. *Equipment*—SS316 Spring used in safety valve (H, 2015).
2. *Environment*—flow line.
3. *Conditions*—fluid: sour gas, pressure: 1440 psig.
4. *Observation and remarks*—when H_2S and stress combine, SCC arises. These fatigues happened because of residual internal metal tension or externally imposed stress. Deformation during manufacture, unequal cooling, excessive temperature, and internal structural arrangements including volume change all contributed to residual internal stress.
5. *Remedy*—by selecting a material that is not prone to SCC and correctly process and fabricating it. To eliminate or minimize stress below the SCC stress threshold.

9.6.2 Case II

1. *Equipment*—a U-bend waste fuel heat exchanger tube made up of 316L SS.
2. *Environment*—shell side: hot waste fuel. Tube side: cooling water.
3. *Conditions*—temperature: 21°C, service life: 5years.
4. *Observation and remarks*—chloride SCC was found in 316 SS heat exchanger tubes in straight portions and U-bends, according to failure analysis. A blockage in the cooling water triggered the hot oil at 21°C, raising the temperature of the water containing around 800 ppm chloride.
5. *Remedy*—the problem was resolved by free circulation of the cooling water.

Short Questions

1. Stress corrosion cracking affects which materials?
2. Which factor is the most important in stress corrosion?
3. Write any three variables which affect SCC.
4. What is the effect of stress on corrosion?

References

Edeleanu, C., & Forty, A. J. (1960). Some observations on the stress-corrosion cracking of α-brass and similar alloys. *Philosophical Magazine, 5*(58), 1029–1040. Available from https://doi.org/10.1080/14786436008235881.

Galvele, J. R. (1987). A stress corrosion cracking mechanism based on surface mobility. *Corrosion Science, 27* (1), 1–33. Available from https://doi.org/10.1016/0010-938X(87)90117-X.

Logan, H. L. (1952). Film-rupture mechanism of stress corrosion. *Journal of Research of the National Bureau of Standards, 48*(2), 99. Available from https://doi.org/10.6028/jres.048.013.

Lou, X. (2010). *Stress corrosion cracking and corrosion of carbon steel in simulated fuel-grade ethanol.* Mech. Eng.

McEvily, A. J., & Bond, A. P. (1965). On the initiation and growth of stress corrosion cracks in tarnished brass. *Journal of the Electrochemical Society, 112*(2), 131. Available from https://doi.org/10.1149/1.2423481.

Natarajan, D., Srinivasan, R., & Shivakumar, M. S. (2014). Phyllanthus wightianus Müll. Arg.: A potential source for natural antimicrobial agents. *BioMed Research International, 2014*, 1–9. Available from https://doi.org/10.1155/2014/135082.

Newman, R. C., & Procter, R. P. M. (1990). *Stress Corrosion Cracking, 25*, 259–270.

Newman, R. C., Corderman, R. R., & Sieradzki, K. (1989). Evidence for dealloying of austenitic stainless steels in simulated stress corrosion crack environments. *British Corrosion Journal, 24*(2), 143–148. Available from https://doi.org/10.1179/000705989798270261.

Pedeferri, P. (2018). *General principles of corrosion* (pp. 1–16). Springer Science and Business Media LLC. Available from https://doi.org/10.1007/978-3-319-97625-9_1.

Pinchback, T. R., Clough, S. P., & Heldt, L. A. (1976). Stress corrosion cracking of alpha brass in a nontarnishing ammoniacal environment:fractography and chemical analysis. *Metallurgical Transactions A, 7*(8), 1241–1243. Available from https://doi.org/10.1007/BF02656612.

Pugh, E. N. (1985). Progress toward understanding the stress corrosion problem. *Corrosion., 41*(9), 517–526. Available from https://doi.org/10.5006/1.3583022.

Schweitzer, P. A. (2006). *Fundamentals of metallic corrosion: Atmospheric and media corrosion of metals* (pp. 1–729). United States: CRC Press. Available from https://www.taylorfrancis.com/books/e/9780849382444.

Sieradzki, K., & Newman, R. C. (1987). Stress-corrosion cracking. *Journal of Physics and Chemistry of Solids, 48*(11), 1101–1113. Available from https://doi.org/10.1016/0022-3697(87)90120-X.

Silcock, J. M. & Swann, P. R. (1979). Environment-sensitive fracture of engineering materials (Z. A. Foroulis, Ed.).

Stress corrosion cracking: theory and practice. Elsevier (2011).

Zaferani, H. S. (2015). *Failure analysis of corrosion case histories.* OMICS Group eBooks, 978–1.

10 Corrosion fatigue

10.1 General conditions, occurrence, and causes

The reduction in fatigue resistance caused by the presence of a corrosive medium is referred to as corrosion fatigue. The mechanical degradation that occurs due to the combination of joint action and cyclic loading can lead to corrosion fatigue (Forrest, 1962). Although corrosion fatigue is different from stress corrosion cracking, due to the nature of the applied load. For instance, while cyclic and repetitive loading can cause corrosion fatigue, static tensile loading can also lead to SCC (Callister Jr & Rethwisch, 2014). While rotating machinery, deaerators, and cycle boilers are frequently linked to corrosion fatigue, this problem can impact any piece of equipment that operates in a corrosive environment and has enough cyclic stresses and stress raisers. In concentrated areas of a component, such as pits, notches, surface flaws, or metallurgical alterations, stress raisers are frequently encountered (Forsyth, 1969). The surface of the component typically develops numerous parallel cracks because of corrosion fatigue. But certain corrosion fatigue fractures can also be rounded, particularly if they develop close to welded connections. The simultaneous action of cyclic stress, tensile stress (directly applied or residual), or plastic strain results in fatigue damage. A fatigue crack will not start and spread if any one of these three conditions is missing. Using efficient nondestructive testing methods, cracks can be found. Two of the most popular techniques for checking for corrosion fatigue are ultrasonic testing and magnetic particle testing (Schijve, n.d.).

10.2 Mechanism

Stage 1: Crack initiation—When fatigue cracks appear at a free surface, they usually start near a stress riser, that is, the toe of a weld. The formation of these cracks is caused by the plastic deformation of the metal. Through plastic deformation, discontinuities are created and moved, strengthening the metal. Work hardening is the term for this occurrence. The metal loses its ability to deform plastically when it gets work hardened, and this condition causes the discontinuity to turn into a crack after reaching its plastic deformation limit (Pao, 1996).

Stage 2: Crack propagation—When a crack initially appears, the constant cyclic stresses cause it to grow until it eventually affects the material's structural integrity.

Stage 3: Failure—Depending on the properties of material, such as its temperature, thickness, and applied stress, a catastrophic fracture might occur. Usually, these failures happen suddenly even without caution. A component that has been fractured due to fatigue may have smooth fracture surfaces without signs of plastic deformation.

Handbook of Corrosion Engineering. DOI: https://doi.org/10.1016/B978-0-323-95185-2.00010-1

10.3 Factors affecting corrosion fatigue

Three categories can be used to group the factors influencing corrosion fatigue:

1. *Mechanical.* It is evident that the intensity of the alternating tensile/pressure load affects how quickly a corrosion fatigue fracture grows. Although the frequency of the alternating load has no impact on ordinary fatigue, it is crucial in the case of corrosion fatigue. Prolonged contact of the fracture surface with the electrolyte at low frequencies increases the effect of corrosion on the crack growth rate. The frequency increases will result in a slower rate of corrosion fatigue fracture growth.
2. *Metallurgical.* Unalloyed steel can be less susceptible to corrosion fatigue even if adding little amounts of Cr and Ni to steel boosts both its general corrosion resistance and tensile strength. Moreover, it is important to consider the electrode potential of metal in the aggressive media and, more importantly, how well it resists pitting corrosion. High tensile strength does not stop crack initiation in typical fatigue, but corrosion fatigue does because corrosion is the one that initiates the cracking process in this case. Additionally, as tensile strength rises, a metal crack that has already begun to propagate will continue to propagate more quickly.
3. *Environmental.* In corrosion fatigue, a corrosive media will accelerate crack initiation and crack propagation. Nearly every fatigue crack begins at the bottom of a pit. This type of corrosion proliferate in situations with lots of oxygen and chloride. Therefore corrosion fatigue will occur sooner in aircraft operating often on transoceanic flights and exposed to the salty maritime environment than in aircraft operating on continental routes. The resistance of the metal to corrosion fatigue typically diminishes as the chemical activity of the medium rises. Increases in temperature will also hasten the development of cracks.

The temperature of the crack can affect the rate of its propagation and the life of the crack. An increase in the temperature can cause the active species to move to the tip of the crack, and this accelerates the crack's propagation. It can also increase the corrosion process by decreasing the hydrogen overvoltage.

10.4 Forms of fatigue

Although there are many different types of fatigue failures, they can generally be divided into two categories: mechanical fatigue and thermal fatigue.

1. *Mechanical fatigue*—As a result of prolonged exposure to cyclic loads, fatigue damage forms. It is important to note that certain classes of components, such as rotating machinery and pressure reduction valves, created for alternating mechanical stresses, rarely fail from fatigue unless another process such as corrosion damages them. The following types of fatigue can be categorized as mechanical in general.
   a. *Vibration fatigue*—Mechanical fatigue, known as vibration fatigue, is carried on by the vibration of pipework or other machinery while it is in use. For instance, using

equipment outside of approved integrity operating windows may lead to vibration fatigue. The main causes of vibration-induced fatigue damage are often bad design, a lack of support, too much support, or stiffness. For vibration fatigue damage to occur, which causes crack initiation and crack propagation, the amplitude and frequency of the vibration are crucial variables.

  b. *Corrosion fatigue*—It occurs because of mechanical fatigue and a chemical attack both occurring at the same time. Metal is known to degrade in corrosive situations. The damage serves as a location of stress concentration and the start of a crack as corrosion progresses. Equipment is coated with thin films and other materials to prevent corrosion, but these materials are regularly damaged by mechanical wear, leaving the equipment vulnerable to the environment.

2. *Thermal fatigue*—It is simply a failure caused by cyclic thermal fluctuations. This mechanism is most frequently found in the tube assemblies of fired heaters. Typically, service starts and stops result in thermal fatigue. Thermal shock is the term for abrupt temperature fluctuations that lead to quick failure. Thermal fatigue is more susceptible to starting and stopping of the turbine/engine. Additionally, rapid heating and cooling rates raise susceptibility.

By using proper design and operation to reduce thermal loads and thermal cycling, the risk of thermal fatigue can be reduced. Designing to reduce stress concentration sites, blending weld profile grinding, and mandating progressive and smooth thickness transitions are a few examples of such approaches. During start-up and shutdown, use regulated heating and cooling rates; reduce the number of starts and stops, especially "trips"; reduce differential thermal expansion in metal junctions of various compositions and; include flexibility that will allow for differential expansion whenever possible (Broek, 1972).

10.5 Corrosion fatigue assessment methods

When components are subjected to cyclic loading, fatigue damage happens in stress/strain concentration points/zones, even when loads are substantially lower than the material stress strength. After some time, fatigue may eventually cause an abrupt fracture failure without warning. To prevent unanticipated loss, various significant fatigue studies have been conducted, and a variety of strategies for fatigue prediction have been put forth (Andresen, 1996). From a microscopic perspective, crack formation and propagation cause fatigue. Generally speaking, there are three ways to characterize fatigue:

10.5.1 Stress/life (S/N) characterization

The fatigue limit or the S−N curve, also known as the fatigue curve or Wöhler curve, is frequently used to characterize the fatigue properties of materials such as technical alloys and polymers. The maximum number of cycles is determined using fatigue curves. A typical S−N curve is shown in Fig. 10−1. The number of cycles before a failure is shown on the

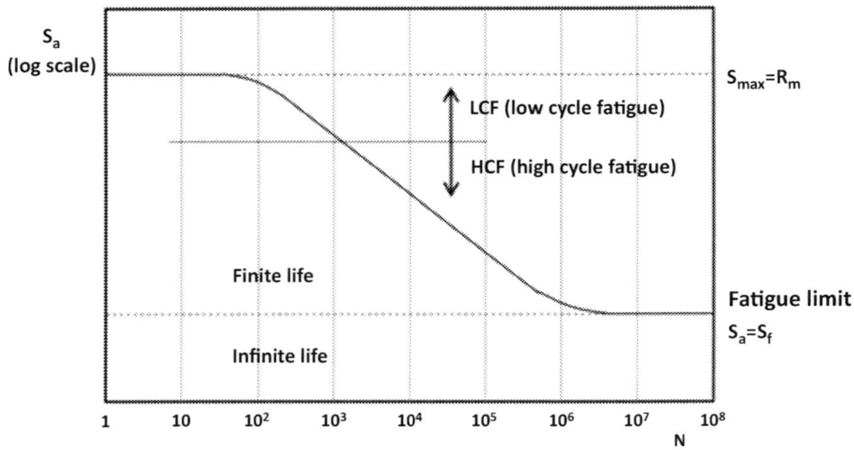

FIGURE 10–1 Typical S–N curve.

horizontal axis using a logarithmic scale. The stress amplitude of each cycle is shown on the vertical axis, which is either linear or logarithmic. A fatigue test is then performed to produce S–N curves, which are derived from the results of the stress tests. Each kind of material has a unique fatigue curve. An atmosphere that is corrosive encourages fracture initiation and reduces fatigue life. The various factors that can affect a material's fatigue life are its internal flaws, fatigue power, temperature, surface quality, design geometry, and residual stresses. For most materials, such as titanium and steel, the theoretical stress amplitude value is below which material will not fail during any given cycle. This value is referred to as a fatigue limit or endurance limit. The curve shows the endurance limit at the asymptotic part (Ferdinand et al., 1992). Tests are conducted by subjecting specimens to cyclic stress with constant amplitude (CA) until the specimen fails. Occasionally, the test is terminated after doing $N > 10^6$ cycles. The results are then interpreted as infinite life.

The following terms are defined for the S–N curve:

Fatigue limit. The fatigue limit, also known as the endurance limit, is the stress level below which fatigue failure does not occur. This limit, where the S–N curve becomes horizontal at higher N values, is only attained by a selected few ferrous (iron-based) and titanium alloys. Other structural metals, such as copper and aluminum, lack a definite point of failure and degrade over time even in the presence of minor stresses. Standard limits for steel range are half of the ultimate tensile strength, to a maximum of 290 MPa (42 ksi) (Hou & Lin, n.d.).

Fatigue strength. The maximum stress that a material can endure for a predetermined number of cycles without breaking is known as fatigue strength. According to the American society for testing and materials (ASTM), fatigue strength, or S_{Nf}, is the stress level at which failure occurs after a predetermined number of cycles (e.g., 10^7 cycles). For instance, the annealed Ti–6Al–4V titanium alloy has a fatigue strength of approximately 240 MPa at 10^7 cycles, and the stress concentration factor is equal to 3.3.

Fatigue life. The fatigue behavior of a material is defined by its fatigue life. According to the S–N plot, it is the number of cycles necessary for failure to occur at a given stress level.

There are two types of fatigue: high-cycle fatigue and low-cycle fatigue. The chief difference between high-cycle and low-cycle fatigue is the number of cycles to failure. The transition between these two types of fatigue is influenced by the stress level, that is, the transition between plastic and elastic deformations.

High-cycle fatigue requires more than 10^4 cycles to failure where stress is low and primarily elastic.

Low-cycle fatigue has low cycle counts due to the frequent plastic deformation that it experiences (i.e., within each cycle). Small changes in stress in the plastic region can result in substantial variations in strain. Low-cycle fatigue, as demonstrated by experiments, is also crack initiation. The relationship between the cyclic crack development rate ($\Delta \mathrm{d}a/\mathrm{d}N$) and the stress intensity range (ΔK) as stated in Eq. (10.1) can be used to explain the behavior of crack growth.

$$\frac{\mathrm{d}a}{\mathrm{d}N} = C(\Delta K)^m \tag{10.1}$$

where C and m are material scaling constants.

Eq. (10.1) is frequently used to estimate the lifetime of fractured structures from material knowledge, particularly in its integral version. For many technical materials, the value of m typically ranges between 2 and 4. Ultimately a region known as Kth is achieved for extremely low K values, below which long cracks do not propagate. When stress is consistently lower than the elastic limit, the stress-life approach (Lin, 2013) is used to predict the life span of high-cycle fatigue. The strain-life approach should be used instead of stress-life approach if the critical zone's stress exceeds the elastic limit.

10.5.2 Strain/life (ε/N) characterization

The low-cycle fatigue life prediction cases, in which components are loaded beyond yield strength, are the main emphasis of the strain-life approach. According to the approach, elastic strain and plastic strain have differing impacts on fatigue life and strain is the primary cause of fatigue damage. The method determines the number of reversals necessary for each reversal to fail as well as the strain history and mean stress of each reversal. The strain cycle curve (also known as the ε–N curve) is then used to calculate material damage. The ideal tests, which record sets of nominal stresses (ε) and life (N), yield the ε–N curve. The Morrow correction is used to calculate the effect of mean stress when it is non-zero. Damage and load time can then be used to compute the fatigue life (Gangloff, 1990).

10.5.3 Fatigue crack growth

The technique used to predict crack growth life is fracture mechanics. It is especially beneficial for parts with flaws or cracks. The main uses of fracture mechanics-based fatigue analysis are establishing inspection intervals, calculating crack development life given a known or assumed crack size, and determining the largest initial flaw size that can be tolerated for design life (frequently incorporating welds). It is crucial to conduct experiments to determine the fatigue crack growth rate (FCGR) of the material under suitable environmental conditions to accurately anticipate crack growth using this method (Gangloff, 1990).

There are two well-recognized standards for conducting FCGR tests: ASTM E647 (E647-13a, 2013) and ISO 12108 (Metallic Materials - Fatigue Testing - Fatigue Crack Growth Method', International Standard Organization, 2018).

10.6 Prevention

Corrosion fatigue can be prevented through:

1. Reducing fatigue by minimizing vibration and pressure fluctuation.
2. Fatigue properties can also be improved by increasing the thickness of the component. However, the increased thickness can result in the final failure being brittle rather than ductile.
3. If the initial flaws are relatively minor, cathodic polarization offers sufficient defense against the environmental contribution to corrosion fatigue of common structural steels. Due to their increased hydrogen sensitivity and usual higher stress loading, high-strength steels are more susceptible to hydrogen-assisted failures.
4. One of the most effective ways to reduce the stress levels in a weld is by stress annealing and good design. This can be done by reducing the stress concentrations in the weld. In addition to this, it is also important to localize the weld in order to minimize the effects of the stress on the other components.
5. By using anodic coatings.
6. CF damage can be prevented or minimized by using smart material selection in conjunction with a suitable design. It should be emphasized that the corrosion fatigue strength of high-strength steel is not higher than steel with lower tensile strength.
7. A change in the environment, such as deaeration or the application of inhibitors, is occasionally necessary. Although deaeration will not always prevent corrosion fatigue, it can at least stop it from starting on steel surfaces without surface flaws in neutral waters.
8. Anodic protection has been shown to increase the fatigue strength of carbon steel and stainless steel in oxidizing environments (it improves passivation).

Short questions

1. What causes fatigue corrosion?
2. What is the difference between SCC and CF?
3. List the factors that affect corrosion fatigue.
4. What is the role of S–N curve in fatigue mechanism?

References

Andresen, P. L. (1996). *Corrosion fatigue testing* (pp. 193–209). ASM International. Available from https://doi.org/10.31399/asm.hb.v19.a0002362.

Broek, D. (1972). *Fatigue crack growth and residual strength of aluminium sheet at low temperatures*. Nat. Aerospace Lab. NLR.

Callister Jr, W. D., & Rethwisch, D. G. (2014). *Materials science and engineering*. John Wiley & Sons.

E647-13a (2013). *ASTM. E647-13a, Standard test method for measurement of fatigue crack growth rates*. ASTM International, 2013.

Ferdinand, P., Beer, E., & Johnston Jr, R. (1992). *Mechanics of materials*. McGraw-Hill, Inc.

Forrest, P. G. (1962). *Fatigue of metals*. Pergamon Press.

Forsyth, P. J. E. (1969). *The physical basis of metal fatigue*. Blackie and Son.

Gangloff, R.P. (1990). *Corrosion fatigue crack propagation in metals*. NASA CR.

Hou, Y. & Lin, H. (n.d.). *Theodore unpublished content fatigue life prediction based on AMPS stress-life approach fatigue FEA*.

Lin, H.T. (2013). *Sefea (Strain-Enriched FEA) theory, benchmark and applications*. 12.

Metallic materials - Fatigue testing - Fatigue crack growth method', International Standard Organization (ISO) 12108. (2018).

Pao, P. S. (1996). *Mechanisms of corrosion fatigue* (pp. 185–192). ASM International. Available from https://doi.org/10.31399/asm.hb.v19.a0002361.

Schijve, J. (n.d.) *Corrosion fatigue*. Springer Science and Business Media LLC, pp. 457–479. Available from https://doi.org/10.1007/978-1-4020-6808-9_16.

PART 2

Corrosion in different environments

11

Corrosive electrolytes

11.1 Introduction

Corrosion problem in the presence of electrolyte is long established and most metallic structures are highly prone to corrosion in these environments. Through chemical and/or electrochemical reactions, the metallic species form relatively more stable compounds such as carbonates, sulfides, hydroxides, and oxides (Verma, Ebenso, & Quraishi, 2017; Izquierdo, Santana, González, & Souto, 2010). Therefore in nature, most of the metals except noble metals exist in their carbonates, sulfides, hydroxides, oxides, etc. form. The carbonates, sulfides, hydroxides, oxides, etc. present over the exterior parts of the metallic structures are called impurities. To get pure metals, the surface impurities should be removed using different types of electrolytes. The electrolyte to be used to clean a particular impurity is selected based on the nature of surface impurities. The solutions used to clean or remove the surface impurities are called electrolytes, aggressive solutions, or cleaning solutions (Muldoon et al., 2013; Wu, Zhang, Pickering, & Allara, 1993; Azevedo, Allély, Ogle, & Volovitch, 2015). Different types of neutral, basic, and acidic solutions are used to clean the surface impurities. However, during the cleaning process, some metallic component is also lost along with the impurities (Stratmann, Streckel, Kim, & Crockett, 1990; Zaferani, Sharifi, Zaarei, & Shishesaz, 2013; Hegazy, Badawi, Abd El Rehim, & Kamel, 2013). Therefore generally, corrosion inhibitors are used to avoid the undesirable loss of metallic components (Schmitt, 1984). Organic compounds serve as the best corrosion inhibitors, and they adsorb on the metal surface electron-rich rich sites (Verma, Olasunkanmi, Ebenso, Quraishi, & Obot, 2016). Their adsorption results in the formation of protective film. Their adsorption on the metal surface mostly involves a physiochemisorption mechanism for which the value of Gibb's free energy of adsorption (ΔG_{ads}) lies between -20 and -40 kJ/mol (Verma et al., 2016).

In this chapter, a collection of different types of commonly used electrolytes and the mechanism of metal dissolution in these electrolytes are described. Particular emphasis is paid to H_2SO_4 (sulfuric acid), HCl (hydrochloric), H_3PO_4 (phosphoric acid), $C_6H_8O_7$ (citric acid), HF (hydrofluoric acid), HNO_3 (nitric acid), HCOOH (formic acid), CH_3COOH (acetic acid), $HClO_4$ (perchloric acid), and H_3NSO_3 (sulfamic acid) based electrolytes. Along with that corrosion mechanism in basic solutions including NH_4OH (ammonium hydroxide), KOH (potassium hydroxide), and NaOH (sodium hydroxide) is also covered. In the last sections, corrosion in neutral solutions containing different concentrations of sodium chloride (NaCl) is also described. These solutions are called brines or synthetic brines. The effect of the presence of different impurities and gases is also designated herein.

Handbook of Corrosion Engineering. DOI: https://doi.org/10.1016/B978-0-323-95185-2.00011-3

11.2 Metallic corrosion in different electrolytes

11.2.1 Acidic electrolytes

11.2.1.1 Hydrochloric acid

Hydrochloric acid or hydronium chloride (HCl) is a well-known and commonly used mineral acid and inorganic chemical. This is also called as muriatic acid. HCl is a colorless inorganic chemical species that bears a characteristic smell. An alchemist J. ibn Hayyan discovered its existence surround CE 800. Hydrochloric acid is also called acid salt or spirits of salts as later (in the 15th century) it was mined from rock salt and "green vitriol" (Iron(II) sulfate). HCl is highly soluble in water and is used as a highly acidic alternative. The acidic solution of HCl is widely used as electrolytes for different industrial and academic purposes (Ahamad, Prasad, & Quraishi, 2010; Motamedi, Tehrani-Bagha, & Mahdavian, 2013; Lebrini, Lagrenée, Vezin, Traisnel, & Bentiss, 2007). Generally, analytical grades of HCl are supplied in 33%–37% HCl form. The highly concentrated HCl solutions are slightly yellow in color. The highly concentrated HCl solutions ($\sim$15%–18% HCl) are useful for the industrial pickling process (Lu & Wang, 2008; Maanonen, 2014). Generally, the acidic pickling process is performed at around 60°C for a specific time interval. A very high or too-low immersion time is avoided. Noticeably, a very long immersion time facilitates the unwanted dissolution of metallic components. In many cases, HCl-based pickling is performed at relatively low temperatures (below 15°C) to minimize metallic losses (Lu & Wang, 2008; Maanonen, 2014). Pickling practice is adopted in specific equipment called a pickling bath. Noticeably, every 10°C increase in temperature increases pickling efficiency by two times (Lu & Wang, 2008; Maanonen, 2014). Some metallic alloys such as stainless steel are highly susceptible to pitting corrosion in a pickling environment therefore, acid pickling cannot be performed in such conditions. The pickling cannot be zinced parts as zinc is relatively vulnerable to corrosion. The EU 1272/2008 divides pickling solutions into three different concentration ranges (Table 11–1). The pickling conditions for different alloys in HCl solutions are presented in Table 11–2.

A literature survey showed that the use of 15%–22% HCl solutions for 7–10 minutes of immersion time in the temperature range of 30°C–40°C represents the best picking condition

Table 11–1 EU 1272/2008 classification of HCl (Regulation No. 1272/2008 of the European Parliament, & of the Council of 16 December 2008 on classification, labelling & packaging of substances and mixtures, amending and repealing Directives 67/548/EEC and 1999/45/EC, and amending Regulation EC No 1907/2006, 2008).

S. no.	HCl concentration	Specification
1	≥ 25%	Hazardous statement code: H 314 Adverse effect: eye damage and skin burns.
2	10%–25%	Hazardous statement codes: H315 and H319 Adverse effect: serious skin and eye irritations
3	≥ 10%	Hazardous statement code: H335 Adverse effect: respiratory irritation.

Table 11–2 Conditions for HCl-based pickling for different alloys (Konlechner, 2014a).

S. no.	Pickling bath temperature	HCl concentration	Example
1	Ambient to 80°C	10%–25%	Carbon steel
2	Ambient to 80°C	Various	Stainless steel
3	Ambient to 60°C	5%–20%	Cu and Cu alloys
4	Ambient	1%–5%	Steel deoxidation
5	Ambient	3%–8%	Neutralization of alkali residues

(Lu & Wang, 2008; Maanonen, 2014). It is important to notice that lower concentrations of HCl, especially 1–3 M HCl are employed by academicians to determine the corrosion inhibition efficiency of organic compounds. The corrosion inhibitors used during pickling processes should be associated with the following properties:

1. Effective retardation of metal dissolution,
2. Lack of overpickling,
3. Effective at low concentration,
4. Effective at high temperatures,
5. Thermally and chemically stable,
6. Minimize hydrogen entry into the metal,
7. Excellent surfactant properties,
8. Good foaming properties.

11.2.1.2 Sulfuric acid (H_2SO_4)

Sulfuric acid (H_2SO_4) is also known as the oil of vitriol. This is one of the most frequently available and commonly used mineral acids. The sulfuric acid solution of different concentrations is widely used for pickling processes, especially in the United States. Like HCl-based picking, pickling efficiency in H_2SO_4 solutions is highly dependent upon the temperature. Based on EU regulation 1272/2008 guidelines, H_2SO_4-based solutions can be classified into the three-concentration range as illustrated in Table 11–3. The temperature and H_2SO_4 concentration ranges for the pickling of different metals and alloys are summarized in Table 11–3. The different pickling conditions for different alloys are presented in Table 11–4. Inspection of the table shows that 5%–25% H_2SO_4 solutions are mostly employed for the pickling of steel and copper alloys. Noticeably, different concentrations of H_2SO_4 concentrations are also used by numerous academicians to demonstrate the anticorrosive effectiveness of newly discovered inhibitor molecules.

11.2.1.3 Phosphoric acid

Phosphoric acid is also called orthophosphoric acid or phosphoric (V) acid. This is a colorless syrupy liquid and has a molecular formula of H_3PO_4. In its pure form, it exists as a colorless solid, and compared to mineral acids, this is a relatively weaker acid. It can exist in its

Table 11–3 EU 1272/2008 regulation classification of H_2SO_4 (Regulation No. 1272/2008 of the European Parliament, & of the Council of 16 December 2008 on classification, labelling & packaging of substances and mixtures, amending and repealing Directives 67/548/EEC and 1999/45/EC, and amending Regulation EC No 1907/2006, 2008).

S. no.	H_2SO_4 concentration	Hazardous statement code	Adverse effect
1	≥ 15%	1A	Skin corrosion
		H314	Eye damage and skin burns
2	5%–15%	–	Skin irritation 2
		–	Eye irritation 2
		H315	Serious skin irritation
		H319	Serious eye irritation

Table 11–4 Temperature and H_2SO_4 concentration range for pickling of different metals and alloys in HCl solution (Konlechner, 2014b).

S. no.	Pickling bath temperature	H_2SO_4 concentration (%)	Example
1	50°C–100°C	10–25	Carbon steel
2	60°C–100°C	5–25	Steel alloys
3	50°C–90°C	10–60	Cu and Cu alloys
4	Ambient	2–10	Steel deoxidation
5	Ambient	3–6	Neutralization of alkali residues
6	Ambient to 80°C	96	Glass and dull burning in a mixture of sulfuric and nitric acid

different deprotonated forms such as $H_2PO_4^-$ (dihydrogen phosphate ion), HPO_4^{2-} (hydrogen phosphate ion), and $PH_3O_4^-$ (phosphate ion). Like HCl and H_2SO_4 solutions, phosphoric acid-based solutions are highly effective oxide removers. The 15%–70% solutions of phosphoric acids are highly used to remove the surface oxide impurities. At its low concentrations, that is, 1%–2%, phosphoric acid forms metal phosphate that provides anticorrosive protection up to some extent. As per the EU 1272/2008 regulation, the classification of H_3PO_4 is presented in Table 11–5, and different pickling conditions in its solutions are given in Table 11–6.

11.2.1.4 Nitric acid (HNO_3) and miscellaneous

Nitric acid (HNO_3) is also called aqua fortis or spirit of niter. This is an extremely corrosive and strong mineral acid. The aqueous solutions of HNO_3 turn yellow due to the formation of nitrogen oxides. The commercial and fuming nitric acid grades are available in 68% and 86%-95%, respectively. The HNO_3 solutions are very rarely used for the pickling process as

Table 11–5 EU 1272/2008 regulation classification of H_3PO_4 (Regulation No. 1272/2008 of the European Parliament, & of the Council of 16 December 2008 on classification, labelling & packaging of substances and mixtures, amending and repealing Directives 67/548/EEC and 1999/45/EC, and amending Regulation EC No 1907/2006, 2008).

S. no.	H_3PO_4 concentration	Hazardous statement code	Adverse effect
1	≥ 15%	1A	Skin corrosion
		H314	Eye damage and skin burns
2	5%–15%	–	Skin irritation 2
		–	Eye irritation 2
		H315	Serious skin irritation
		H319	Serious eye irritation

Table 11–6 Temperature and H_3PO_4 concentration range for pickling of different metals and alloys in H_3PO_4 solution (Konlechner, 2014c).

S. no.	Pickling bath temperature	H_3PO_4 concentration (%)	Example
1	40°C–80°C	2–20	Steel
2	40°C–80°C	10–40	Pickling of Cu and Cu alloys
3	40°C–80°C	15–60	Burning of Cu and Cu alloys
4	40°C–80°C	5–30	Al and Al alloys
5	Variable	2–20	Cleaning solutions
6	Variable	2–10	Phosphating solution
7	40°C–80°C	10–60	H3PO4 mixture with other acids

they are highly corrosive. For example, pure copper and its alloys completely dissolve in HNO_3 solutions. However, some of the metals show slight resistance due to the formation of nitrates, for example, silver forms silver nitrate in HNO_3. The pickling conditions for some common alloys in HNO_3 solutions are shown in Table 11–7. Apart from the above electrolytes, the sulfamic acid ($H_3NSO_3$and perchloric acid ($HClO_4$) solutions are also widely used for different cleaning practices and corrosion inhibition studies of various organic compounds. More so, natural and synthetic crude oils are also used as electrolytes in many corrosion inhibition studies. These electrolytes contain various corrosive gases and chemicals such as organic amines and organic acids. These species are corrosive either by themselves or by combining with other species including water. Other acidic electrolytes include HF however it is extremely hazardous if spilled. Acetic acid (CH_3COOH), citric acid ($(COOH)_2$), and formic acid (HCOOH) and their HCl and an H_2SO_4-based mixtures can also be used as electrolytes. $(COOH)_2$ is cheaper and safer to use than HCOOH and CH_3COOH. H_3NSO_3 (sulfamic acid) is mostly used to remove iron oxide and carbonate scales. Perchloric acid ($HClO_4$) is mainly employed to clean stainless steel surfaces. The mixture of acidic solutions can be used to remove complex impurities.

Table 11–7 Temperature and HNO_3 concentration range for the pickling of different metals and alloys (Konlechner, 2014d).

S. no.	Pickling bath temperature	HNO_3 concentration (%)	Example
1	RT to 60°C	60–65	Heat treatment of Cu and Cu-based alloys
2	RT to 60°C	60–65	Matte and gloss burning in H_2SO_4 and HNO_3 mixture
3	RT to 60°C	3–25	Still alloys and other metals in a mixture of HF and HNO_3
4	RT to 60°C	1–30	Al, Mg, Zn, Cu, and other metals
5	RT to 50°C	25–35	Passivation
6	RT to 60°C	10–65	Etching of different metals
7	RT to 60°C	5–35	Other substances

11.2.2 Corrosion in acidic electrolytes

In aqueous solutions, metals undergo corrosive degradation by two reactions, namely anodic oxidation, and cathodic hydrogen evolution reactions. The anodic and cathodic reactions are presented below:

1. Anodic oxidation

$$M \rightarrow M^{n+} + ne^- \tag{11-1}$$

2. Cathodic reduction

$$H^+ + e^- \rightarrow \frac{1}{2}H_2 \tag{11-2}$$

The metallic corrosion in different acidic electrolytes can be presented as follows:

1. *In the HCl medium*: Metal (M) and metal oxide (MO and M_2O_3) can dissolve in the HCl medium using the following mechanisms:

$$2HCl + M \rightarrow H_2 + MCl_2 \tag{11-3}$$

$$3HCl + 2M' \rightarrow 3H_2 + 2M'Cl_3 \tag{11-4}$$

$$2HCl + MO \rightarrow H_2O + MCl_2 \tag{11-5}$$

$$6HCl + M_2O_3 \rightarrow 3H_2O + 2MCl_3 \tag{11-6}$$

where M and M′ are the di- and trivalent metals, respectively.

2. *In the H_2SO_4 medium*: Metal (M) and metal chloride (MCl_2) can dissolve in the H_2SO_4 medium using the following mechanisms:

$$H_2SO_4 + M \rightarrow H_2 + MSO_4 \tag{11-7}$$

$$H_2SO_4 + MCl_2 \rightarrow 2HCl + MSO_4 \quad (11-8)$$

where M represents a divalent metallic element. Similarly, metallic corrosion also takes in other acidic solutions.

11.2.3 Corrosion in neutral electrolytes

11.2.3.1 Corrosion in NaCl medium

The NaCl-based electrolytes can be considered neutral solutions and their different concentrations, mainly 3% and 3.5% NaCl solutions, are very commonly used for metals and alloys. The mechanism of corrosion of different metals in such electrolytes is presented as follows:

1. *Corrosion of iron*

$$Fe \rightarrow 2e^-/3e^- + F^{2+/3+} \text{Anodic reaction} \quad (11-9)$$

$$2H_2O + O_2 + 4e^- \rightarrow 4OH^- \text{Cathodic reaction} \quad (11-10)$$

The ions so formed react with chloride and produce ferrous and ferric chloride as shown below:

$$Fe^{2+} + 2Cl^- \rightarrow FeCl_2 \quad (11-11)$$

$$Fe^{3+} + 3Cl^- \rightarrow FeCl_3 \quad (11-12)$$

The formation of hydroxides can be illustrated as per Eqs. (11–13)–(11–16):

$$2OH^- + Fe^{2+} \rightarrow Fe(OH)_2 \quad (11-13)$$

$$3OH^- + Fe^{3+} \rightarrow Fe(OH)_3 \quad (11-14)$$

$$2H_2O + FeCl_2 \rightarrow 2HCl + Fe(OH)_2 \quad (11-15)$$

$$3H_2O + FeCl_3 \rightarrow 3HCl + Fe(OH)_3 \quad (11-16)$$

2. *Corrosion of aluminum*: Aluminum is also liable to corrode in NaCl media. Dissolution of Al can be presented below:

$$Al \rightarrow Al^+ + 3e^- \text{ Anodic reaction} \quad (11-17)$$

$$2H_2O + O_2 + 4e^- \rightarrow 4OH^- \text{ Cathodic reaction} \quad (11-18)$$

The Al ion so formed reacts with water and forms different hydroxide complexes immediately. The formation of complexes can be presented as follows:

$$H_2O + Al^{3+} \Leftrightarrow H^+ + [Al(OH)]^{2+} \quad (11-19)$$

$$2H_2O + Al^{3+} \Leftrightarrow 2H^+ + [Al(OH)_2]^+ \quad (11-20)$$

$$3H_2O + Al^{3+} \Leftrightarrow 3H^+ + [Al(OH)_3] \quad (11-21)$$

$$4H_2O + Al^{3+} \Leftrightarrow 4H^+ + [Al(OH)_4]^- \quad (11-22)$$

$$5H_2O + Al^{3+} \Leftrightarrow 5H^+ + [Al(OH)_5]^{2-} \quad (11-23)$$

$$6H_2O + Al^{3+} \Leftrightarrow 6H^+ + [Al(OH)_6]^{3-} \quad (11-24)$$

The development of the hexa-coordinated complex takes place in segments of a microsecond. Some polynuclear complexes also form as per the following mechanisms:

$$2H_2O + 2Al^{3+} \rightarrow 2H^+ + [Al_2(OH)_2] \quad (11-25)$$

$$4H_2O + 2Al^{3+} \rightarrow 4H^+ + [Al_2(OH)_4] \quad (11-26)$$

Aluminum ions can also react with chloride ions to form several reactive intermediates as shown in Eq. (11–27):

$$Cl^-_{excess} + Al^{3+} \rightarrow AlCl_3/AlCl^{2+}/AlCl_2^+ \quad (11-27)$$

The mono-, di-, and trichloro-compounds are exceedingly sensitive and effortlessly liquify in H_2O, HCl, and NaCl. Consequently, the development of these intermediates indorses the Al corrosion in NaCl medium.

3. *Corrosion of aluminum*: Copper corrosion in NaCl is also ahead precisely courtesy because of its feasting in several industries together with marine fields. The existence of water formation of CuCl, $CuCl_2$, and Cu_2O can be presented as follows:

$$H_2O + 2Cu \rightarrow 2H^+ + Cu_2O + 2e^- \quad (11-28)$$

$$2H^+ + 2Cu + \frac{1}{2}O_2 \underset{-H_2O}{\longrightarrow} 2Cl^- + 2Cu^+ \rightarrow CuCl \quad (11-29)$$

$$2H^+ + Cu + \frac{1}{2}O_2 \underset{-H_2O}{\longrightarrow} 2Cl^- + Cu^{2+} \rightarrow CuCl_2 \quad (11-30)$$

Thus, CuCl, $CuCl_2$, and Cu_2O are formed as corrosion products. However, the formation of $Cu_2(OH)_3Cl$ (atacamite) is also possible.

$$2H_2O+Cu_2O+Cl^- \rightarrow H^+ + Cu_2(OH)_3Cl+e^- \quad (11-31)$$

11.2.3.2 Corrosion in NaCl medium saturated with CO_2

The crude oils contain NaCl solution and CO_2 as an impurity. The anodic and cathodic reactions (for iron) in such conditions can be written as follows:

1. Anodic reactions

$$Fe \rightarrow Fe^{2+} + 2e^- \quad (11-32)$$

However, the above reaction proceeds in several steps which initiates through the adsorption of water as presented below:

$$H_2O + Fe \rightarrow Fe \cdot H_2O_{ads} \quad (11-33)$$

$$OH^- + Fe \cdot H_2O_{ads} \rightarrow H_2O + FeOH^-_{ads} \quad (11-34)$$

$$FeOH^-_{ads} \rightarrow e^- + FeOH_{ads} \quad (11-35)$$

$$FeOH_{ads} \rightarrow e^- + FeOH^+ \quad (11-36)$$

$$FeOH^+ \rightarrow e^- + Fe^{2+} \quad (11-37)$$

However, water easily replaces with chloride ions as presented below:

$$Cl^- + Fe \cdot H_2O_{ads} \rightarrow H_2O + FeCl^-_{ads} \quad (11-38)$$

$$OH^- + FeCl^-_{ads} \rightarrow Cl + ^-FeOH^+ + 2e^- \quad (11-39)$$

In an aqueous medium, CO_2 converts into carbonic acid (H_2CO_3). In such medium major cathodic reactions can be a reduction of water, carbonic acid, and H^+.

$$2H_2CO_{3(aq)} + 2e^- \rightarrow H_{2(g)} + 2HCO^-_{3(aq)} \quad (11-40)$$

$$2H^+_{(aq)} + 2e^- \rightarrow + H_{2(g)} \quad (11-41)$$

$$2H_2O_{(L)} + 2e^- \rightarrow + H_{2(g)} + 2OH^- \quad (11-42)$$

11.2.4 Corrosion in oil and gas industries (industrial electrolytes)

The oil and gas industry is badly affected by corrosion failure assisted by a few corrosive gases. CO_2 and H_2S are the major corrosive gases, and they are responsible for sweet and sour corrosion, respectively. Most of the oil and gas industries contain both gases. The mechanism and rate of corrosion greatly depend upon the partial pressure of CO_2 and H_2S, represented by P_{CO2} and P_{H2S}, respectively. The mechanism of sweet and sour corrosion is depicted below:

11.2.4.1 Sweet corrosion

The dry CO_2, that is, in the absence of water, CO_2 is not corrosive. However, in the presence of water, it dissolves and gets converted into carbonic acid. The anodic and cathodic reactions for a divalent metal such as iron in a CO_2 medium can be written as follows:

1. Anodic reactions

$$M \leftrightarrow 2e^- + M^{2+} \tag{11-43}$$

$$CO_3^{2-} + M^{2+} \leftrightarrow MCO_3 \tag{11-44}$$

$$2HCO_3^{2-} + M^{2+} \leftrightarrow M(HCO_3)_2 \tag{11-45}$$

$$M(HCO_3)_2 \leftrightarrow H_2O + CO_2 + MCO_3 \tag{11-46}$$

In the above equations, carbonate and bicarbonate form by the dissociation of carbonic acid (H_2CO_3).

2. Cathodic reactions

$$H_2O + CO_2 \leftrightarrow H_2CO_3 \tag{11-47}$$

$$2HCO_3^{2-} + 2e^- \leftrightarrow 2CO_3^{2-} + H_{2(g)} \tag{11-48}$$

$$HCO_3^{2-} \leftrightarrow CO_3^{2-} + H^+ \tag{11-49}$$

The overall reaction can be presented as follows:

$$M + CO_{2(g)} + H_2O_{(l)} \rightarrow H_2O_{(g)} + MCO_3 \tag{11-50}$$

Several factors affect the metallic corrosion in CO_2 medium. A detailed description can be found elsewhere.

11.2.4.2 Sour corrosion

The metallic corrosion in H_2S is known as sour corrosion and it can take place through the development of metal sulfide (Ms) as per the following steps:

$$H_2S + M \leftrightarrow H^+ + MSH^- \tag{11-51}$$

$$MSH^- \leftrightarrow MSH + e^- \tag{11-52}$$

$$MSH \leftrightarrow MSH^{+} + e^{-} \quad (11-53)$$

$$MSH^{+} \leftrightarrow MS + H^{+} \quad (11-54)$$

11.2.5 Corrosion in basic electrolytes

The NaOH, KOH, and NH_4OH solutions are sometimes used to clean the metal surface. The dissolution of metals in these solutions can be illustrated as follows:

1. Anodic reactions

$$M \rightarrow M^{2+} + 2e^{-} \quad (11-55)$$

$$M' \rightarrow M'^{3+} + 3e^{-} \quad (11-56)$$

$$2H_2O + M^{2+} \rightarrow 2H^{+} + M(OH)_2 \quad (11-57)$$

$$3H_2O + M'^{3+} \rightarrow 3H^{+} + M'(OH)_3 \quad (11-58)$$

2. Cathodic reactions

$$2H_2O + MCl_2 \rightarrow 2HCl + M(OH)_2 \quad (11-59)$$

$$3H_2O + M'Cl_3 \rightarrow 3HCl + M(OH)_3 \quad (11-60)$$

$$H^{+} + e^{-} \rightarrow \frac{1}{2}H_2 \quad (11-61)$$

$$\frac{1}{2}O_2 + H_2O + 2e^{-} \rightarrow 2OH^{-} \quad (11-62)$$

M and M′ signify the di- and trivalent metal, respectively. Cathodic reactions accessible in Eqs. (11–59) and (11–60) are possible only in the presence of chloride ions.

Short questions

1. What is the significance of electrolytes in metallurgy?
2. Write the chemical name and formula of some common electrolytes and describe the mechanism of corrosion.
3. What does "sweet and sour corrosion" refer to?
4. How impurities present in corrosive electrolytes affect the corrosion rate?
5. Is there any correlation between corrosion rate and aggressiveness of an electrolyte?
6. Describe the effect of immersion time on corrosion rate in an electrolyte with and without a corrosion inhibitor.

7. Describe the corrosion rate phenomenon at the interface of an electrolyte and air (differential aeration corrosion cells).
8. What are the most common electrolytes for descaling and surface cleanings processes?
9. What are the most common electrolytes for acidizing/acid pickling process?
10. Describe the effect of different parameters on sweet corrosion.

References

Ahamad, I., Prasad, R., & Quraishi, M. (2010). *Corrosion Science, 52,* 933–942.

Azevedo, M. S., Allély, C., Ogle, K., & Volovitch, P. (2015). *Corrosion Science, 90,* 482–490.

Hegazy, M., Badawi, A., Abd El Rehim, S., & Kamel, W. (2013). *Corrosion Science, 69,* 110–122.

Izquierdo, J., Santana, J. J., González, S., & Souto, R. M. (2010). *Electrochimica Acta, 55,* 8791–8800.

Konlechner, D. (2014a). *Metal pickling by inorganic acids 1 – hydrochloric acid HCl.* https://www.kon-chem.com/inorganic-acids-for-pickling-of-metals-1-hydrochloric-acid-hcl/?lang = en

Konlechner, D. (2014b). *Metal pickling by inorganic acids 2 – sulfuric acid H2SO4.* https://www.kon-chem.com/metal-pickling-by-inorganic-acids-2-sulfuric-acid-h2so4/?lang = en

Konlechner, D. (2014d). *Metal pickling by inorganic acids 3 – nitric acid HNO3.* https://www.kon-chem.com/metal-pickling-by-inorganic-acids-nitric-acid-hno3/?lang = en

Konlechner, D. (2014c). *Metal pickling by inorganic acids 4 – phosphoric acid H3PO4.* https://www.kon-chem.com/metal-pickling-by-inorganic-acids-4-phosphoric-acid-h3po4/?%20lang = en

Lebrini, M., Lagrenée, M., Vezin, H., Traisnel, M., & Bentiss, F. (2007). *Corrosion Science, 49,* 2254–2269.

Lu, B.-n & Wang, S.-w (2008). Zhejiang Chemical Industry.

Maanonen, M. (2014). *Steel Pickling in Challenging conditions.* Materials Technology and Surface Engineering – Thesis. Helsinki Metropolia University of Applied Sciences.

Motamedi, M., Tehrani-Bagha, A. R., & Mahdavian, M. (2013). *Corrosion Science, 70,* 46–54.

Muldoon, J., Bucur, C. B., Oliver, A. G., Zajicek, J., Allred, G. D., & Boggess, W. C. (2013). *Energy & Environmental Science, 6,* 482–487.

Regulation No. 1272/2008 of the European Parliament and of the Council of 16 December 2008 on classification, labelling and packaging of substances and mixtures, amending and repealing Directives 67/548/EEC and 1999/45/EC, and amending Regulation (EC) No 1907/2006*, EC, and amending regulation, L: Legislation (2008) 353.*

Schmitt, G. (1984). *British Corrosion Journal, 19,* 165–176.

Stratmann, M., Streckel, H., Kim, K., & Crockett, S. (1990). *Corrosion Science, 30,* 715–734.

Verma, C., Ebenso, E. E., & Quraishi, M. (2017). *Journal of Molecular Liquids, 233,* 403–414.

Verma, C., Olasunkanmi, L. O., Ebenso, E. E., Quraishi, M. A., & Obot, I. B. (2016). *The Journal of Physical Chemistry C, 120,* 11598–11611.

Wu, Y., Zhang, P., Pickering, H., & Allara, D. (1993). *Journal of the Electrochemical Society, 140,* 2791.

Zaferani, S. H., Sharifi, M., Zaarei, D., & Shishesaz, M. R. (2013). *Journal of Environmental Chemical Engineering, 1,* 652–657.

PART 3

Electrochemistry basics & principles of corrosion

12 Principles of corrosion

Although the corrosion process is considered complex and complicated to understand, the detailed explanation is even more complicated. If the basic principle involved in the process is understood, it will be easier to find a way to control the corrosion. So, in this chapter, the outline of some basic principles of the corrosion process is illustrated. This chapter covers the basic thermodynamic and kinetic principles involved in the corrosion process.

12.1 Corrosion reactions

Metal atoms oxidize and exit the metal lattice as ions, which causes metallic corrosion. The anodic half-reaction is the name given to the metal's oxidation process. Anodes are the places on a metal surface where oxidation takes place. The cathodic half-reaction, in which electrochemically active species in the electrolyte are reduced, is known as reduction. The regions on the metal's surface where reduction occurs are known as cathodes. Anodes and cathodes may be as small as atoms or as big as to be visible to the unaided eye.

As long as positive and negative ions travel through the electrolyte (environment on the metal surface) in the direction of the anodes and cathodes, respectively, to preserve the electrical charge neutrality of the metal and electrolyte, anodes and cathodes can be separated by finite distances. Metallic corrosion must be initiated and maintained by both cathodic and anodic processes. Electrons, reactive elements (ions or molecules), and ions or molecules generated by the corrosion process are all represented by symbols in electrochemical corrosion reaction equations. For instance, the anodic electrochemical equation depicts the corrosion of iron as follows:

$$Fe^{\circ} \rightarrow Fe^{2+} + 2e^{-} \qquad (12-1)$$

where Fe° represents iron atoms at the metal surface, Fe^{2+} represents iron ions, and $2e^{-}$ represents the two electrons produced by the anodic reaction. Eq. (12–1) produces free electrons, which is why it is known as an anodic half-reaction. The electron reaction in Eq. (12–1) does not have a matching equation. A cathodic half-reaction must also be stated to account for the reduction of electrochemically active species with the electrons produced in the anodic half-reaction. The reduction of hydrogen ions is an illustration of a cathodic half-reaction:

$$2H^{+} + 2e^{-} \rightarrow H_2 \qquad (12-2)$$

The overall corrosion reaction is a combination of the anodic and cathodic half-reactions:

$$Fe^{\circ} + 2H^{+} \rightarrow Fe^{2+} + H_2 \qquad (12-3)$$

Handbook of Corrosion Engineering. DOI: https://doi.org/10.1016/B978-0-323-95185-2.00012-5

According to Eq. (12–3), hydrogen ions in the metal's surroundings (i.e., the electrolyte) at the metal surface are reduced by iron atoms to make iron ions, which produce electrons that reduce hydrogen ions to form hydrogen molecules. The theory that corrosion may be halted by stopping one of the half-reactions from happening and/or eliminating electrochemically active species from the electrolyte is supported by Eq. (12–3) and it's component anodic and cathodic equations (Eqs. 12–1 and 12–2). However, corrosion is brought on by a wide variety of electrochemically active substances.

12.2 Thermodynamic state functions

A state function is one whose properties depend only on the system's present state and not on the path taken to get there. (By the "state"in thermodynamics, we mean having a particular set of descriptors rather than the specific physical state of the system). There are five standard state functions:

1. the internal energy E,
2. the entropy S,
3. the enthalpy H,
4. the Helmholtz free energy F, and
5. the Gibbs free energy G.

12.2.1 Internal energy

The internal energy, *E,* of a system is the total energy contained within the system. Internal energy includes contributions from both kinetic energy and potential energy sources. These include the translational, rotational, and vibrational motion of atoms. Forces within an atom, such as the attraction of an electron to the positively charged nucleus, and intermolecular forces between molecules (e.g., Van der Waals forces) are included. Internal energy also includes gravitational forces, magnetic forces, and electrical forces. The fact that E is a state function means that when the system changes from some condition (condition 1) to another condition (condition 2), the change in internal energy is given by:

$$E = E_2 - E_1 \tag{12-4}$$

Because both E_2 and E_1 are state functions, their change ($E_2 - E_1$) does not depend on the path taken between 1 and 2.

It is important to note that while the internal energy E is a state function, the quantities q and w are not. Heat and work are not contained in a system but are observed when a system changes from one state to another.

The first law of thermodynamics states that small changes in internal energy:

$$dE = q + w \tag{12-5}$$

where q is the heat absorbed by the system and w is the work done on the system. It is significant to remember that while the values q and w are not state functions, the internal energy E is.

12.2.2 Entropy

The entropy S is defined by:

$$dS = \frac{q}{T} \tag{12-6}$$

where T is the Kelvin temperature (°C + 273.1).

The entropy S is also a state function and has the properties:

1. $dS = S_2 - S_1$
2. $dS = 0$ for a reversible system
3. $dS > 0$ for an irreversible system

Entropy is a measure of the disorder of a system, with the entropy increasing for an irreversible system. This observation is the basis for the familiar statement that the entropy of the universe is increasing.

12.2.3 Enthalpy

The enthalpy H is defined by:

$$H = E + PV \tag{12-7}$$

where P and V are the pressure and the volume of gases involved.

The change in enthalpy (dH) between two states measures the heat of a reaction at constant pressure. That is, at constant pressure, the heat q_p (the subscript indicates constant pressure) is given by:

$$q_p = H \tag{12-8}$$

Corrosion reactions usually occur under conditions of constant pressure (i.e., atmospheric pressure).

12.2.4 Helmholtz and Gibbs free energies

The Helmholtz free energy F is defined by:

$$F = E - TS \tag{12-9}$$

And the Gibbs free energy G is given by:

$$G = H - TS \tag{12-10}$$

Both A and G are state functions. Their significance is as follows:

- At constant temperature T and volume V, the system is at equilibrium when $dF = 0$.
- At constant temperature T and pressure P, the system is at equilibrium when $dG = 0$.

Most corrosion reactions occur at constant temperature and pressure, so the Gibbs free energy G is the appropriate equilibrium measure for corrosion scientists and engineers. At constant temperature and pressure, Eq. (12–10) gives:

$$dG = dH + TdS \tag{12-11}$$

where dG, dH, and dS are the changes in Gibbs free energy, enthalpy, and entropy, respectively, when the system undergoes a change from one state to another.

12.3 Free energy and spontaneity

The notions of reversibility, equilibrium, and spontaneity are related. A reaction is reversible if it can proceed in either direction. A reaction is at equilibrium if the forward reaction rate is the same as the reverse reaction rate. A reaction is spontaneous if it proceeds in a given direction without external influence.

For the dissolution of iron in acids at room temperature, the overall reaction, Fe(s) + 2H + (aq) → Fe_2 + (aq) + H_2(g), is spontaneous at 25°C because it proceeds as written above if a piece of iron is immersed in a hydrochloric acid solution. In addition, the reaction reaches an equilibrium in which the anodic half-cell reaction, Fe(s) → Fe_2 + (aq) + $2e^-$, proceeds at the same rate as the cathodic half-cell reaction, 2H + (aq) + 2e − → H_2(g). However, the overall reaction is not reversible because we cannot treat a ferrous salt with hydrogen gas at 25°C and produce solid iron. The reduction of Fe_2O_3 can produce solid iron under special conditions and in a process that requires intensive energy input. The reduction of iron ore (Fe_2O_3) at elevated temperatures over 1000°C and in the presence of carbon (as coke) and $CaCO_3$ (limestone) is the basis for the production of iron from iron ore.

Another way of viewing corrosion is that it is the thermodynamic process by which metals revert to their natural form as ores. The change in Gibbs free energy (dG) is a powerful indicator of spontaneity. Fig. 12–1 illustrates the behavior of the change in Gibbs free energy G for the case of equilibrium, spontaneous, and nonspontaneous reactions.

12.4 Change of Gibbs free energy (dG)

As discussed above, the tendency for any chemical reaction to go, including the reaction of a metal with its environment, is measured by the Gibbs free-energy change, dG. The more negative the value of dG, the greater the tendency for the reaction to go. For example, consider the following reaction at 25°C:

$$Mg + H_2O(l) + \tfrac{1}{2}O_2(g) \rightarrow Mg(OH)_2(s)\ (dG^\circ = -596,600\ \mathrm{J/mole}) \tag{12-12}$$

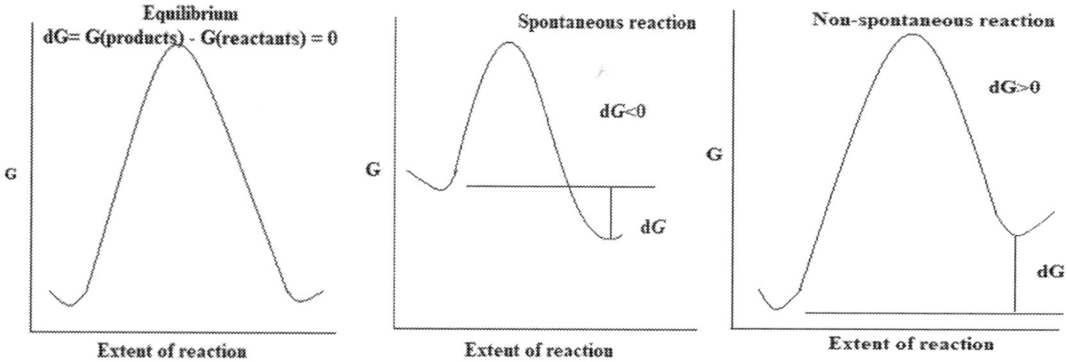

FIGURE 12–1 The change in Gibbs free energy for the case of equilibrium, spontaneous, and nonspontaneous reactions (1) is at equilibrium if the free energy change *G* is zero, (2) proceeds spontaneously if *G* is negative, (3) is not spontaneous if *G* is positive.

The large negative value of dG° (reactants and products in standard states) indicates a pronounced tendency for magnesium to react with water and oxygen. On the other hand for copper:

$$Cu + H_2O(l) + {}^1/_2O_2(g) \rightarrow Cu(OH)_2(s)\ \left(dG^\circ = -119,700\ J/mole\right) \qquad (12-13)$$

The reaction tendency is less or we can say that the corrosion tendency of copper in aerated water is not as pronounced as that of magnesium. Finally, for gold:

$$Au + 3/2H_2O(l) + 3/4O_2(g) \rightarrow Au(OH)_2(s)\ (dG^\circ = +65,700\ J/mole) \qquad (12-14)$$

The free energy is positive, indicating that the reaction has no tendency to go at all; and gold, correspondingly, does not corrode in aqueous media to form $Au(OH)_3$.

It should be emphasized that the tendency to corrode is not a measure of reaction rate. A large negative dG may or may not be accompanied by a high corrosion rate, but, when dG is positive, it can be stated with certainty that the reaction will not go at all under the particular conditions described. If dG is negative, the reaction rate may be rapid or slow. In view of the electrochemical mechanisms of corrosion, the tendency for a metal to corrode can also be expressed in terms of the electromotive force (emf) of the corrosion cells that are an integral part of the corrosion process. Since electrical energy is expressed as the product of volts by coulombs (joules, J), the relation between dG in joules and emf in volts, E, is defined by:

$$dG = -nFE \qquad (12-15)$$

where n is the number of electrons (or chemical equivalents) taking part in the reaction, and F is the Faraday (96,500 C/eq). The term dG can be converted from calories to joules by using the factor 1 cal = 4.184 absolute joules. Accordingly, the greater the value of E for any cell, the greater the tendency for the overall reaction of the cell to go.

12.5 The Nernst equation

E^o only applies when a metal is submerged in a solution of it's own ions at unit activity, as normal electrode potentials do. Corrosion processes rarely encounter this state. The half-cell potential at a different concentration may be calculated using the Nernst equation in terms of the potential of the standard electrode. The Nernst equation essentially represents a cell's emf in terms of the activity of the products and reactants involved in the cell process. Consider a general cell reaction:

$$M_1 + M_2{}^{n+} \leftrightarrow M_2 + M_1{}^{n+} \qquad (12-16)$$

For the following reaction Eq. (12–16), M_1 and M_2 represents metal electrodes in a cell, such as Cu and Zn. Now the above reaction can be written as follows:

$$Cu + Zn^{2+} \leftrightarrow Zn + Cu^{2+} \qquad (12-17)$$

The difference between the molar free energies of the products and reactants determines the free energy change (dG) of a process.

$$dG = \sum_{ni} dG(i) - \sum_{ni} dG(j) \qquad (12-18)$$

As a result, the reaction's free energy change may be written as:

$$dG = (G_{M_2} + G_{M_1{}^{n+}}) - (G_{M_1} + G_{M_2{}^{n+}}) \qquad (12-19)$$

If the substances are in their standard states the above expression becomes:

$$dG^\circ = (G^\circ{}_{M_2} + G^\circ{}_{M_1{}^{n+}}) - (G^\circ{}_{M_1} + G^\circ{}_{M_2{}^{n+}}) \qquad (12-20)$$

The following equation relates the free energy of any metal, such as M_1 in its standard state, with any other state:

$$G_{M1} - G^\circ{}_{M1} = RT\ln a_{M1} \qquad (12-21)$$

where a_{M1} is the activity of metal M_1, R is the universal gas constant, and equal to 8.314 joules/degree/mole, and T is the absolute temperature in Kelvin, which is (273.16 + °C). Subtracting Eq. (12–20) from Eq. (12–19), we get:

$$dG - dG^\circ = \left(G_{M_2} - G_{M_2}{}^\circ\right) + \left(G_{M_1{}^{n+}} - G^\circ_{M_1{}^{n+}}\right) - \left(G_{M_1} - G^\circ_{M_1}\right) + \left(G_{M_2{}^{n+}} - G^\circ_{M_2{}^{n+}}\right) \qquad (12-22)$$

Substituting from Eq. (2–51), we obtain:

$$dG = dG^\circ = RT\ln a_{M2} + RT\ln a_{M1} - RT\ln a_{M2} + RT\ln a_{M2^{n+}} \qquad (12-23)$$

$$dG - dG^{\circ} = RT\ln\left[\frac{a_{M_2}a_{M^{n+}_1}}{a_{M_1}a_{M^{n-}2}}\right] \quad (12-24)$$

where the "*a*" values are any arbitrary activity values, for which dG is the corresponding free energy of that reaction. dG° is the free energy change for the reaction involving standard states (unit activities). As $dG = -nFE$, and also $dG^{\circ} = -nFE^{\circ}$, Eq. (12–24) becomes:

$$E - E^{\circ} = \frac{RT}{nF}\ln\left[\frac{a_{M_1}a_{M^{n-}_2}}{a_{M_1}a_{M_1^{n-}}}\right] \quad (12-25)$$

$$E - E^{\circ} = \frac{RT}{nF}\ln\left[\frac{a_{product}}{a_{reactant}}\right] \quad (12-26)$$

Eq. (12–26) is a general form of the Nernst equation.

suppose,

$$M \rightarrow M^{z+} + ze^{-}$$

$$E_M = E_M{}^{\circ} - \frac{RT}{nF}\ln\left[a_M{}^{z+} \times za^{z}{}_{e-}\right] \quad (12-27)$$

$$E_M = E_M{}^{\circ} - \frac{RT}{nF}\ln\left[a_M{}^{z+}\right] \quad (12-28)$$

As a_M and $a_e = 1$.

at 25°

$$E_M = E_M{}^{\circ} - \frac{0.059}{2}\ln\left[a_M{}^{z+}\right] \quad (12-29)$$

It expresses the emf involved in an electrochemical reaction in terms of *R, T, F, n* activities of products and activities of reactants. However, the IUPAC sign convention requires the above minus sign to be replaced by a plus sign. The Nernst equation is re-written to comply with the IUPAC sign convention, as:

$$E = E^{\circ} + \frac{RT}{nF}\ln\frac{\text{activity product of oxidized state}}{\text{activity product of reduced state}} \quad (12-30)$$

12.6 Pourboix diagram or stability diagram

Potential-pH diagrams are commonly referred to as Pourbaix diagrams after its originator, Belgian corrosion scientist and electrochemist Pourbaix (1963) (Portero and Ant ó n, 2007; Thompson et al., 2000). These graphs demonstrate how voltage and pH affect a metal's stability. They are analogs of phase equilibrium diagrams, which show how temperature and the proportion of the metal in each phase affect the stability of various phases.

The diagram's axes represent the main variables that the corrosion may affect. The amount of oxidizer in the solution may be changed, and the metal can also be electrically potentialized, to change the potential between the metal and the solution, which is shown on the vertical axis. On the horizontal axis, the solution's pR is shown. The dominant species in each location is shown together with the zones of stability on the diagrams. In order to demonstrate the corrosion limits where soluble corrosion products would be scarcely detectable, the boundaries are frequently drawn at equilibrium concentrations of 10^{-6}M for locations where metal ions are stable. When solution concentrations are known to be greater than 10^{-6}M or the temperature is not 25°C (77°F), the figure can be altered to represent the real circumstances in a given corrosion scenario.

12.6.1 Pourbaix diagram for water

The conditions of thermodynamic equilibrium for each reaction are represented by a line in a Pourbaix diagram. Fig. 12–2 shows the Pourbaix diagram for water. Oxygen is produced above line b. According to the Nernst equation, potential and pH have the following relationship for this equilibrium:

$$\phi = \phi^{\circ} - 2.303\frac{RT}{2F}\log\frac{1}{(\mathrm{H}^{+})^{2}} \quad (12-31)$$

With

$$\varphi^{\circ} = 1.229, \mathrm{T} = 298.2\mathrm{K}, \mathrm{R} = 8.314\mathrm{J}/\mathrm{deg} - \mathrm{mole}, \text{and } \mathrm{F} = 96,500C/\mathrm{eq},$$

$$\varphi = 1.229 - 0.0592\mathrm{pH} \quad (12-32)$$

Above line b, which is specified by this equation, oxygen is evolved at the surface of an immersed electrode. Below this line, the water is stable.

Hydrogen is developed below the line and in accordance with the reaction $2H^{+} + 2e^{-} \rightarrow H_2$. The link between potential and pH for this equilibrium is given by the Nernst equation:

$$\phi = \phi^{\circ} - 2.303\frac{RT}{2F}\log\frac{1}{(\mathrm{H}^{+})^{2}} = -0.0592\mathrm{pH} \quad (12-33)$$

With

$$\varphi^{\circ} = 0$$

$$\varphi = -0.0592 \text{ pH}$$

A submerged electrode's surface releases hydrogen gas below line a, which is represented by this equation. Water is steady between lines a and b.

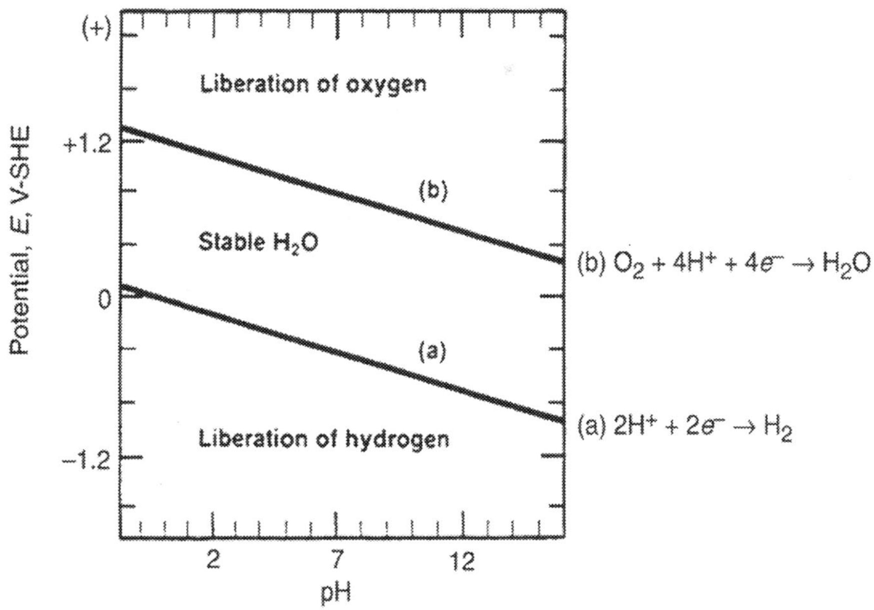

FIGURE 12–2 Pourbaix diagram for water at 25°C, showing the oxygen line, b, above which oxygen is evolved, and the hydrogen line, a, below which hydrogen is evolved, from the surface of an immersed electrode. Between these two lines, water is stable.

12.6.2 Pourbaix diagram for iron

Fig. 12–3 displays the Pourbaix diagram for iron. A horizontal line denotes a reaction without pH; in other words, one in which neither H^+ nor OH^- are present, as in the case of the reaction $Fe^{2+} + 2\ e^- \rightarrow Fe$.

Using the Nernst equation for this equilibrium:

$$\phi = \phi^\circ - 2.303\frac{RT}{2F}\log\frac{1}{(Fe^{2+})} \tag{12-34}$$

$$\phi = -0.440 + 0.0296\log(\mathrm{Fe}^{2+}) \tag{12-35}$$

If (Fe^{2+}) is assumed to be 10^6, then the horizontal line on the Pourbaix diagram, = 0.617 V, results. H^+ or OH^-, but not electrons, are involved in a vertical line, as in $2Fe^{3+} + 3H_2O \rightarrow Fe_2O_3 + 6H^+$. This reaction is shown by the vertical line separating Fe^{3+} from Fe_2O_3 in Fig. 12–3. According to this equilibrium:

$$K = \frac{(\mathrm{H}^+)^6}{(\mathrm{Fe}^{3+})^2} \tag{12-36}$$

$$\log K = 6\log(H^+) - 2\log(Fe^{3+}) \quad (12-37)$$

$$\log K = 6pH - 2\log(Fe^{3+}) \quad (12-38)$$

Since $dG^o = -RT\ln K$ and $dG^o = -8240 j/mole$

$$\log K = 1.43$$

$$\log(Fe^{3+}) = -0.72 - 3\,pH \quad (12-39)$$

Taking $(Fe^{3+}) = 10^{-6}$, we have $pH = 1.76$.

The equilibrium reaction, $2Fe^{3+} + 3H_2O \rightarrow Fe_2O_3 + 6H^+$, is represented by the vertical line at pH 1.76 in the Pourbaix diagram for iron (Fig. 12–3). Fe_2O_3 is the stable phase to the right of this line (i.e., at pH > 1.76), and as a protective film, one would anticipate that it would offer some defense against corrosion. Since ferric ions are stable in solution to the left

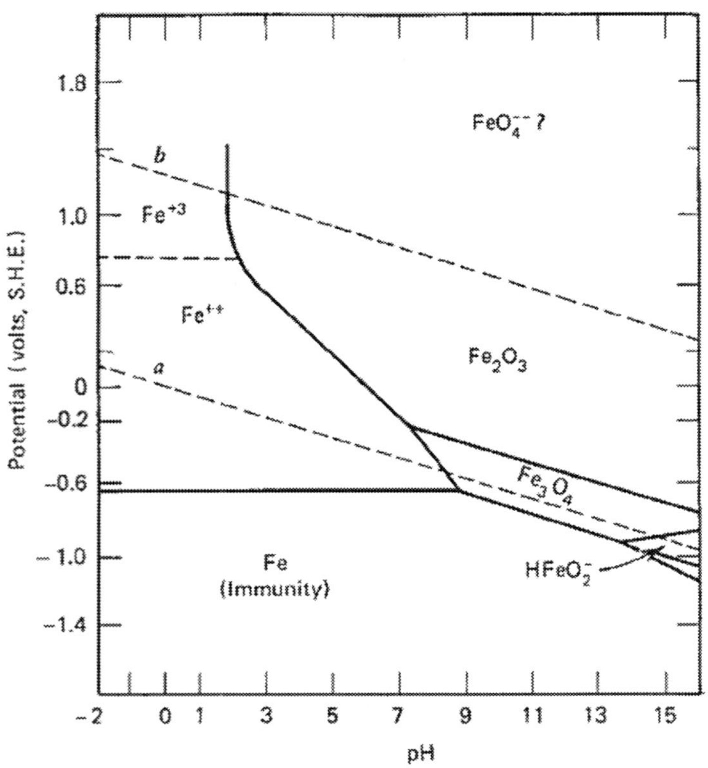

FIGURE 12–3 Pourbaix diagram for the iron–water system at 25°C.

of this line (i.e., at pH 1.76), corrosion is anticipated to occur without the aid of a surface oxide deposit.

A sloping line involves H^+, OH^-, and electrons.

The equilibrium in Fig. 12–3 is shown by this line, which reads $Fe_2O_3 + 6H^+ + 2e^- \rightarrow 2Fe^{2+} + 3H_2O$. Fe_2O_3 is a stable phase to the right of this line and is anticipated to create a surface oxide film that shields the underlying metal from corrosion. Fe^{2+} is a stable species that are present in the solution to the left of this line. The pH values of the solutions that are in direct contact with the metal surface are displayed in Pourbaix diagrams. This value can occasionally be different from the bulk solution's value, like when Fe is present in aerated water. In extremely alkaline solutions, soluble hypoferrites ($HFeO_2$) can develop within a constrained active potential range. At highly noble potentials, soluble ferrates (FeO_4^{2-}) can form in alkaline solutions, although the stable field is not well defined. When any reaction involves ions other than H^+ or OH^-, it is assumed, in general, that the activity equals 10^{-6}. Thus the horizontal line at -0.62 V means that iron will not corrode below this value to form a solution of concentration $>10^{-6}$M Fe^{2+} in accord with $Fe \rightarrow Fe^{2+} + 2e$-, $\varphi = -0.44 + (0.059/2)\log(10^{-6}) = -0.62$ V. If a value other than 10^{-6} is used, the lines separating the phases are shifted. The fields marked Fe_2O_3 and Fe_2O_3 are sometimes labeled "passivation" on the assumption that iron reacts in these regions to form protective oxide films. This is correct only insofar as passivity is accounted for by a diffusion-barrier oxide layer.

12.6.3 Pourbaix diagram of aluminum

According to the Pourbaix diagram of aluminum (Fig. 12–4), hydrargillite, $Al_2O_3 \cdot 3H_2O$, is the stable phase between pH values of 4 and 9. In fact, it is believed that this film is to blame for the widespread success of aluminum in structural applications. The amphoteric character of aluminum is also predicted by this figure, with corrosion occurring in extremely acidic and alkaline environments and being protected by an oxide sheet between pH 4 and 9.

12.6.4 Pourbaix diagram of magnesium

In contrast to aluminum, Mg^{2+} is the stable species in solution for the majority of the potential—pH range, according to the Pourbaix diagram of magnesium (Fig. 12–5). Magnesium decreases water because the potential—pH range in which it is stable is much lower than the range in which water is stable. Hydrogen evolution is the cathodic reaction. A protective coating of oxide or hydroxide may offer some protection between pH values of around 8.5 and 11. Magnesium is protected from corrosion by a stable $Mg(OH)_2$ coating at pH values higher than roughly 11.5. Magnesium cannot be produced by electrolysis of it's aqueous solutions because magnesium would not be produced due to the extremely active magnitude of the equilibrium potential of the reaction, $Mg^{2+} + 2e^- \rightarrow Mg$.

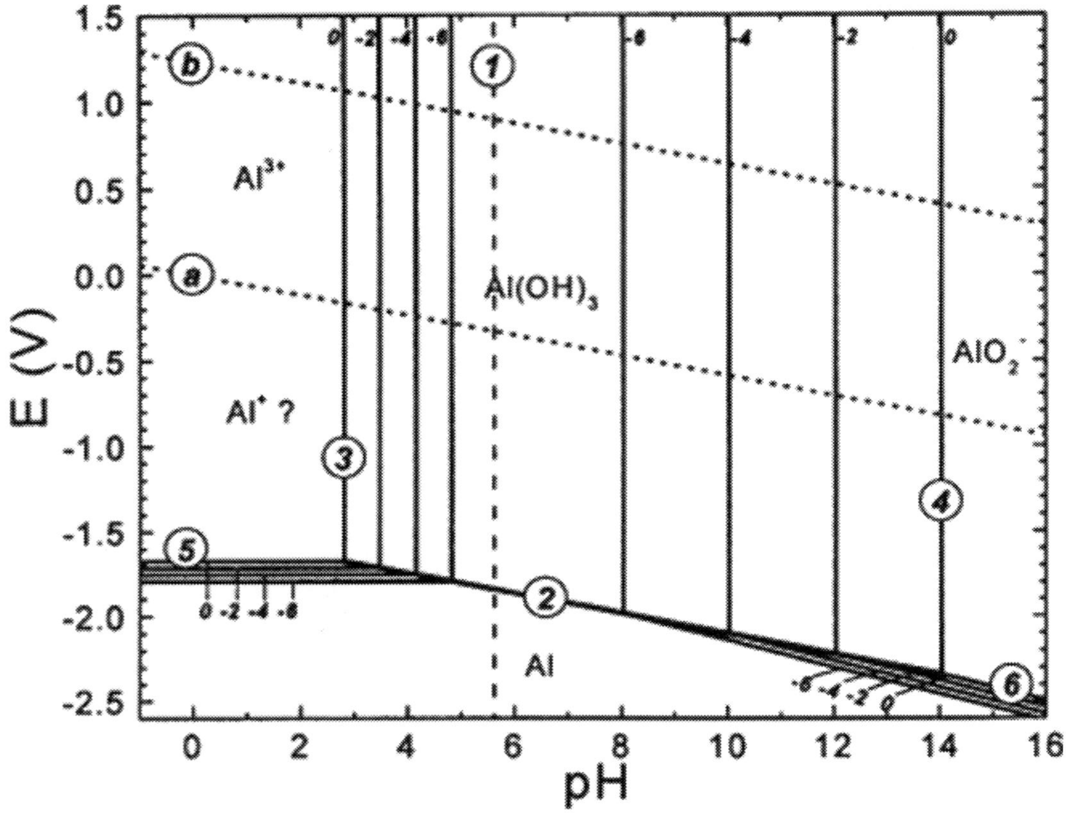

FIGURE 12–4 Pourbaix diagram for the aluminum–water system at 25°C.

12.6.5 Pourbaix diagram for copper

Fig. 12–6 depicts a Pourbaix diagram for a copper/water system. While copper is more noble than iron ($E^{\circ}_{Cu} = 0.337$ V vs $E^{\circ}_{Fe} = 0.444$ V), iron is less stable in water. Copper in acidic electrolytes is not inert. Copper oxide, Cu^{+}, and Cu^{2+} are only protective in electrolytes that are only mildly acidic and alkaline. The area of immunity in the diagram extends over the line (a) which indicates the development of oxygen. If copper was given a higher noble potential, it would corrode in both weak and severely acidic environments. Additionally, it would corrode under more oxidizing potentials and severely alkaline environments. Controlling the corrosion of metals in aquatic environments is feasible by determining the pH and potential ranges. Due to it's propensity to generate protective oxide layers that shield the metal from corrosion, aluminum is extremely active yet may be utilized with little risk of corrosion. Titanium has an excellent tendency for passivation as shown by a highly extended passive zone that extends over the entire pH range, but it corrodes under reducing and

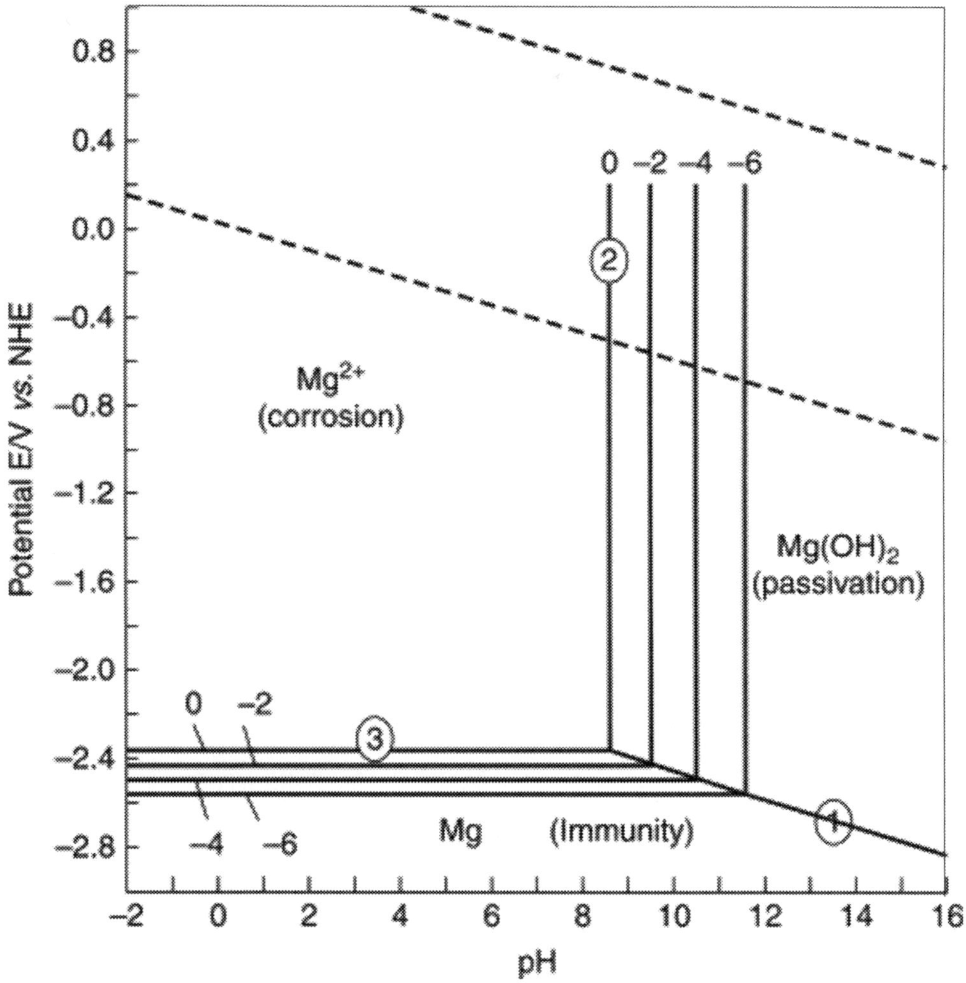

FIGURE 12–5 Pourbaix diagram for the magnesium–water system at 25°C.

highly oxidizing conditions, as shown in the lower left region of the pH-potential diagram, despite having a very narrow immunity zone toward the bottom of the diagram.

12.6.6 Benefits of Pourbaix diagrams

Pourbaix diagrams provide a significant amount of thermodynamic data in a highly effective and condensed style. Pure metal corrosion in an aqueous environment may be prevented by using the information in the illustrations. Corrosion may be managed by adjusting the pH and voltage in the areas of immunity and passivation. For instance, iron corrosion may be controlled by going to slightly alkaline regions and raising the pH of the environment.

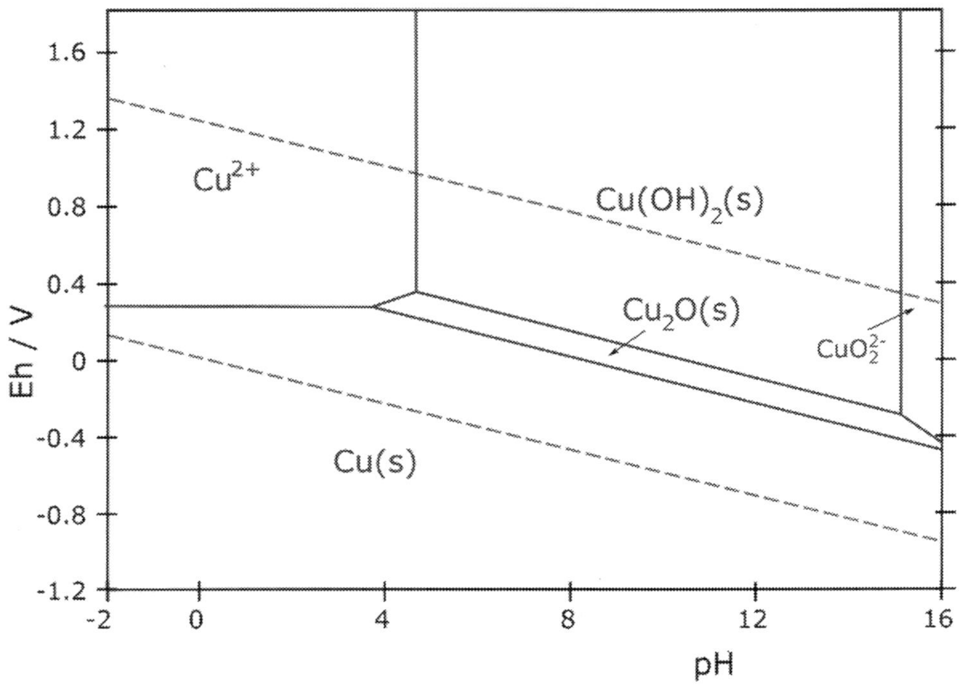

FIGURE 12–6 Pourbaix diagram for the copper–water system at 25°C.

Cathodic protection is a technology that eliminates corrosion by altering the potential of iron to more negative values. Additionally, by causing stable oxide film to develop on the surface of transition metals, increasing the potential to higher positive values slows corrosion. If an alkaline atmosphere is maintained, steel in reinforced concrete does not corrode. On the other hand, if the pH is higher than 8.0, an alkaline environment for aluminum is disastrous. The advantages of Pourbaix diagrams in corrosion prediction and control are clearly seen in the aforementioned case.

12.6.7 Limitations of Pourbaix diagrams

Pourbaix diagrams do not provide information about reaction rates, that is, whether any potential reaction occurs quickly or slowly when the energy changes are favorable, because they are dependent on thermodynamic data. The statistics point out the circumstances under which diffusion-barrier films may develop on an electrode surface, but they do not quantify how effective they might be when certain anions, such as SO_4^{2-} or Cl^-, are present. They also do not specify the precise circumstances under which it is feasible to form nonstoichiometric metal compound films, some of which are crucial for predicting corrosion rates. The stoichiometric compounds that any fewer stable molecules may turn into when equilibrium is reached, however, are clearly depicted in the figures. Standard Pourbaix diagrams do not take into account alloys; for instance, it is impossible to properly estimate the corrosion of iron-nickel alloys by superimposing conventional Pourbaix diagrams for iron and nickel. Despite these drawbacks, the diagrams are nevertheless useful for describing how metal is equilibrated when it is exposed to acids or alkalis or when a certain potential is applied to it.

However, there are several other limitations of these diagrams, which are summarized below:

1. These diagrams do not contain any information on the reactions; they are solely based on thermodynamic data. The kinetics of the process could prevent achieving thermodynamic stability. Regarding response rates, no information is given.
2. Only equilibrium circumstances in the given environment are taken into account; variables like temperature and velocity, which might significantly alter the rate of corrosion, are not.
3. It is not feasible to choose a species' activity as 10^{-6} g/ mol at random.
4. Pourbaix diagrams deal with pure metals, which engineers are not very interested in.
5. Since porosity, thickness, and substrate adhesion are significant elements that affect the protective efficacy of insoluble corrosion products, it is incorrect to assume that all insoluble products are protective.

Even though the aforementioned drawbacks seem to be significant, the Pourbaix diagrams' benefits significantly surpass their drawbacks.

12.7 Kinetics of corrosion reaction

While electrode kinetics focuses on the speeds of such events, thermodynamics provides an indicator of the propensity for electrode reactions to occur. It would be more acceptable to refer to the kinetics of these events as corrosion kinetics, as the reactions in question are mostly corrosion reactions. It is crucial to get a thorough grasp of the kinetics of the reaction occurring on an electrode surface in contact with an aqueous electrolyte in order to comprehend the theory of aqueous corrosion. Determining the number of reactants left in products after a certain period of time is a step in techniques used to analyze the pace of a reaction. Understanding the nature of the irreversible processes that occur on the electrode surface during aqueous corrosion is crucial.

12.7.1 Faraday's law

Metal in a corrosive or potentially corrosive environment develops a quantifiable open circuit potential (OCP) as a result of the transfer of electrons from the metal to electrochemically active ions and molecules. The following factors affect the OCP magnitude:

1. the type of metal,
2. the surface condition of the metal,
3. the type of reference electrode used to measure the potential, and

4. the chemical composition of the environment—and thus the chemical composition of the metal's electrical double layer (EDL).

When the OCP has a negative magnitude, spontaneous corrosion is anticipated. However, when spontaneous corrosion is anticipated by $G = nFE$, the metal OCP for a particular environment does not tell us how quickly the corrosion will progress.

According to Faraday's law, the amount of metal lost or deposited by an electrical current is proportional to the current's intensity and the size of the metal ions' charges (Keenan & Wood, 1966). For instance, a 1 A electrical current creates 6.02×10^{23} electrons and 1 Coulomb/s of charge transfer. The gram molecular weight of iron divided by 2, or 27.56 g of iron, is equivalent to 6.02×10^{23} electrons since iron ions have a charge of 2. In other words, an iron loss of 27.56 g would result from a corrosion current of 1 A. As a result, Faraday's law serves as the foundation for calculating corrosion kinetics utilizing electrical devices.

Corrosion dynamics determine the lifespan of structures constructed from metals and metal alloys. Sensitive electronic devices can estimate corrosion rates thanks to the transfer of electrons from the metal to electrochemically active ions and molecules in the environment.

12.7.2 The Butler–Volmer equation

After the EDL has been formed, corroding metals are often not at equilibrium and metal atoms continue to exit the metal lattice (i.e., the corrosion process moves ahead). Metal corrosion is therefore at a constant state. As a result, steady-state corrosion is not adequately represented by the EDL model. A steady-state EDL model for a corroding metal is shown in Fig. 12–7. It contains electrochemically active species that propel the corrosion process ahead. In other words, corrosion is shown in Fig. 12–7 as the loss of metal atoms from the metal lattice.

Electrical current and variations in metal potential brought on by an external power source are related by the Butler–Volmer (BV) equation (Bard & Faulkner, 1980). The BV equation is

$$\mathrm{I}\ i_{\mathrm{corr}}\left\{\exp-\left(\mathrm{snFh/RT}\right)-\exp\left((1-\mathrm{s})\mathrm{nFh/RT}\right)\right\} \qquad (12-40)$$

where R is the ideal gas constant, 1.986 cal/mole K, T is the temperature in degrees Kelvin, n is the number of electrons in the anodic half-reaction, F is Faraday's constant, 96,500 C/equivalent or 23,060 cal/V, I is the external current density, in A/cm^2, flowing to or from an electrode because of an applied potential, i_{corr} is the corrosion current density, in A/cm^2, that occurs when the electrode is at it's OCP, σ is a coefficient with values ranging from 0 to 1, η is the test electrode over the potential-difference between the electrode OCP and applied potential (Vapplied—OCP), F/RT is 0.039 mV21 at 25°C.

Cathodic current is represented by the exp(nF/RT) component in Eq., while anodic current is represented by the exp((12σ)nFη/RT) component. The external current, I is zero

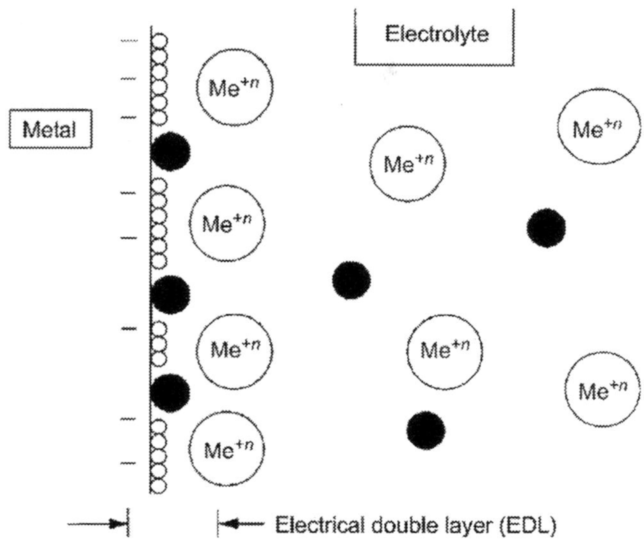

FIGURE 12–7 An equilibrium model for the electrical double layer.

and the anodic and cathodic currents are equal at OCP = 0. (i.e., no current flows to or from the electrode).

When we apply a negative or positive voltage to a test electrode in an electrochemical corrosion test cell, the Butler-Volmer equation demonstrates that an external current develops. An illustration of a potential-current curve for iron produced using the Butler-Volmer equation is shown in Fig. 12–8. The curve contains both cathodic and anodic branches, as can be seen in the figure. The anodic branch oxidizes iron whereas the cathodic branch reduces it. The curve's inflection point, or OCP for iron, is also the potential at which there is no net current flowing to or from the electrode (i.e., $i = 0$). Therefore, extrapolating linear sections of the anodic and cathodic branches in Fig. 12–8 to the OCP, as illustrated by the solid lines, allows one to estimate the corrosion current density (i_{corr}). A completely anodic or cathodic electrode generates external currents that may be detected by electronic instruments. When the overpotential is smaller than around 250 mV, the reverse happens.

12.7.3 Mixed potential theory of corrosion kinetics

By displaying simply the solid lines, as seen in Fig. 12–9, current-potential curves similar to those in Fig. 12–8 can be made simpler. The Evans diagram in Fig. 12–9 is named after U.R. Evans, who initially suggested this simplification. An Evans diagram can be employed for steady state corrosion processes like the iron in acid corrosion reaction.

In addition to an overlay of the Evans diagrams for hydrogen and iron, Fig. 12–9 shows the overall corrosion process for iron in acid. On the current axis, the current densities for hydrogen and iron corrosion are designated, and the proper electrochemical half-reactions

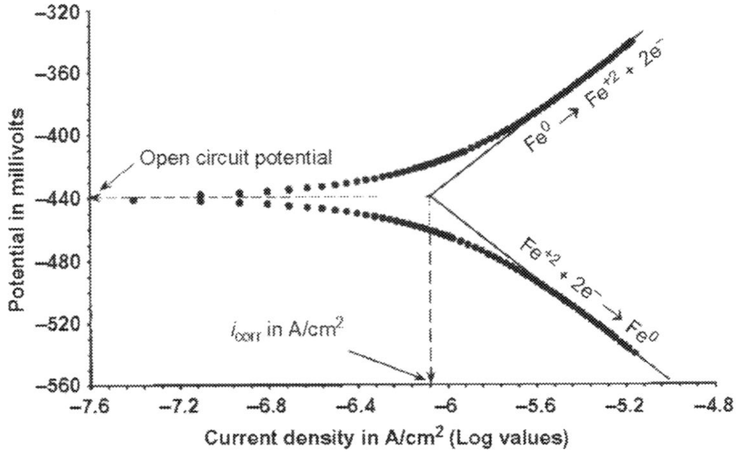

FIGURE 12–8 A Butler–Volmer current–potential curve.

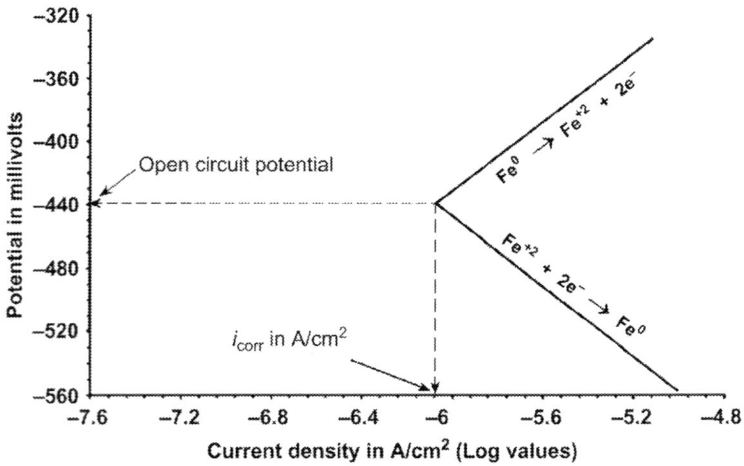

FIGURE 12–9 Evans diagram for iron.

are situated along each branch. Additionally, the Evans diagram in figure shows how hydrogen ions are decreased at the surface of an iron electrode, which causes corrosion to advance. A more detailed discussion of corrosion dynamics is shown in Fig. 12–10 (i.e., corrosion rate).

The OCP for the combined reaction (intersection point of the cathodic and anodic branches) lies between the OCP values for the individual reactions in Fig. 12–10, where the cathodic branch of the hydrogen diagram crosses the anodic branch of the iron diagram. Keep in mind that the overall corrosion current is actually larger than the current densities

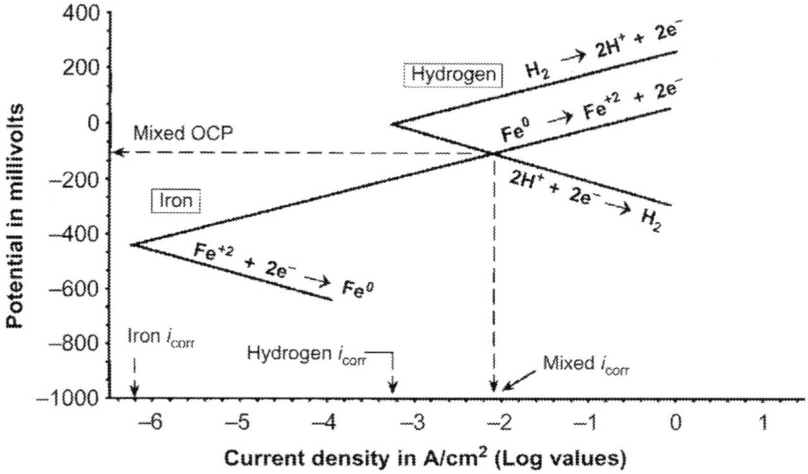

FIGURE 12–10 Evans diagram for corrosion of iron in an acid electrolyte.

for the individual processes, meaning that iron corrosion with hydrogen ions is more rapid (has a higher i_{corr}) than it would be without them. The mixed potential theory combines the electrochemical half-reactions for the metal and the electrochemically active ions or molecules, as seen in Fig. 12–9. In comparison to other theories, the mixed potential theory offers a more comprehensive description of natural corrosion. Compared to the Butler-Volmer equation, the mixed potential theory paints a more detailed picture of natural corrosion. When selecting metals and metal alloys for service in each environment, we must consider the following (Annand et al., 1978):

- the thermodynamics of the metal within a given environment, and
- the corrosion kinetics of the metal in the environment (i.e., electrolyte).

References

Annand R. R., Hilliard, H. M., Tait, W. S. (1978). 641.

Bard, A. J. & Faulkner, L. R. (1980).

Keenan, C. W., & Wood, J. H. (1966). *General College Chemistry.*

Portero, M. J. Mu. ñ oz, & Ant ó n, Garc í a (2007). *Corrosion, 63*(7).

Thompson, W. T., Kaye, M. H., Bale, C. W., & Pelton, A. D. (2000). Uhlig's corrosion handbook. 125–136. In *Wiley Pourbaix diagrams for multielement systems.*

PART
4

Materials for corrosion

13

Metals and alloys susceptible for corrosion

Chemical surface reactions that take place as a result of the presence of aggressive substances in the environment can lead to the degradation of metals. Steel cables of suspension bridges, steel reinforcements in concrete structures, or functional materials like dental alloys or copper leads for printed circuits are a few examples of metals that can be considered damaged. Only noble metals, such as platinum and gold, are able to resist corrosion under normal environmental conditions. This is because the reaction with atmospheric oxygen cannot be performed due to certain thermodynamic principles. For most types of metals, such as aluminum and titanium, the chemical reaction resulting from atmospheric oxygen is spontaneous. It involves a net gain in entropy of the metal/environment system together with its surroundings. The prerequisite for any useful engineering material is, of course, that oxidation is slow under normal service conditions, at least in dry air. Except in moist air, this is usually the result of a microscopically thin oxide film formed by slight initial oxidation, which prevents further corrosion of non-noble metals.

13.1 Carbon steel

In environments where carbon steel (CS) is highly corrosive, such as acidic and seawater, the actions of microorganisms can lead to the dissolution of the corrosion product films of this type of corrosion. In the crystal structure of certain types of metals, the iron atoms are arranged in a manner that allows them to leave space. The spaces between the iron atoms are where the oxygen atoms are located. This interaction between the iron atoms and the water triggers the formation of a brown deposit on the surface of a metal. This phenomenon, which is referred to as "oxidation," is characterized by a layer of brownish deposit on the surface of the metal and the reaction is called "redox."

$$H_2O + 2e^- \rightarrow 2OH^- \tag{13-1}$$

The oxidant undergoes a reduction. It is said to reduce. A reducing agent is a species capable of releasing electrons (electron donor).

$$Fe \rightarrow Fe^{2+} + 2e^- \tag{13-2}$$

The reducing agent undergoes oxidation. It is said to oxidize. The redox equation is:

$$Fe + 2OH^- \rightarrow Fe(OH)_2 + 2e^- \tag{13-3}$$

Handbook of Corrosion Engineering. DOI: https://doi.org/10.1016/B978-0-323-95185-2.00013-7

Molecules of dioxygen dissolved in water attack the iron atoms and tear off two electrons forming Fe^{2+} ions.

$$Fe + O_2 + 2H_2O \rightarrow Fe^{2+} + 4OH^- \qquad (13-4)$$

Simultaneously, the electrons, oxygen, and water result in the formation of hydroxyl ions.

$$2H_2O + O_2 + 4e- \rightarrow 4OH^- \qquad (13-5)$$

This is the first step, but the oxidation continues. Oxygen attacks Fe^{2+} ions, extracts an electron, and transforms them into Iron(III), Fe^{3+}. And the ballet continues, producing new hydroxyl ions. Fe^{3+} ions and hydroxyl ions come together in a precipitate and result in rust.

$$4Fe(OH)_2 + 2H_2O + O_2 \rightarrow 4Fe(OH)_3 \qquad (13-6)$$

The bright orange that spreads on the surface:

$$2Fe(OH)_3 \rightarrow Fe_2O_3 + 3H_2O \qquad (13-7)$$

The situation can be summed up as follows: oxidizing iron releases two electrons which are accepted by oxygen which is reduced, iron

$$Fe \rightarrow Fe^{2+} + 2e - O_2 + 2e - \rightarrow O_2 \qquad (13-8)$$

The redox equation is simpler:

$$Fe^{2+} + O_2 \rightarrow Fe_2O_3 \qquad (13-9)$$

Iron oxide is a brownish layer that develops on the metal's surface. This metal is also wet on its surface, which results in a new redox equation form. The following reaction can be used to describe it schematically:

$$4Fe + 3O_2 + 2H_2O \rightarrow 4FeO(OH) \qquad (13-10)$$

The compound FeOOH, which can be considered as hydrated hematite:

$$2FeO(OH) \rightarrow Fe_2O_3 + H_2O \qquad (13-11)$$

Rust is a complex mixture of different crystalline and amorphous phases of iron oxides and hydroxides, with a stoichiometry that only estimates the overall FeOOH formula. Its three main components are as follows:

- Lepidocrocite γ-FeOOH of rhombohedral structure,
- α-FeOOH goethite with a rhombohedral structure,
- Magnetite Fe_3O_4 of cubic structure.

13.2 Stainless steel

Despite its high mechanical properties, stainless steel (SS) can still be highly resistant to corrosion. This is because its fine layer of chromium oxide can self-repair itself when broken or breached. Unfortunately, excessive corrosion can occur if the damage is too extensive. Most commonly used SSs, such as 316 and 304, are not resistant to HCI at high temperatures or concentrations (Saefuloh et al., 2020). High temperatures can cause severe corrosion in SSs. Due to the presence of various elements such as copper, nickel, and molybdenum, they can resist the effects of dilute acid. However, pitting can result in the failure of certain austenitic steels when subjected to hydrochloric acid. In addition to pitting, type 316 steel can also crack at 0°C in 5% HCl. Although stress corrosion cracking is not expected to occur at high corrosion rates (10 m/year), it can happen due to the presence of corrosion products, such as $FeCl_3$. The most common cause of pitting corrosion is chloride, which can penetrate and destroy the film that is responsible for the resistance of SS to corrosion (Kappes, n.d.). Some types of SSs, such as 20Cb-3, which have a high nickel content (32%−38%), 2%−3%Mo, and 3%−4%Cu can resist the dilute HCI at ambient temperatures. However, it can also be prone to pitting and crevice attacks in acid chlorides. Although the exact mechanisms by which chloride ions damage an oxide film have been studied, there is still no clear understanding as to how they do so. It is generally believed that when pitting corrosion occurs, the chloride ions will penetrate the film and reach the surface of the metal. The incorporation of chloride into the outer layer of the film has been verified (Cobb, 2010). Because oxide films are thinner, their chemical composition is more complicated, and they cannot be reduced cathodically, studying their composition and structure is much more difficult than studying iron films. As a result, chemical and electrochemical methods for studying SS passivity are restricted.

The presence of chromium and molybdenum as alloying elements allows the passive behavior of SS in different environments (Wang et al., n.d.). The surface of these SS is ideal for microbial adhesion, which makes it susceptible to stress crevice corrosion or pitting in solutions containing chlorides. The presence of peroxides in the marine environment can also encourage the development of corrosion in these materials. During bacterial metabolism, the generation of peroxides results in an ennoblement of the SS' sting potential. Although all types of SS have their own unique characteristics, they tend to deteriorate when exposed to colonies of microorganisms. There are several types of corrosion that can develop on these surfaces. The most common types are:

Pitting corrosion: The surface of SS's passive layer can be attacked by various chemical species. One of these is chloride ion, which is commonly found in everyday products such as salt and bleach. Another type of damage that can eat into the steel's surface is called harsh pitting corrosion. It can also be caused by elevated temperatures. To avoid pitting corrosion, make sure that the SS is not exposed to harmful chemicals. You can also choose a more resistant type of steel by avoiding prolonged exposure to harmful substances.

Crevice corrosion: The supply of oxygen is required to ensure that the passive layer can be formed on the surface of SS. However, in tight spaces, it is not always possible for oxygen to get to the surface, which can allow it to be attacked. To prevent this, it is important to seal the crevices with a flexible sealant.

General corrosion: Unlike other types of steels, SS does not usually corrode uniformly. However, due to the presence of certain chemicals, such as acids, the passive layer can be attacked uniformly. This can lead to the metal loss affecting the entire surface of the steel. Some of the most aggressive types of corrosion are hydrochloric acid and sulfuric acid.

Galvanic corrosion: If two different types of metals are in contact with an electrolyte, such as water, then a galvanic cell might be established to accelerate the process of corrosion. However, this method can be avoided by separating the two components using a nonmetallic insulator, such as rubber.

Stress corrosion cracking: The rare form of corrosion known as SCC requires a combination of corrosive species, temperature, and tensile stress to occur. This type of corrosion can usually be encountered in swimming pools or hot water tanks. However, if it does happen, it can cause severe damage to the steel's mechanical properties in just a couple of days. Another type of corrosion that can occur in oil and gas production is sulfide stress corrosion cracking.

13.3 Copper and nickel alloys

Although copper alloys are well known for their ability to resist corrosion in seawater, they are susceptible to corrosion under certain circumstances, such as when they are exposed to water pollutants or when the material's chemical composition is not ideal. Copper alloys are not usually protected with paint coatings. They can additionally be subjected to antifouling activities, which are usually performed through the dosing of Fe^{2+} ions into seawater cooling systems. In mixed metal systems, this may be necessary (Powell & Michels, 2000). Biocides can help protect copper alloys from various types of corrosion. Some of these include selective corrosion, which is caused by the dezincification or denickelification of copper, and stress corrosion, which is caused by the interaction of a more noble material with a copper alloy. Cavitation erosion, corrosion fractures, galvanic corrosion, and pitting corrosion are also common.

Among the various types of Cu alloys are brasses, bronzes, aluminum bronzes, and copper superalloys, such as Monel, Ni-Mo, Ni-Cr-Mo, Ni-Cr-Fe-Mo, and traditional nickel alloys: Ni-Cr-Fe, Ni-Fe-Cr, Fe-Ni-Cr-Mo, and Cu-Ni alloys CuNi\70/30, CuNi\90/10. These have been known to exhibit high-strength corrosion when exposed to seawater. Despite these skills, there have been reports of bacteria colonization after several months of exposure.

Over the years, copper-nickel alloys have been developed for seawater service. Initially, these were used for the manufacturing of piping systems and condensers. Their suitability has been confirmed by their continued use in various marine applications. These alloys are commonly used in the desalination, power, and offshore industries and also continue to be used by many Navies and in merchant shipping. Although they were initially developed for their resistance to macrofouling, it was eventually realized that they also have characteristics that make them ideal for reducing the need for biocides in seawater systems and condensers. This has allowed them to avoid the need for cleaning regimes and drag forces in boats and offshore platforms. As the applications for copper-nickel alloys expand, their knowledge continues to evolve.

An excellent conductor of electrical and thermal energy, high mechanical strength, and low corrosion resistance are some of the characteristics of copper-nickel alloys that are commonly used in marine environments (Metikoš-Huković, Babić, Škugor, & Grubač, 2011; Yuan & Pehkonen, 2007). The 90/10 copper-nickel alloy is commonly utilized in various applications, such as seawater pipes, heat exchangers, condensers in ships, and power plants (Mousavi & Baghgoli, 2016). The high degree of Cu-Ni alloy's corrosion resistance is mainly due to its positive equilibrium potential and the high thermodynamic stability of Cu. The high level of corrosion resistance of Cu-Ni can be attributed to the positive equilibrium potential and high thermodynamic stability of Cu. The positive equilibrium potential makes it difficult for copper to ionize. In order to increase the Cu_2O film's resistance to corrosion, nickel was added. This process causes the Cu^+ ions to disappear, resulting in an increase in the filml's ionic resistance. On the other hand, Ni^{2+} replaces Cu^+, which results in an increase in electronic resistance (North & Pryor, 1970). It is also relevant to the development of a layer of copper oxide on the surface of the copper alloy. This layer, which is composed of CuxO and Cu-hydroxide/oxide layer, is designed to protect the copper matrix. In addition, corrosion products can often deposit on the film and play a protective role (Kong et al., 2018; Metikoš-Huković et al., 2011). Compared to other types of metal alloys, copper and copper alloys are more resistant to corrosion. In marine exposure conditions, the Cu_2O can rapidly form on the surface of the matrix and react with chlorides, which then causes the formation of CuCl, which usually converts to $Cu_2(OH)_3Cl$ as end corrosion products (Zhang, Odnevall Wallinder, & Leygraf, 2014). According to a study conducted by Fuente, these defective components were confirmed in bare copper at atmospheric conditions. They were also observed in laboratory conditions with pre-deposited NaCl and humidified air (de la Fuente, Simancas, & Morcillo, 2008; Strandberg & Johansson, 1998). The failure mainly occurs on the surface of the Cu alloys in marine environments, which severely affects the safety of ship and marine engineering (Xia et al., 2019).

13.4 Aluminum and its alloys

The Latin word for aluminum is alumine, which is derived from the Latin word for potassium alum ($KAl(SO_4)_20 \cdot 12H_2O$). Aluminum compounds are mainly used in the production of various products, such as pottery and dyestuff. With an annual global consumption of over 25 million tons, aluminum is the world's leading nonferrous metal. The production of aluminum has been increasing steadily since 1950 (Brough and Jouhara, 2020). The various advantages of aluminum and its alloys are some of the factors that have been attributed to the increasing demand for these materials. These include their lightweight, electrical conductivity, resistance to corrosion, flexibility, and surface treatments. Some of the other factors that are used by end-users in the construction, transportation, and electrical engineering industries are their ease of recycling and the diversity of Al alloys. Due to its rapid passivated nature, Al is an active metal that can provide excellent corrosion resistance. It is commonly used in the aerospace industry as a component for the manufacture of aircraft fuel tanks. However, there have been instances of biocontamination caused by the presence of microorganisms in the tanks made from Al alloys. This issue can be caused by the presence of water, which can allow fungi and other microorganisms to grow. The fundamental reactions of the corrosion of aluminum in aqueous medium have been the subject of many studies (Lashgari & Malek, 2010). In simplified terms, the oxidation of aluminum in water proceeds according to the equation:

$$Al \rightarrow Al^{3+} + 3e^{-} \qquad (13-12)$$

When aluminum loses three electrons, it goes into solution as trivalent Al^{3+}. This reaction is balanced by the reduction of the solution's ion content, which captures the electrons. It can be shown by the presence of a pH close to neutral in various aqueous media, such as seawater, freshwater, and moisture.

Reduction of H^{+} protons:

$$3H^{+} + 3e^{-} \rightarrow 3/2H_2 \qquad (13-13)$$

protons result from the dissociation of water molecules:

$$H_2O \rightarrow H^{+} + OH^{-} \qquad (13-14)$$

reduction of oxygen dissolved in water:in alkaline or neutral media:

$$O_2 + 2H_2O + 4e^{-} \rightarrow 4OH^{-} \qquad (13-15)$$

in acidic media:

$$O_2 + 4H^{+} + 4e^{-} \rightarrow 2H_2O \qquad (13-16)$$

Under atmospheric pressure and at 20°C, the solubility of oxygen in water is 43.4 mg/kg. It can decrease with increasing temperature, reaching around 30.8 mg/kg at 40°C and 13.8 mg/kg at 80°C. Aluminum's corrosion in aqueous media is caused by two electrochemical reactions. One of these is oxidation, while the other is reduction.

$$Al \rightarrow Al^{3+} + 3e^{-} \qquad (13-17)$$

$$3H^{+} + 3e^{-} \rightarrow \frac{3}{2}H_2 \qquad (13-18)$$

The overall reaction can be written as:

$$Al + 3H^{+} \rightarrow Al^{3+} + \frac{3}{2}H_2 \qquad (13-19)$$

The change in the oxidation state of aluminium, from 0 to +3, initiates the reaction. When three electrons are lost from the solution, this occurs.

The formation of alumina $Al(OH)_3$ by aluminum corrosion is a white gel that forms insoluble precipitate in water. It can also be found in pits that are filled with white gelatinous particles. After the corrosion, 27 g of aluminum will form 33.6 lit hydrogen. The amphoteric nature of aluminum is shown in E–pH diagram (Fig. 13–1). Stable species equilibrium diagrams are used to determine the direction of possible reactions and the domain of stability of a given material. However, they cannot predict the rate of corrosion.

Pourbaix diagrams cover several domains representing three possible situations:

- Soluble corrosion products exist. According to Pourbaix, a concentration of 10^{-6} M (which amounts to 0.027 mg/L for aluminum) is a lower threshold that can be assumed to lead to the formation of corrosion. The upper limits of concentrations of 10^{-5}, 10^{-4}, and 10^{-3} are also defined.
- Passivation is when an insoluble oxide or hydroxide is formed on the metal's surface.
- Immunity, if the concentration of M^{n+} ions is less than 10^{-6} M.

The diagrams used to describe the ideal liquid, such as water at 25°C, are limited due to how they refer to a metal as pure as possible, which is not an alloy. They do not take into account chloride, which can be used in pitting corrosion. Also, they do not take into account the base and acid components' pH values. The method does not indicate the risk of cathodic corrosion in the form of immunity when the potential of a highly electronegative element is present. This is because the method relies on the data collected solely on thermodynamics (Scully, 1975). Since the E–pH diagram of an aluminum alloy 5086 is dependent on the

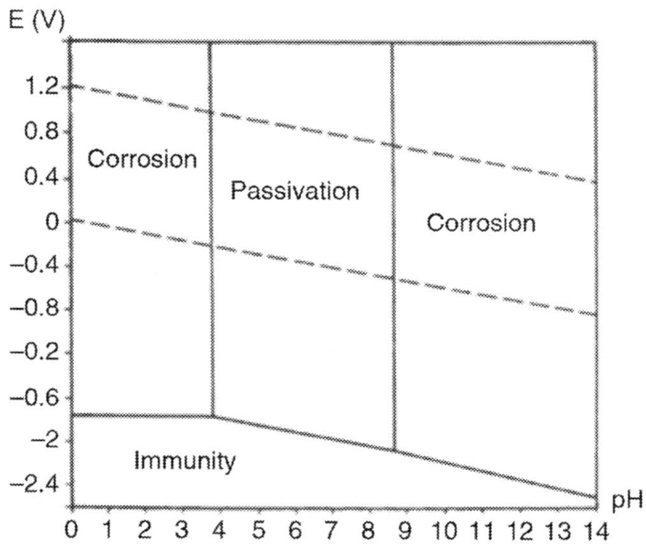

FIGURE 13–1 Pourbaix diagram (E–pH) of aluminum . *From Deltombe, E. & Pourbaix, M. (n.d.).* Comportement électrochimique de l'aluminium, diagramme d'équilibre tension pH du système Al–H2O à 25 8C, *Cebelcor.*

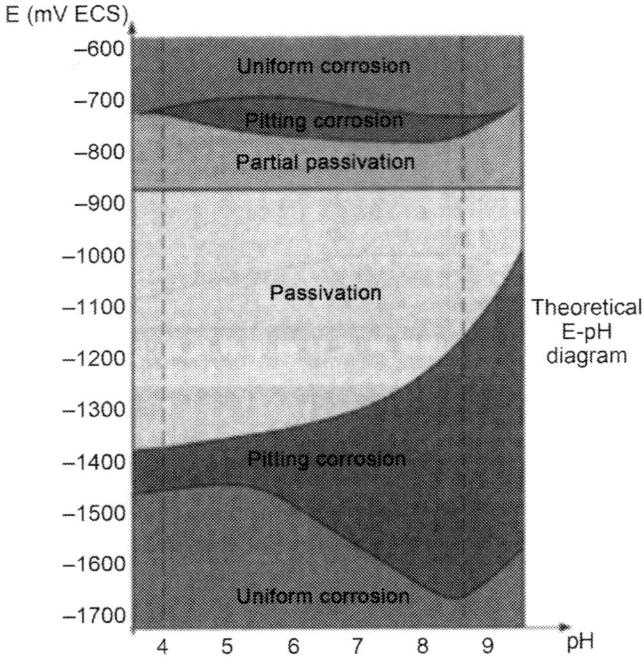

FIGURE 13–2 Experimental E–pH diagram of 5086 in the presence of chloride. *From Ph, G., Rameau, J. J. & Reboul, M. (1982). Diagramme experimental potentiel pH de l'aluminium pour l'eau de mer, Revue de l'aluminium (pp. 261–272).*

precautions taken (Fig. 13–2), it is also interesting to plot it in seawater at a pH adjustment of 30 g/L of NaCl. The graph shows the various potentials of the aluminum alloy. These include the corrosion potential E_0, which is a mixed metal–water mixture; the pitting potential E_c, which is the most electronegative at the moment of pitting; the passivation potential E_p, which is the minimum that has to be set to passivate the existing pits; and the anodic attack E_{ga}, which is the most likely scenario of pitting corrosion spreading over the entire surface. Under the influence of cational polarization, the Ecc is the most probable. When anodic polarization occurs, pitting corrosion may become uniform. Although there is no reduction in AlO^{2-} or Al^{3+} ions, the reduction in water under local alkalinization results due to consumption of H + . This natural oxide layer will then dissolve in an alkaline medium.

Figs. 13–1 and 13–2 show the diagrams that were made based on the measurement of the potentials. The following statements should be made: The pitting potential and the corrosion potential are very close between pH 4 and 9. But, the pH does not depend on it.

- Pitting corrosion takes place only in the range of pH in which the oxide layer is totally insoluble.
- Uniform cathodic attack is catastrophic corrosion that may dissolve up to 10 μm/h under cathodic polarization.
- Immunity is theoretically inaccessible at pH values below 9, because at potentials low enough to enter the domain of immunity of aluminum, water is no longer stable, and hydrogen will be released. Aluminum alloys are subject to the following kinds of

corrosion: pitting corrosion; exfoliation corrosion; galvanic corrosion occurring when in contact with more noble alloys (Gimenez, Rameau, & Reboul, 1981).

13.5 Titanium

Due to its reactive nature, titanium is often exposed to air and other environments that have oxygen. This allows it to form a thin surface film of oxide, which contributes to its excellent corrosion resistance (Zhang et al., 2015). Compared to other structural metals, titanium is more resistant to corrosion in various chemical environments. It is also less prone to failure due to galvanic, erosion, and stress-induced corrosion. The most effective titanium films are usually developed when water, in trace amounts, is present. However, if the environment is heavily oxidizing, the film may not be as protective. Rapid oxidation can also occur due to pyrophoric conditions (Cramer & Covino, 2005). In environments where the protective films are broken down, titanium and its alloys can corrode rapidly. Some of the most important factors that can contribute to this are hydrochloric acid, hydrofluoric acid, oxalic acid, and sulfuric acid (Fekry, 2009). Although titanium can be attacked by various media, such as hydrofluoric acid, it can be reduced by adding certain types of inhibitors, such as acid salts and oxidizing acids. Dry chlorine can also attack titanium, though it is more resistant to other oxidizing agents, such as CO_2, SO_2, and wet chlorine. Although titanium exhibits excellent corrosion resistance when subjected to nitric acid at up to 350°F, it can be attacked at higher concentrations, such as at 375°F. The rate of corrosion can be as high as 100% at higher levels of nitric acid. Even at 550°F, the rate of attack in 20% HNO_3 is only 12%. Although titanium alloys can be safely used in an anhydrous HNO_3 without experiencing pyrophoric reactions, caution should still be taken. Titanium's resistance to chromic acid and aqua regia is good. The increase in H_2SO_4 concentration in nitric and sulfuric acid mixtures can lead to higher corrosion rates. Like other metals, titanium also has good resistance to dilute alkali solutions. Hot, strong, caustic solutions will attack unalloyed titanium and titanium alloys. There is no proof that titanium alloys are prone to caustic embrittlement, similar to that of SS and CS. Compared to SS, titanium is more resistant to pitting and corrosion in neutral chloride solutions (Boulton & Betts, n.d.). Certain types of boiling solutions, such as stannic chloride, aluminum chloride, magnesium chloride, calcium chloride, and cupric chloride, can cause pitting of titanium alloys. However, titanium is not affected by these solutions. It is not subjected to highly corrosive sodium hypochlorite and ferric chloride solutions (Han, Jiang, Shi, Deng, & Li, 2012).

Studies have shown that titanium is resilient to various types of corrosion, including stress corrosion, pitting, galvanic corrosion, erosion, and crevice corrosion. It is not recommended for use in liquid or gaseous oxygen due to the possibility of a violent reaction. Although titanium can resist stress corrosion in various media, such as methanol or fuming HNO_3 or methanol containing HCI, H_2SO_4, or Br^-, it is not always possible to prevent it from cracking under plane strain conditions. This is because titanium with high oxygen levels can rapidly propagate in seawater at low-stress levels (Wang, Su, & Szklarska-Smialowska, 1988).

This phenomenon is often referred to as stress corrosion cracking. Although the common solutions used for corrosion tests do not affect titanium alloys, recent studies have shown that stress corrosion can occur in certain types of titanium. These include titanium alloys exposed to high ambient temperature and aqueous media. It has also been shown that certain types of aqueous stress corrosion solutions can affect the fatigue life of Sharp-notch test specimens. These include distilled water, 3.5% NaCl, and tap water (Efird & Moller, 1979).

References

Boulton, L. H., & Betts, A. J. (n.d.). Corrosion performance of titanium and titanium stabilised stainless steels. *British Corrosion Journal*. 26.

Brough, D., & Jouhara, H. (2020). The aluminium industry: A review on state-of-the-art technologies, environmental impacts and possibilities for waste heat recovery. *International Journal of Thermofluids*.

Cobb, H. M. (2010). *The history of stainless steel "The Material Informations Sociaty"* (pp. 59–91). ASM International.

Cramer, S. D., & Covino, B. S. (2005). *Corrosion of titanium and titanium alloys* (pp. 252–299). ASM International. https://doi.org/10.31399/asm.hb.v13b.a0003822.

de la Fuente, D., Simancas, J., & Morcillo, M. (2008). Morphological study of 16-year patinas formed on copper in a wide range of atmospheric exposures. *Corrosion Science, 50*(1), 268–285. Available from https://doi.org/10.1016/j.corsci.2007.05.030.

Efird, K. D., & Moller, G. E. (1979). Electrochemical characteristics of AISI 304 and 316 stainless steels in fresh water as functions of chloride concentration and temperature. *Materials Performance, 18*(7), 34–38.

Fekry, A. M. (2009). The influence of chloride and sulphate ions on the corrosion behavior of Ti and Ti-6Al-4V alloy in oxalic acid. *Electrochimica Acta, 54*(12), 3480–3489. Available from https://doi.org/10.1016/j.electacta.2008.12.060.

Gimenez, Ph, Rameau, J. J., & Reboul, M. C. (1981). Experimental pH potential diagram of aluminum for sea water. *Corrosion, 37*(12), 673–682. Available from https://doi.org/10.5006/1.3577557.

Han, D., Jiang, Y. M., Shi, C., Deng, B., & Li, J. (2012). Effect of temperature, chloride ion and pH on the crevice corrosion behavior of SAF 2205 duplex stainless steel in chloride solutions. *Journal of Materials Science, 47*(2), 1018–1025. Available from https://doi.org/10.1007/s10853-011-5889-6.

Kappes, K. M. A. (n.d.). Localized corrosion and stress corrosion cracking of stainless steels in halides other than chlorides solutions: A review. *Corrosion Reviews*. 38.

Kong, D., Dong, C., Zheng, Z., Mao, F., Xu, A., Ni, X., ... Li, X. (2018). Surface monitoring for pitting evolution into uniform corrosion on Cu-Ni-Zn ternary alloy in alkaline chloride solution: Ex-situ LCM and in-situ SECM. *Applied Surface Science, 440*, 245–257. Available from https://doi.org/10.1016/j.apsusc.2018.01.116; http//www.journals.elsevier.com/applied-surface-science/.

Lashgari, M., & Malek, A. M. (2010). Fundamental studies of aluminum corrosion in acidic and basic environments: Theoretical predictions and experimental observations. *Electrochimica Acta, 55*(18), 5253–5257. Available from https://doi.org/10.1016/j.electacta.2010.04.054.

Metikoš-Huković, M., Babić, R., Škugor, I., & Grubač, Z. (2011). Copper-nickel alloys modified with thin surface films: Corrosion behaviour in the presence of chloride ions. *Corrosion Science, 53*(1), 347–352. Available from https://doi.org/10.1016/j.corsci.2010.09.041.

Mousavi, M., & Baghgoli, T. (2016). Application of interaction energy in quantitative structure-inhibition relationship study of some benzenethiol derivatives on copper corrosion. *Corrosion Science, 105*, 170–176. Available from https://doi.org/10.1016/j.corsci.2016.01.014.

North, R. F., & Pryor, M. J. (1970). The influence of corrosion product structure on the corrosion rate of Cu-Ni alloys. *Corrosion Science, 10*(5), 297–311. Available from https://doi.org/10.1016/S0010-938X(70)80022-1.

Powell, Carol A., & Michels, Harold T. (2000). Copper-nickel alloys for seawater corrosion resistance and anti-fouling - a state of the art review. *Corrosion*.

Saefuloh, I., Kanani, N., Gumelar. Ramadhan, F., Rukmayadi, Y., Yusuf, Y., Abdullah, S., & Susilo, S. (2020). The study of corrosion behavior and hardness of AISI stainless steel 304 in concentration of chloride acid solution and temperature variations. *Journal of Physics: Conference Series, 1477*(5), 052058. Available from https://doi.org/10.1088/1742-6596/1477/5/052058.

Scully, J. C. (1975). *International serious of monographs on materials science and technology, 17.*

Strandberg, H., & Johansson, L. G. (1998). Some aspects of the atmospheric corrosion of copper in the presence of sodium chloride. *Journal of the Electrochemical Society, 145*(4), 1093–1100. Available from https://doi.org/10.1149/1.1838422; http://jes.ecsdl.org/content/by/year.

Wang, J. H., Su, C. C., & Szklarska-Smialowska, Z. (1988). *Effects of Cl− Concentration and temperature on pitting of AISI 304 stainless steel.* 44.

Wang, Z., Paschalidou, E.-M., Seyeux, A., Zanna, S., Maurice, V., & Marcus, P. (n.d.). Mechanisms of Cr and Mo enrichments in the passive oxide film on 316L austenitic stainless steel. *Frontiers in Materials* 6.

Xia, T., Zeng, L., Zhang, X., Liu, J., Zhang, W., Liang, T., & Yang, B. (2019). Enhanced corrosion resistance of a Cu−10Ni alloy in a 3.5 wt% NaCl solution by means of ultrasonic surface rolling treatment. *Surface and Coatings Technology, 363*, 390–399. Available from https://doi.org/10.1016/j.surfcoat.2019.02.039; http://www.journals.elsevier.com/surface-and-coatings-technology/.

Yuan, S. J., & Pehkonen, S. O. (2007). Surface characterization and corrosion behavior of 70/30 Cu-Ni alloy in pristine and sulfide-containing simulated seawater. *Corrosion Science, 49*(3), 1276–1304. Available from https://doi.org/10.1016/j.corsci.2006.07.003.

Zhang, R., Ai, X., Wan, Y., Liu, Z., Zhang, D., & Feng, S. (2015). Surface corrosion resistance in turning of titanium alloy. *International Journal of Corrosion, 2015*, 1–8. Available from https://doi.org/10.1155/2015/823172.

Zhang, X., Odnevall Wallinder, I., & Leygraf, C. (2014). Mechanistic studies of corrosion product flaking on copper and copper-based alloys in marine environments. *Corrosion Science, 85*, 15–25. Available from https://doi.org/10.1016/j.corsci.2014.03.028.

PART

5

Corrosion investigation & analysis

14

Laboratory-controlled tests

The kind of corrosion test that is conducted on a metal depends on the environmental factors to which it is exposed. Although there are no standardized tests for corrosion, the most reliable method to ensure its reliability is by reviewing a service history (Meikle et al., 2017). Unfortunately, since the service history is rarely accessible, additional testing is often required to confirm the component's authenticity. This is because the information that is collected outside the service history is not always reliable. In most cases, materials' corrosion properties are required for use in certain applications. Having the necessary expertise in corrosion testing can help identify potential materials for use in these areas. Aside from long-term exposure, service experience can also help identifying potential issues. Due to the complexity of a system's corrosion mechanism, it can be difficult to determine which type of test is most appropriate for a given application. Also, localized corrosion tests are not ideal.

Before a test program can be conducted, it is important to consider the various factors that can affect its design. This list aims to provide the most common factors that can affect a test program's success:

- What are the objectives of the test?
- How should the results be interpreted?
- Is the sample homogeneous?
- How representative is the sample?
- Are the tests destructive?
- How can the information be integrated with earlier or other tests?
- How many factors control the specimen's behavior?
- How many factors are to be included in the tests?
- Which of these factors interact and which have negligible interaction?
- How many specimens are available, and how are they produced (batch or sequentially)?
- How much control is there over testing?
- What type of data is to be measured?
- How expensive are the tests and/or specimens?
- How difficult would it be to plan for various types of human errors?

It is also important to develop a strategy that addresses the various requirements of the test program. Having a well-defined testing strategy can help minimize the number of questions that the test team has to answer.

Handbook of Corrosion Engineering. DOI: https://doi.org/10.1016/B978-0-323-95185-2.00014-9

The test methods have been broadly grouped into three main classifications in the following sections (International, 2012):

- cabinet tests
- immersion tests
- high temperature/high-pressure (HT/HP) testing

14.1 Cabinet tests

A cabinet test is a type of laboratory procedure that involves placing specimens in a device that has been exposed to controlled exposure conditions (Brossia, 2018). It is commonly used to evaluate the performance of materials in harsh environments, such as industrial areas and coastal regions. Aside from temperature, various other factors such as humidity, salt, temperature, UV radiation, and corrosive gases can also be included in the tests. The objective of a cabinet test is to determine the corrosion resistance of various materials that are intended for outdoor use. Aside from determining the materials' properties, the test also aims to establish the acceleration factors that are contributing to the degradation of the materials.

The main categories of cabinet tests are given below:

14.1.1 Humidity test

A humidity test is a type of corrosion analysis that involves exposing a material to varying environmental factors and corrosive agents to determine the rate of corrosion. This procedure can be used to evaluate the performance of various industrial materials. Unlike salt spray testing, humidity testing does not require salt. It creates a controlled environment that allows the specimens to be exposed to varying environmental conditions. Parts or panels are exposed to various types of chemicals and temperatures for a certain amount of time to test their corrosion resistance and other properties. Fifteen standards have been established by the American Society of Testing and Materials (ASTM) to address the various methods involved in the production and managing humidity and fog in cabinets. These standards are used to test various types of products, such as copper tubing systems. The basic humidity test is mainly used to evaluate the properties of materials and determine their corrosivity.

A solid-state humidity sensor detecting the present humidity situation and a feedback controller should be included in the test cabinet. The humidity control system uses a blower motor to circulate chamber air across a heater coil in the chamber's bottom, with an atomizer nozzle fogging into the airstream.

Some common humidity standards are given below:

ASTM D2247 Standard Practice for Testing Water Resistance of Coatings in 100% Relative Humidity

ASTM D4585 Cleveland Condensing Humidity

Daimler Chrysler LP-463-PB-09-01 Condensing Humidity Resistance Test

General Motors GM4465P Water Fog Humidity Test

14.1.2 Corrosive gas test

The use of corrosion gas testing is also beneficial for identifying potential issues with components that will be used in harsh industrial environments. To simulate harsher conditions, the levels of corrosive gases are added to the humidity. Some of these tests are also designed to reveal certain material features. To test the presence of pores in certain types of gold and platinum coatings, ASTM B775 (Test Method for Porosity in Gold Coatings on Metal Substrates by Nitric Acid Vapor) and B799 (Test Method for Porosity in Gold or Palladium Coatings by Sulfurous Acid/Sulfur-Dioxide Vapor) are commonly used. These two tests are carried out using extremely high concentrations of nitric acid and sulfuric acid vapor. The moist sulfur dioxide test (ASTM G 87) is commonly used in industrial settings to mimic the corrosion that can occur in certain materials. In addition, a flowing gas test (ASTM B827) is also commonly performed. This type of test involves introducing various components, such as hydrogen sulfide, nitrogen dioxide, and chlorine, into a chamber with controlled humidity and temperature. This type of test is ideal for the electronics sector.

14.1.3 Salt spray testing

A salt spray test in line with DIN EN ISO 9227 can be used to measure the corrosion resistance of a material or a protective anti-corrosion coating on it. It is the most widely used and oldest cabinet test. Salt spray (Fog) testing is another name for it (n.d.). It is commonly used to compare expected and actual corrosion resistance. The duration of the coating in the test and its actual projected life have just a weak association. Because corrosion is not a straightforward process, it is influenced by various external influences. Since salt spray testing is relatively simple and quick to perform, it is commonly used to determine if a certain material's pass-or-fail grade is acceptable. This is usually done as part of a quality control process or to evaluate the effectiveness of a certain production method. The first known international salt spray test was conducted in 1939.

In order to test the effects of saltwater on various specimens, a salt spray experiment is performed in a sealed chamber. The solution used is composed of NaCl, and the concentration varies. A corrosive experiment is then performed to simulate the effects of saltwater fog. The appearance of the oxides is then studied to determine the product's corrosion resistance. The longer it takes for the oxides to emerge, the more resistant the product is. Depending on the type of material being tested, the typical salt spray corrosion resistance test duration can range from 96 to 480 hours. This test is performed as part of ISO 12944-9, which was previously known as ISO 20340. Although both ISO 12944-9 and B-117 are used to measure the protection of a certain type of material against corrosion, the results of the two tests have significant differences. Unlike the B-117 test, which only involves static salt spray, ISO 12944–9 is a cyclic test that involves exposing panels to UV exposure for 72 hours, followed by 72 hours of salt spray. This cycle, which lasts for 25 weeks for a total of 4200 hours, involves repeated exposure to varying levels of salt spray, freezing, and ultraviolet radiation. Although this test is not an actual outdoor environment, it is like the weather conditions that we experience in the summer. The experiments can be performed in three steps

1. The first step in the process of evaluating a coating is to make small panels, usually 3 × 5 inches or 4 × 6 inches wide. The back and edges of the panels are then taped to protect them. The front surface of the panels is then scratched using a sharp instrument.
2. After the panels have been prepared, they are placed inside a test cabinet. According to ASTM B-117, the panels should be tilted back 15 to 30° from their vertical position to allow the vapor to flow through the chamber. The test cabinet is then closed. An atomizer then delivers 5% NaCl salt solution into the chamber at a steady rate and temperature.
3. The salt fog will cause droplets to fall onto the panels as the temperature of the solution rises. Following this, the test panels will be periodically checked, and the salt spray cabinet will be monitored to ensure that it performs properly throughout the test. It is established whether the test is continuing as planned or if the coating on the test panels is deteriorating prematurely during these brief inspections of the panels.

After the salt spray testing has been completed, the test specimens are washed with a deionized solution to remove any loose corrosion products. The results of the examination are then analyzed using either electrical or microscopic methods.

The salt spray test also has a few drawbacks. Aside from not providing the necessary results, it can also only provide results that are applicable to the life expectancy of coatings. It is also destructive.

14.2 Immersion testing

The simplest form of immersion testing is to immerse a material in a solution or an interesting environment. This test determines a material's resistance to harsh environmental conditions. In this process, a specimen is placed inside a container that is filled with a solution. The weight loss that a material can experience due to corrosion is measured using a standard laboratory coupon. The ASTM G31 test method converts a material's corrosion rate to a temporal unit, which is usually such as millimeters per year or mils per year (mpy). The duration of the test can vary, but findings can be obtained after a 24-hour exposure period, a 10-day exposure period, or a 20-day exposure cycle.

There are numerous evaluation criteria for immersion testing that are frequently used to quantify the effect of corrosion. The most utilized criteria are appearance (i.e., how the specimen appears after testing) and weight change. The standard technique for administering the test, for the most part, also includes information on evaluation and pass/fail criteria. There are also a variety of relevant standard evaluation processes, such as:

- ASTM G1
- ASTM G46

Some examination methodologies are tailored to a particular type of corrosion. ASTM G31 (weight change or weight loss), for example, can be applied to any mode/form of corrosion and transformed to a corrosion rate using the equation:

$$\text{corrosion rate} = \frac{8.76 \times 10^4 W}{At\rho}$$

where *CR*, corrosion rate; *W*, mass loss in g; *A*, an exposed area in cm^2; *t*, time in hours; ρ, material density in g/cm^3. If localized corrosion is the main cause of the corrosion, then the weight loss test should not be used to determine the true corrosion rate. This occurs because the rate at which corrosion occurs at the site is much higher than the rate at which corrosion occurs on the surface. One of the most challenging factors in determining the performance of corrosion-resistant materials is the small changes in the specimen weight. This makes it difficult to interpret and reproduce the results. ASTM G31 suggests a minimum test period of the following expression that can help overcome this issue:

$$\text{test time(h)} = \frac{787.4}{\text{corrosion rate(mmpy)}}$$

If a 1 mm/year corrosion rate is measured, the suggested minimum test period is 787 hours or approximately 32 days. As a result, relatively modest corrosion rates necessitate longer test times to ensure accuracy. As a result, even modest changes in the observed corrosion rate might have a significant impact. Such changes could result from posttest cleaning methods and possibly ambient species deposition (e.g., scale or deposits). NACE test method TM0169-95 (281-228-6200) offers recommendations for determining optimum test times for immersion service testing.

The fundamental benefit of immersion tests is that a large range of circumstances and materials may be tested simultaneously with little effort and equipment. This has the benefit of establishing "safe operating" conditions in which a specific material can be utilized. At temperatures, less than 120°C, the corrosion rate of Type 303 stainless steel in 65% nitric acid has been observed to be 0.5 mm/year. When the temperature is raised to 135°C, the measured corrosion rate rises to 5 mm per year, with further temperature increases resulting in accelerated corrosion (International, 2012). The study also indicated that Type 303 stainless steel can be used if the operating temperature is below 120°C. However, it could not be used if the operating temperature is predicted to exceed this value. This could also lead to the use of other materials. To help in the selection process, similar studies are performed on a wide range of environmental variables. These studies can help determine the ideal material for various applications. Variables that are frequently examined include temperature, pH, halide concentration (chloride, fluoride, bromide, etc.), the concentration of other inorganic and organic species, inhibitors, acidic gases (e.g., CO_2, H_2S), pressure, flow rate, type (e.g., single-phase, multiphase), and stress level.

14.3 High-temperature/high-pressure testing

High temperatures and high pressures (HPHT) testing, also known as autoclave testing, is performed in a controlled environment that replicates the high temperatures (HT) and high pressures (HP) ($>$ 0.07 MPa or 10 psig) found in many commercial and industrial operations (Roberge, 2012). There are numerous types of commercial HT/HP environments that are important to various industries, such as petroleum, nuclear, chemicals, aircraft, and transportation. These environments require special testing to evaluate their dependability, serviceability, and corrosion concerns. In the oil and gas industry, the pitting, corrosion cracking, and crevice resistance of various corrosion-resistant materials are evaluated using HP/HT testing. These are commonly subjected to harsh service conditions such as H_2S and sweet service. Moreover, for various applications, such as tanks and vessels, internal coatings and linings are subjected to HP/HT testing to imitate the conditions in a laboratory environment. The coating or lining in question is evaluated by exposing it to a specified composition of liquids and gases (both sweet and sour), often in combination with temperature and pressure. This testing is frequently used to qualify new coatings for a particular environment or to determine their resistance to process circumstances like steam out, gas pressure changes, rapid decompression, and so on. NACE TM 0185 and client-specific standards and bespoke test methodologies are all covered. Element also offers a variety of pre- and post-exposure coating characterization tests to assess the film's performance.

Salient features for performing high-temperature/high-pressure testing

- Volumes from 500 mL to 25 L
- Design pressures up to 350 bar and temperatures up to 650°C
- Available in different materials of construction – SS 316, Hastelloy C, Inconel, Titanium, etc.
- Stirred/non-stirred autoclave systems
- Recirculating loop and electrochemical corrosion testing
- For static, dynamic, loop, and electrochemical studies

There will be a brief description of four versions of conventional HT/HP test procedures that have been proven to be beneficial in materials evaluation involving corrosion phenomena. These types of evaluations, on the other hand, can be achieved with proper planning and test vessel design. Among them are:

14.3.1 Windowed test vessels

To make visual measurements or observations within the boundaries of test containers, unique transparent windows, and other fittings such as fiber optics have been employed. These windows may have to perform various activities linked to the entry of light or other radiation if these are part of the test variables, in addition to being able to tolerate pressures, temperatures, and corrosion conditions.

14.3.2 Electrochemical measurements

For tests conducted inside HT/HP vessels, most traditional electrochemical techniques were applied. The reference electrode has traditionally been the most important electrochemical component in these investigations. The reference electrode's design and construction are critical since they must offer a stable and consistent reference potential. Various test vessels have been modified to allow an external reference electrode to reduce the effects of temperature, pressure, contamination, or a combination of these factors in many applications.

14.3.3 Hydrogen permeation

Hydrogen permeation in materials that are subjected to high temperature and high pressure conditions can pose a threat to their safety. In this case, diffusion constants and hydrogen permeation rates are required to determine the potential risk of hydrogen assault. Solid-state devices may be required for hydrogen permeation measurements at high temperatures.

14.3.4 Mechanical property testing

Slow strain rate, fracture, and fatigue testing can all be performed using HT/HP vessels. The key issue is always choosing fixtures that can tolerate the corrosive conditions created during HT/HP tests.

Short questions

1. What is salt spray testing and how does the process work?
2. How do you conduct a humidity test for corrosion testing?
3. Is salt spray a destructive test?
4. What test methods are available for corrosion evaluation?
5. What are the two common testing methods used to study the corrosion on a piece of steel?

References

Brossia, S. (2018). Laboratory assessment of corrosion. In *Handbook of Environmental Degradation of Materials* (3rd ed., pp. 27–50). Elsevier.

International, NACE. (2012). International, standard guide for laboratory immersion corrosion testing of metals. *NACE International.*

Meikle, T., Tadolini, S. C., Sainsbury, B.-A., & Bolton, J. (2017). Laboratory and field testing of bolting systems subjected to highly corrosive environments. *International Journal of Mining Science and Technology, 27*(1), 101–106. Available from https://doi.org/10.1016/j.ijmst.2016.11.017.

Roberge, P. R. (2012). In *Handbook of Corrosion Engineering*, (2nd ed.). New York, Chicago, San Francisco, Lisbon, London, Madrid, Mexico City, Milan, New Delhi, San Juan, Seoul, Singapore, Sydney, Toronto: McGraw-Hill Education.

15

Field tests methods

Although field tests are generally conducted in different environments, such as the atmosphere, soils, and water, and they are generally not under the control of investigators. This means that they are the most reliable way of evaluating a material's performance in a certain service. This chapter provides an overview of the various types involved in conducting these tests, including the planning, and designing of the test program. Conducting field corrosion testing involves preparing materials with known properties and composition. They are then placed into an existing environment of interest. The specimens are then removed and evaluated. This kind of testing differs from laboratory testing in that it is not conducted under an investigator's supervision. In the case of chemical processing operations, the environment is created by the chemical operation. Various tests methods are discussed below:

15.1 Corrosion coupons testing

One of the most common methods of monitoring corrosion is by installing corrosion coupons (Speight, 2014). These are typically inserted into equipment or machinery that has a corrosion coupon holder. They are then used to monitor the condition of the pipelines, tanks, and heat exchangers. Even though different factors might affect the location of the coupons, they are usually placed in areas where severe corrosion is expected. After the coupons have been installed, they will be considered to see if there are any modifications in size and weight, or the visual inspection will be considered. The disadvantage of this approach is that results cannot be obtained by accelerating the environmental circumstances (n.d.-a). Coupons are basically a sample of metal that is weighed, exposed to a particular environment, removed, cleaned, and then reweighed to calculate the weight loss due to corrosion over a certain time. The amount of weight loss in mils per year (MPY), which is the most often used unit of measurement, is determined using the NACE's standard methodology. Most other corrosion monitoring techniques also use this reference. Although MPY is by far the most popular unit of measurement, MM per year and other units of measurement are also utilized. Corrosion coupons for weight loss come in a wide variety of sizes and materials. Once one understands that the more surface area exposed to the environment, the more accurate the readings will be, and the quicker a reading may be taken, the size becomes less significant. The most popular shape, known as a stripped coupon, is frequently offered in the following sizes: 1/2″ × 1/16″ × 3″ or 7/8″ × 1/8″ × 3″ (Singh, 2014). Additionally, other configurations are also employed, including rod or pencil-shaped coupons, prestressed

Handbook of Corrosion Engineering. DOI: https://doi.org/10.1016/B978-0-323-95185-2.00015-0

coupons in a variety of forms, and "lifesaver"-shaped coupons that frequently have electronic probes attached. Coupons for weight loss corrosion are available in any material needed. However, the material is typically either that which contains the environment (i.e., the pipe, vessel, or tank) or low-carbon steel, which is highly corrosive.

The sample weight loss can then be converted to an equivalent uniform corrosion rate by using an analytical calculation using Eq. (15–1) (ASTM G31-72, 2004):

$$V = -\left(\frac{K.\Delta m}{\rho St}\right) \tag{15-1}$$

where V is corrosion rate (μm/year), ΔM is sample weight variation (mg), ρ is alloy density (g/cm^3), S is an area of the sample exposed to the medium (cm^2), t is exposure duration (h), and K is a nondimensional coefficient (8.76×10^6, with the units used for the other terms).

Coupons are typically exposed for 90 days as part of a monitoring program before being removed for lab analysis. This provides fundamental corrosion-rate measurements four times annually. Any single coupon exposure that results in weight loss gives the "average" value of corrosion that takes place during that exposure. The drawback of the coupon technique is that if corrosion upset happens during the exposure period, the coupon alone will not be able to pinpoint when it happened, and depending on the upset's peak value and duration, it might not even be able to detect a statistically significant increased weight loss.

Online corrosion coupons are commonly used for the evaluation of fluid corrosion in critical process circuits. These are made of either carbon or alloy steel, and they can be fixed or retractable. In high-temperature environments, such as those involving molten rock, the coupons should be sealed with an appropriate seal system. Also, different types of coupons can be used, such as circular, rectangular, or helical. It can be either regular or prestressed. The level of corrosion can be determined by the number of additives that are added to the solution. This can be done by modifying the coupon's corrosion rate.

The data collected from corrosion coupons should be analyzed to determine the optimal strategy for analyzing them. Some of the factors that should be considered include the location of the facility, the exposure time, and the product/scale composition.

Coupons should be distributed throughout the system's many components where the worst-case corrosion rates are anticipated to enable effective monitoring in each system. This could include low spots, sluggish regions, and regions with high speeds or turbulence. Additionally, pipelines and piping should be kept an eye out at both the 6 and 12 o'clock positions. To assess corrosion rates, the service fluid must be exposed for a long period of time (3–6 months). It should be remembered that some issues, such as bacterial fouling, take longer to manifest on the surface of coupons.

How should the coupons be oriented and positioned?

To evaluate and contrast various data sets, the orientation of the corrosion coupons should be uniform. A corrosion coupon should typically be placed with its orientation parallel to the process flow. It is crucial to keep an eye on both the 6 and 12 o'clock positions in systems with multiphase/stratified flow or in systems where bottom-of-line deposits may form. To accurately depict the flow regime of the vessel or pipe, the corrosion coupon should be positioned in the plane of the wall. The use of flush-mounted coupons rather than invasive coupons will result in the best representation of flow in each system.

How many coupons should be installed?

Typically, more than one coupon is put at any coupon monitoring site for the following two reasons: The accuracy of a single coupon may be affected negatively by numerous uncontrollable circumstances. The dependability and precision of the data are greatly enhanced when paired coupons are used, with an average value taken. Coupons for corrosion reveal the corrosion that has already taken place. Multiple corrosion coupons might be inserted at the same spot and retrieved at various intervals if it was necessary to determine if the corrosion rate was constant or fluctuating over a specific period.

Types of corrosion coupons

There are many different coupon types. The most common coupon types are described below:

Strip coupons: These are the most common and economical type.

Disc coupons: these are helpful for multiphase flows, particularly when combined with ladder coupon holders to assess the rates of corrosion along the pipeline or piping's diameter. Coupons for flush discs can be installed so that they line up with the interior of the piping, pipeline, or vessel. This offers two benefits: The flow in the vessel will be shown on the coupon more precisely. This will make it possible for pipelines to be pigged. No prior knowledge of product flow is necessary.

Corrosion coupon holders: Two or more coupons can be fixed at one location with the help of a coupon holder. The coupon holder is chosen based on the operational conditions, the number and type of coupons, the configuration, and the system entry method. The most typical types of holders are:

Fixed coupon holder: It may be constructed from a range of materials and is the most economical. It is necessary to depressurize the system before inserting and removing objects. It is utilized in systems that frequently shutdown or in systems that are simple to isolate. It is easy to install.

Adjustable coupon holders: These coupon holders are utilized in systems that demand pressures of up to 125 psi. In the water treatment business, this arrangement is frequently used for inserting coupons through a full port valve. In a municipal water pumping station, for instance, a leak during test coupon removal would not be a serious issue.

Retractable coupon holders: Extensively employed in refineries. These are utilized in environments where the leakage is not permitted and ratings up to 2000 psi are necessary. To insert or withdraw coupons from these systems, a line (process) shutdown is not necessary. The assembly is used to insert coupons through valves with a minimum port size of 1". To prevent unintentional discharge, a safety cord is offered.

Retrievable coupon holders: When pressure ratings of up to 3600 psi are necessary, these coupon holders are employed with high-pressure access systems (HPTM or MHTM). Widely employed in the extraction of oil and gas. There is no need to stop any processes.

The advantages of this test include its affordability, positive weight loss results, the ability to collect samples of bacteria or corrosion products from the surface for further examination and testing, and the fact that coupons are not prone to instrument failure like electronic methods of corrosion monitoring. The drawback of corrosion coupons is that they can only provide average readings and that they take a long time to produce findings.

15.2 Ultrasonic thickness monitoring

The use of ultrasonic method is commonly used in the inspection of industrial equipment to determine the thickness of the components that are being tested (Song, Guo, Jia, & Tu, 2022). It is very important that the metal loss caused by corrosion or erosion is detected before it can affect the operation of the equipment. A Polish engineer from Katowice named Werner Sobek developed the first ultrasonic thickness gauge in 1967 (n. d.-b). This first ultrasonic thickness gauge assessed the speed of the waves it released in certain test materials, and from this speed measurement, it computed the thickness in micrometers using a mathematical calculation. The use of the ultrasonic method can also help determine if the item or structure needs to be repaired or replaced. An instrument known as an ultrasonic thickness gauge is used to check the internal corrosion of storage tanks and pipelines. The ultrasonic sound wave created by a transducer passes through the back wall and reflects its source. This allows one to measure the thickness of the material by considering the velocity and reflection time of the sound waves. The instrument is useful when there is no way to access both sides of a test specimen, as it allows one to calculate the thickness of the material. The data collected by an ultrasonic thickness gauge helps customers determine if the item they are testing has the necessary metal thickness. An ultrasonic thickness gauge measures the exact amount of time needed for a sound pulse produced by a tiny probe known as an ultrasonic transducer to pass through a test material and reflect off the interior surface or far wall. The measured thickness of the test piece is displayed on a digital screen, allowing the user to instantly inspect the results. The portability of the instrument makes it ideal for on-site inspection.

Additional nondestructive testing techniques can be employed to further examine the results if the approach has identified an issue.

When calibrated correctly, measurements may typically be made with an accuracy of +/− 0.001" or 0.01 mm, and in some circumstances, accuracy can even approach 0.0001" or 0.001 mm (n.d.-c). The best way to assess accuracy in a particular application is to utilize reference standards with known thicknesses. The ability to measure the thickness of an object's surface is carried out with the help of the UT probe. With its ultra-high accuracy, the measurement can be performed globally or locally. The compact and lightweight design of the UT probe makes it ideal for measuring the thickness of an object's surface. It can also be used to record the results in a data logger. Without the need for surface smoothing on inaccessible locations, very thin plates, high-temperature conditions, etc., various probes can be used to measure thickness above paint (through-coat or echo-to-echo functions) and on surfaces with severe local corrosion (use of pencil-type probes).

15.3 Advantages of the method

1. This is a high-precision quick measurement.
2. The inspected object can be in-service.
3. No surface preparation is required; minor wire brushing is needed.
4. Measurements can be made without the removal of the paint. Measurements can be made in inaccessible regions using suitable probes.
5. Corrosion rate calculations can be made with thickness reduction observations with repeated measurements at the same points.
6. Measurements on high-temperature surfaces are possible.

15.4 Electrical resistance testing

By using electrical resistance (ER) probes and instruments, it is possible to quantify metal loss due to corrosion or erosion (Xu et al., 2020). Apart from liquid metals and a few conductive molten salts, this approach can be applied almost anywhere. When the corrosion rate of coupons must be determined online, ER probes are employed to assess the rate of corrosion. The decrease in thickness or surface area of a cross-section caused by corrosion in an environment causes a metal's ER to rise. The rate of metal dissolution can be measured over time by observing changes in ER, and the corrosion rate can be computed in mpy or mm/year. Any environment, including aqueous solutions, oil (hydrocarbons), soil, gas, and the atmosphere, can be employed with the ER probes. Depending on the kind of metals, system, and testing environment, the probe can be made in a variety of geometries. The sensing components of ER probes are manufactured of the metal or alloy for which corrosion data is needed. The probe bodies could be higher, less corrosive alloys, or the same alloy but thicker.

15.4.1 Principles of operation

The ER of a metal or alloy element is given by Eq. (15−2):

$$R = r.\frac{L}{A} \tag{15-2}$$

where L is element length, A is cross-sectional area, r is specific resistance reduction (metal loss) in the element's cross-section due to corrosion will be accompanied by a proportionate increase in the element's ER.

ER probes with an element that is openly "exposed" to the corrosive fluid and a "reference" element sealed inside the probe body enable practical measurement. Measuring the resistance ratio minimizes the impact of changes in the ambient temperature because temperature changes affect the resistance of both the exposed and shielded elements equally. Therefore after thermal equilibrium is reached, any net change in the resistance ratio can only be attributed to metal loss from the exposed element. A portable instrument can be used to measure the ER probe periodically, or a unit that is permanently mounted can measure it continuously. In either situation, ER equipment will generate a linearized signal that is inversely correlated to the metal loss of the exposed element. The instrument's output change rate serves as a gauge for the rate of deterioration. Continuously collected data is typically sent to a computer or data logger where it is processed to provide information on the rate of direct corrosion.

15.4.2 ER sensing elements

Sensing components come in a range of geometrical arrangements, alloy compositions, and thicknesses. These ER sensing components are described below:

Wire loop: The most used type of element is a wire loop. This kind of element is a solid option for most monitoring systems because of its high sensitivity and low susceptibility to system noise. Typically, wire loops are glass-sealed into an endcap before being welded to the probe body. A glass seal is a good option for most applications since it is often chemically inert and has good pressure and temperature ratings. Carbon steel, AISI 304, and 316 stainless sheets of steel are alloys that are frequently glass sealed. Teflon-sealed elements are also available in places where glass may be prone to corrosion issues. To safeguard the element from floating debris in the pipe system, probes with wire loop elements are typically fitted with a flow deflector (also known as a velocity shield).

Tube loop: Where great sensitivity is required to quickly identify low corrosion rates, tube loop elements are advised. A small bore, hollow tube is used to make tube loop elements, which are then shaped into the aforementioned loop configuration. The most prevalent alloy is carbon steel. It's also possible to use tube loops that are pressure-sealed with Teflon into the probe. To reduce potential distortion in rapidly flowing systems, flow deflectors can be added to probes using the tubular loop element.

Strip loop: Strip loop components resemble wire and tube loop arrangements. A flat element created in a loop geometry is a strip loop. Depending on the application, the strip loop may be glass or epoxy sealed into the endcap. The strip loop is a particularly delicate component. Strip loops should only be used in extremely low-flow situations since they are quite brittle.

Cylindrical elements: A reference tube is welded inside of a tube element to create cylindrical elements. This component is entirely welded together before being attached to the probe body. Exotic alloy elements can be made very readily because this element is entirely welded together. If a glass-sealed element is not an option, this probe is best suited for tough environments like high-velocity and high-temperature systems. They offer a huge measurement surface, have the quickest dynamic response to temperature transients, and are least affected by any conductive deposits (such as iron sulfide) that may be present. These probes can be utilized in environments with pH values between 0 and 14 and at temperatures up to 500°F because of their design. For temperatures exceeding 50°F, a high-temperature model is also offered. To avoid the effects of potential galvanic corrosion between the probe body and probe element, the cylindrical element probe's body is often supplied in the same alloy as the element.

Spiral loop: A thin metal strip produced on an inert basis makes up the spiral loop element. The material is especially tough and perfect for high-flow programs. The element is extremely sensitive due to its relatively high resistance, which also results in a high signal-to-noise ratio.

Flush mount: Elements with flush mounts are made to fit flush against the vessel wall. This component is particularly good at replicating the actual corrosion along the vessel walls inside surfaces. Due to its flush nature, this element can be employed in pipeline systems that are subject to pigging operations and is less likely to be damaged in high-velocity systems. For erosion-corrosion monitoring, the flush-mounted type is acceptable, and the wire loop design offers greater sensitivity.

Surface strip: These elements are thin rectangular components with a relatively wide surface area that are used in atmospheric probes to produce more accurate results in nonhomogeneous corrosive conditions. To assess the efficacy of cathodic protection currents applied to the outside surfaces of subterranean structures, strip elements are frequently utilized in underground probes.

15.4.3 Corrosion rate calculation

The device generates a linearized signal (S) while measuring the ER probe that is proportionate to the total metal loss of the exposed element (M). The element thickness and geometry affect the true numerical value. These geometrical and dimensional variables are considered while calculating the "probe life" (P), and the metal loss is given by:

$$M = \frac{S \times P}{100} \tag{15-3}$$

Metal loss is conventionally expressed in mils (0.001 inches), as is element thickness.

Corrosion rate (C) is derived by:

$$C = \frac{P \times 365(S_2 - S_1)}{\Delta T \times 100} \tag{15-4}$$

T is the lapse time in days between instrument readings S_1 and S_2.

However, ER probes do have some restrictions. First off, the ER technique's sensitivity is typically too low to detect sudden changes in corrosion conditions. The response time for detecting a corrosion rate on the order of 1 mpy (25 μm/year) is between 1 and 10 days. On the other side, the response time is significantly shortened to a matter of a few hours if the anticipated corrosion rate is 100 mpy (2.5 μm/year). Thinner/smaller elements can be used to increase sensitivity, but the probe life is sacrificed in the process. The ER method has a drawback in that it can only quantify homogeneous corrosion and cannot detect localized corrosion processes. When the likelihood of conductive corrosion products or scales precipitating on the probe exists, further restrictions come into play. For instance, because the iron sulfide corrosion product that formed is conductive and tends to underestimate the amount of metal loss experienced, sulfide environments are particularly challenging to interpret with ER probes. In certain instances, the amount of iron sulfide precipitation was high enough to show that metal gain had occurred. Therefore, this type of response slightly restricts the use of ER probes in sour oil and gas systems.

Short questions

1. Which instrument is used for monitor rate of corrosion?
2. What kind of technique is employed to measure thickness?
3. What is the maximum thickness that can probably be tested with ultrasound?
4. What are the advantages of ultrasonic testing?
5. How does an electrical resistance probe work?
6. What is the maximum thickness for ultrasonic testing?

References

ASTM G31-72. (2004). *Standard practice for laboratory immersion corrosion testing of metals*. American Society for Testing and Materials.

Singh, Ramesh (2014). *Chapter Five - Hazards and Threats to a Pipeline System Pipeline Integrity Handbook Risk Management and Evaluation* (pp. 45–120). Elsevier.

Song, J., Guo, D., Jia, J., & Tu, S. (2022). A new on-line ultrasonic thickness monitoring system for high temperature pipes. *International Journal of Pressure Vessels and Piping, 199*, 104691. Available from https://doi.org/10.1016/j.ijpvp.2022.104691.

Speight, J. G. (2014). *Oil and gas corrosion prevention: From surface facilities to refineriesoil and gas corrosion prevention: From surface facilities to refineries* (pp. 1–151). United States: Elsevier Inc. Available from http://www.sciencedirect.com/science/book/9780128003466.

Xu, Y., Liu, L., Zhou, Q., Wang, X., Tan, M. Y., & Huang, Y. (2020). An overview of major experimental methods and apparatus for measuring and investigating erosion-corrosion of ferrous-based steels. *Metals, 10*(2), 180. Available from https://doi.org/10.3390/met10020180.

16 Electrochemical test methods

16.1 Introduction

The explosion of electrochemical methods has greatly benefited the field of corrosion engineering and science. These tools have allowed scientists and engineers to investigate the various aspects of corrosion, such as its kinetic and thermodynamic properties (Pourbaix, 1984). Over the years, the development of effective methods for preventing and understanding corrosion has been carried out in both the laboratory and the field. One of the advantages of electrochemical methods overexposure techniques is their ability to provide fast results, nondestructiveness, and high resolution. These methods are relatively cheap to perform and can be used quickly. These methods also have the advantage of being able to provide mechanistic information. Electrochemical methods for corrosion are mainly based on the response of the charge-to-discharge process to a controlled electrical disturbance (Pourbaix, 1984; Rechnitz, 1963; Silverman & in, 2000).

For many years, the use of various electrochemical techniques like linear polarization (LPR), potentiodynamic polarization, and electrochemical impedance spectroscopy (EIS) have been regarded as a powerful complementary tool for the traditional weight loss approach (Geethanjali & Subhashini, 2015; Njoku et al., 2016; Zand et al., 2013).

16.1.1 Electrochemical impedance spectroscopy

The advantages of this technique are its fast and flexible nature, high efficiency, and nondestructive nature. It can be used in various applications such as biomedical devices, sensors, and fuel cells (Huet, 1998; Kelly et al., 2002; Koutsouras et al., 2017; Li et al., 2021; Wagner et al., 1998). It is a transfer function method that can be applied to the processes.

Electrochemical techniques have been successfully utilized in the field of corrosion protection and inhibition. They have been shown to be useful in the analysis of the passive layers and the formulation of effective corrosion inhibitors. In addition, they have been able to provide the details of the protection mechanisms. The ability to analyze the effects of an adsorbed or coating film on different materials has been shown to be useful in the development of effective corrosion protection strategies. From the EIS data mechanistic information, surface properties, and electrode kinetics can be predicted (Macdonald, 2001). The increasing number of applications of EIS has greatly affected the field of electrochemistry. Its potential to study the various physical and electronic properties of various types of electrochemical systems such as the diffusion coefficient, adsorption processes, resistances to charge transfer, capacitances, etc. has led to the development of new tools for the evaluation of their inhibitory action. One of these is the ability to calculate corrosion-related parameters.

Handbook of Corrosion Engineering. DOI: https://doi.org/10.1016/B978-0-323-95185-2.00016-2

The three-electrode cell used for performing an EIS measurement consists of a reference electrode, a counterelectrode, and a working electrode (WE). The working is referred to as samples of the metal being examined. To achieve a stable-state condition, the WEs are placed in a testing electrolyte. This ensures that the corrosion potential of the metal is computed based on the other measurements. The small-amplitude signal, that is, 5–50 mV is applied to a sample over a frequency range of 0.001 to 100,000 Hz to make an accurate measurement of the system's resistance and capacitance. The instrument then records the real and imaginary components of the system's response. The instrument uses an algorithm to analyze the data and determine the system's initial circuit parameters. The computer then uses the data collected by the system to calculate the appropriate parameters for a fitting. The fitting curve of the spectrum is used to determine the quality of the steel substrate. It is also possible to connect the data to the corrosion conditions of the steel.

16.2 Nyquist plots

A frequency response plot, also known as a complex plane plot or a Cole–Cole plot, is used in control engineering and signal processing. A Nyquist plot is a widely used tool to evaluate a feedback system's stability. The imaginary part of the transfer function is shown on the *x*-axis, while the real part is displayed on the *y*-axis in Cartesian coordinates (Brett et al., 1993) as shown in Fig. 16–1A. The frequency is represented by a frequency-based graphic. The function is derived from the polar coordinates, and its phase is the corresponding coordinate. In polar coordinates, the radial coordinate is the transfer function's gain, and the phase is its corresponding angular coordinate making the Nyquist plot unique. The appropriate impedance at a certain frequency is defined by any location on the map. In the electrochemistry group, plotting $-Z_{imag}$ (also written as $-Z$ "or $-Zj$") on the *y*-axis so that the detail lies inside the first

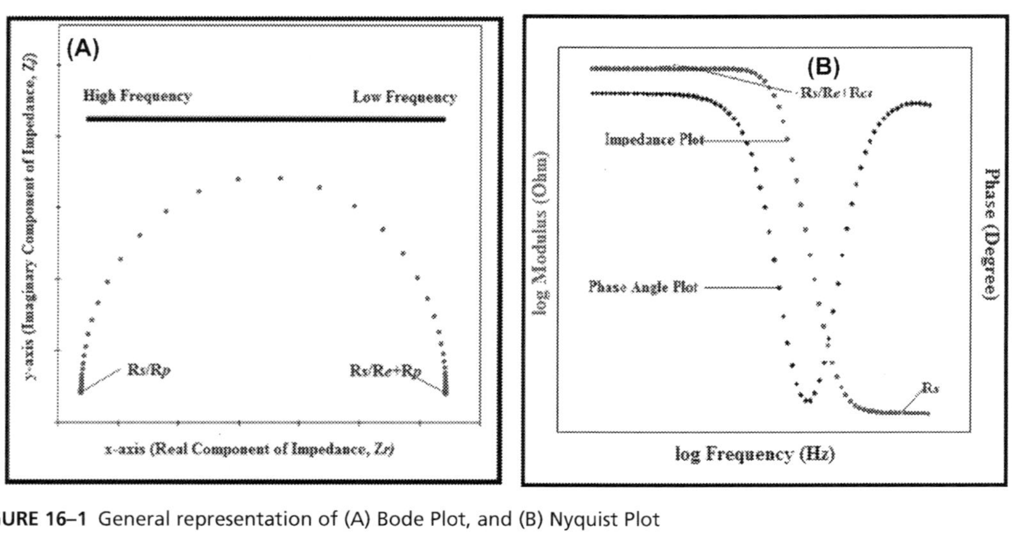

FIGURE 16–1 General representation of (A) Bode Plot, and (B) Nyquist Plot

quadrant of a graph is a common norm. Nyquist charts are useful for obtaining details about a system's attributes, but they always lose frequency information. The Nyquist map has one fundamental flaw: the frequency cannot be calculated just by looking at the graphs. The frequency of crucial data points, such as high and low real axis intercepts and the characteristic frequency of the arc, was annotated to overcome this constraint (Cole & Cole, 1941).

16.2.1 Bode plot

In the 1930s, Hendrik Wade Bode was the first to introduce the concept of a Bode plot to the world of control system engineering. A Bode plot is a graph that shows the stability of a system. The magnitude and phase plots are used to map the system's frequency response. The Bode plot is shown in Fig. 16–1B as a two-in-one spectrum plot that combines the phases and the logarithm of impedance. It shows the system's stability by considering the logarithm of impedance (log $|Z|$) and phase shift at the first and second ordinates, respectively.

The phase diagram also shows the relationship between the magnitude of the impedance vector as well as the phase angle. It can be represented using the following equation that is derived from the real and imaginary elements.

$$z = \sqrt{Zr^2 + Zj^2} \tag{16-1}$$

where Zr = the real component of the impedance and Zj = the imaginary component of the impedance.

An electrical circuit model is used to analyze the data collected from the EIS. It features a variety of conventional components such as capacitors, inductors, and resistors. The parts shown in Fig. 16–2 can be joined in a network to create an equivalent model.

The ideal impedance of various components such as capacitors, inductors, and resistors are determined by the following equations (Macdonald, 2006):

$$Z_R = R \tag{16-2}$$

$$Z_c = \frac{-j}{\omega C} \tag{16-3}$$

$$Z_L = j\omega L \tag{16-4}$$

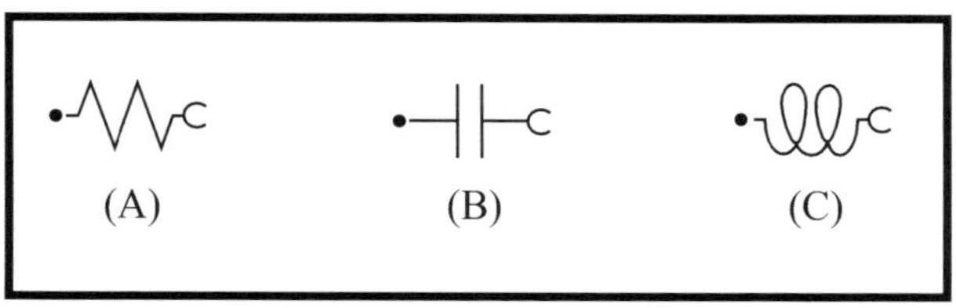

FIGURE 16–2 Passive elements of the Equivalent electrical circuits: (A) resistor, (B) capacitor, and (C) inductor.

A perfect resistor's impedance will be fully real, and the current flowing through it will always be in phase with the voltage, showing a phase change of 0 (in-phase), whereas an ideal inductor and capacitor's impedance will be purely imaginary. In terms of frequency response, an inductor's impedance is the opposite of capacitors. The impedance of an inductor grows as the frequency rises, whereas the capacitor's impedance lowers as the frequency rises. The capacitors have a -90-phase shift, while the inductors have a 90-phase shift (out of phase) (Bard & Faulkner, 2001).

The simplified version of Randle's circuit shown in Fig. 16–3 is a well-known cell model for developing complex circuits and a starting point for other complex models (Macdonald, 2001; Raja & Sethuraman, 2008).

The simplified version of the circuit shown in Fig. 16–3 is composed of a solution resistance, a charge transfer resistance, and a double-layer capacitance. The two layers of the double-layer capacitance are parallel to each other, and this combination is in series with the circuit's solution resistances. The Nyquist plot in Fig. 16–1A shows that a semicircular arc crosses the real axis (Z') at two points. The value of the solution resistance at the origin is determined by the intercept closest to the source, while the value of the overall resistance ($Rs + Rct$) is determined by the intercept farther away (Macdonald, 2011).

More complex models are generated based on the electrode's physical features, such as a partially blocked electrode, an electrode covered with an inert porous layer, a completely porous electrode, an electrode coated with many layers of inert porous layers, and so on. Combining the fundamental circuits yields the relevant impedance for a highly nested electrical circuit. Other modeling components utilized in electrochemical circuits for diverse complex systems include Warburg impedance (W) and constant phase elements (CPEs) (see Fig. 16–4).

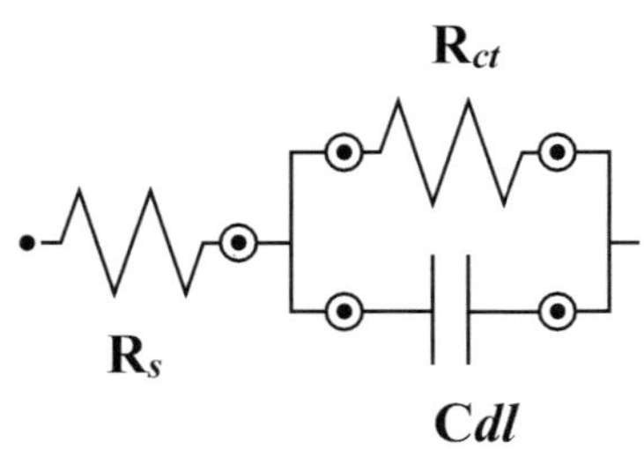

FIGURE 16–3 Simple randles equivalent circuit used for the fitting of EIS data for a corrosion system under charge transfer control, with the one-time constant feature

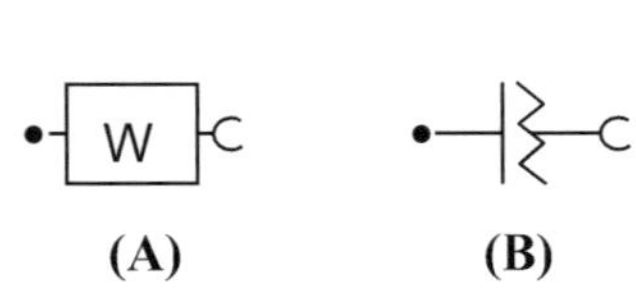

FIGURE 16–4 (A) Warburg Impedance, (B) constant phase element, and other circuit elements used for the fitting of EIS data.

W is the general impedance associated with mass transfer processes that result in a 45 degrees phase shift, such as diffusion control reactions. In contrast, CPE is a generic element that can be used to simulate nonideal or faulty capacitors. It causes a phase shift of 80–90 degrees. The Warburg impedance and the CPEs are calculated using the equations below (Warburg, 1899):

$$Z_w = \frac{1}{Y_0\sqrt{j\omega}} \quad (16-5)$$

$$Z_Q = \frac{1}{Y_0\,(j\omega)^n} \quad (16-6)$$

On a microscopic level, the specific electrochemical interface is not the smooth, uniform, and defect-free surface that is usual for a pure capacitor (Sastri, 1998). It has a wide range of structural properties, including porosity, surface inhomogeneities, and surface roughness. As can be observed in Table 16–1, the semicircle looks to be flattened in the Nyquist plot (point 2). Due to the high current flow generated by surface inhomogeneities, CPEs are commonly used to replace the double-layer capacitors in electronic circuits. The impedance of a device is computed by considering the CPEs as follows:

$$Z_{CPE} = \frac{1}{Y}(j\omega)\alpha \quad (16-7)$$

where *Y* is a coefficient between 0 and 1 that is related to electrode capacitance.

A straight line with a slope of 45 degrees is identified in the Nyquist plot when the corrosion mechanism is regulated not only by a charge-transfer process but also by the diffusion of

Table 16.1 Examples of maximum scan rates for performing valid polarization plots.

Maximum scan rate (mV/s)	Capacitance (μF/cm²)	Polarization resistance (kΩ cm²)	Solution resistance (Ω cm²)
5.1	100	1	10
0.51	100	10	10
0.05	100	100	10
0.005	100	1000	10
6.3	100	1	100
0.51	100	10	100
0.05	100	100	100
0.005	20	1000	100
25	20	1	10
2.5	20	10	10
0.25	20	100	10
0.025	20	1000	10
50	20	1	100
2.6	20	10	100
0.25	20	100	100
0.025	20	1000	100

charged species through the corrosion products (Parveen et al., 2016) (Table 16–1, point 3). A converted Randle's-type circuit model with a semi-infinite Warburg diffusional impedance element (*W*) is utilized to mimic the EIS results. Low-frequency tails were thought to be closely related to the diffusion coefficient of electroactive substances across corrosion layers. Only a few corrosion EIS experiments, to the author's knowledge, have produced Nyquist plots using a single high-frequency capacitive loop and a low-frequency inductive loop (Table 16–1, point 4). The existence of the inductive ring could be owing to pitting corrosion and associated adsorption/desorption of intermediates on the electrode surface, or due to fast anodic dissolution.

Although the inductive loop may eventually disappear as the protective layer is produced on the surface, larger inductive loops can still be considered more serious cases of pitting corrosion. To understand the effect of this phenomenon on the circuit, an inductor and resistance are added. An electrochemical impedance study is also performed to investigate the growth of protective coating layers on the surface of various metals. The continuum in this case is composed of two low-frequency loops and two high-frequency loops, without an inductive loop (Table 16–1, point 5). The protection layer's resistance and the charge transfer resistance are R1, and the electric double layer's capacitance by C2 and the charge transfer resistance are R2. In the literature about the use of the EIS technique to study corrosion, it is often noted that the diagrams show complicated diagrams with multiple arcs at varying frequencies. These diagrams can also include an inductive loop at high frequencies, a second capacitive arc at medium frequencies, and an inductive loop at the lowest frequencies (Table 16–1, point 6).

The understanding of the two capacitive arcs at high and medium frequencies, respectively, is a contentious and ambiguous problem. According to some studies, the capacitive arc of high frequency is caused by charge transfer and the capacity effect of the electrochemical double layer, whereas the capacitive arc of medium frequency is caused by mass transport processes related to metal ion diffusion via the corrosion products layer. Other investigations, on the other hand, link high-frequency capacitive arc resistance to passivating coatings or corrosion-resistant layers, while medium-frequency capacitive arc resistance is linked to the charge transfer reaction (Feliu Jr, 2020).

16.2.2 Advantages and disadvantages of using EIS

Compared to other methods, such as those commonly used in assessing corrosion rates, an electrochemical analysis system can provide valuable information on the unique kinetics of certain electrochemical reactions. This method can also distinguish between different types of electrochemical reactions based on their own relaxation durations and provide exact information on their distinct kinetics in the form of corrosion rates. The different processes that occur in different types of electrochemical reactions, such as the formation of double electric layers, the presence of ohmic resistance, and charge transfer resistance in electrochemical reactions, occur at high frequencies. On the other hand, slower processes, such as the formation of adsorbed layers by intermediates in complex electrochemical reactions or diffusion (transport phenomena) in solutions, occur at low frequency (Amirudin & Thieny, 1995). This can be used to measure corrosion resistance (which is inversely proportional to corrosion rate) and examine the formation of corrosion or passivation layers on the metal surface, showing whether corrosion occurs by activation, concentration, adsorption, or diffusion. It also includes information on the double layer at the metal/electrolyte interface and the dielectric properties of surface oxides (Melo et al., 2016).

EIS uses a tiny excitation amplitude in comparison to polarization curves, resulting in little corrosion potential perturbation. Because EIS is nondestructive, it can be used to evaluate corrosion rates in situ for lengthy periods of immersion, minimizing the number of samples required for measurements. EIS permits far lower corrosion rate values to be determined compared with nonelectrochemical approaches (weight loss or hydrogen evolution). EIS provides a consistent and repeatable estimate of the metal's immediate corrosion rate. The EIS technique can be used to investigate high-impedance systems such as coatings and linings, as well as high-purity water, organic coating/metal systems, and corrosion in a low-conductive solution.

Despite the obvious benefits of utilizing EIS, the technique has certain drawbacks, including the following:

1. One of the most common issues with an analysis of an electric system is that there are several comparable models with low chi-squared. This can lead to errors that can be used to match the same data (Kirkland et al., 2012). The selection of the correct electric model is also important for assessing the system's corrosion resistance and capacitance.
2. Because the impedance is averaged over the entire macroscopic electrode surface, global EIS measurements have poor lateral resolution (Moreto et al., 2014). This makes studying localized electrochemical corrosion responses, in which the electrochemical characteristics of the microdefect are averaged over the entire electrode surface, more difficult (Cesiulis et al., 2016).
3. One of the significant drawbacks of AC impedance measurements is that the polarization resistance cannot be converted into a corrosion rate without first knowing the Tafel slope values and the Stern–Geary coefficient (Lorenz & Mansfeld, 1981).
4. Unlike polarization curves, the EIS method is not able to determine the variations in the system's corrosion potentials and anodic and cathodic reaction kinetics due to the different alloying components, microstructural characteristics (e.g., secondary phases), or solutions (Kirkland & Birbilis, 2014).

16.3 Polarization methods

For conducting laboratory corrosion tests, various polarization techniques such as cyclic voltammetry (CV), potentiodynamic polarization, and potentiostaircase are commonly used. These techniques can provide valuable information about the various factors that influence the mechanism and rate of corrosion and the susceptibility of certain materials to corrosion. Even though these methods are well-established, their results can sometimes be misleading.

The goal of these methods is to change the potential or the current produced by a working electrode as a function of time or potential. Anodic polarization involves shifting the potential in an anodic direction, which then causes the working electrode to become an anode and electrons to be removed. The working electrode becomes more negative during cathodic polarization, and electrons are added to the surface, causing electrodeposition in some situations. Both anodic and cathodic polarizations are done in a cyclic manner for cyclic polarization. The following items make up the polarization testing equipment:

1. A potentiostat keeps the working electrode's potential close to a predefined value.
2. A current tester measures the current generated by an applied voltage. Some potentiostats directly output the current logarithm, allowing for the charting of current vs. potential curves. It is also crucial that the current-measuring gadget can autorange or change the scale automatically.
3. The ability to immediately store data in a computer or plot it out. This is also critical.
4. Polarization cells. Several test cells for making polarization measurements are available commercially.
5. Whether testing small coupons or sheet materials, or testing inside autoclaves, polarization cells can be configured to meet individual testing requirements. The electrodes can be put directly into a process stream in a plant setting. A cell's characteristics include the following.
   a. The working electrode, that is, the sample for testing or analysis, may be accompanied by one or more auxiliary or counterelectrodes.
   b. A solution bridge and Luggin probe are frequently used to isolate the reference electrode from the solution. This setup eliminates solution exchange with the reference electrode while allowing it to be positioned extremely close to the working electrode's surface to reduce the influence of solution resistance.
   c. A thermometer to determine temperature.
   d. An inlet and outlet for gas to allow deaeration, aeration, or introduction of specific gases into the solution.
   e. An electrical connection can be made directly with the working electrode, which will not be affected by the solution.
   f. Unless this is the desired result, totally immerse the working electrode in the solution to eliminate any crevices at the solution interface.
   g. The test cell is made of a material that will not corrode or deteriorate during the test and will not contaminate the test fluid. The cell's volume must be large enough to allow corroding ions to be removed from the working electrode's surface without altering the solution potential.
   h. If necessary, a stirring device, such as a stirring bar or bubbling gas, guarantees that the solution chemistry is homogeneous.

In ASTM G3 (Standard Practice for Conventions Applicable to Electrochemical Measurements in Corrosion Testing, n.d.), there are several examples of polarization measurements which are briefly described below:

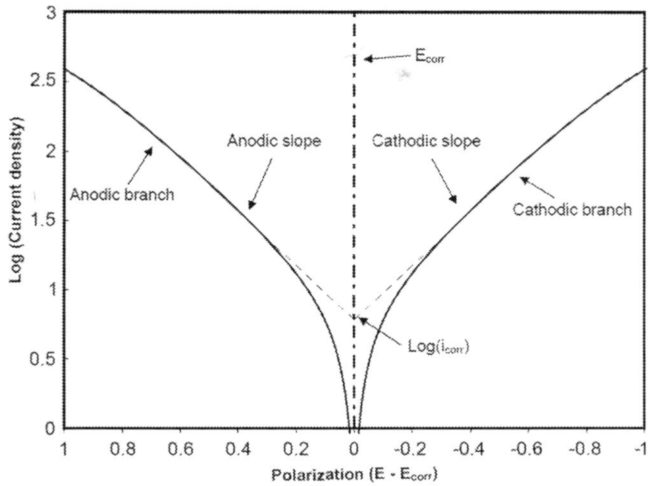

FIGURE 16–5 Hypothetical polarization diagram for an active system with anodic and cathodic branches.

Potentiodynamic polarization is a process in which a current through the electrolyte changes the electrode's potential at a predetermined rate. It is the most widely used polarization testing method for determining corrosion resistance, and it is utilized for a range of purposes (2000). Tafel analysis is a common dc technique that employs bigger, applied potentials. This results in larger, quantifiable currents, but because the current–potential relationship is nonlinear, a semi-log plot known as a "Tafel" plot is utilized. Fig. 16–5 depicts hypothetical curves for active and active–passive behavior, respectively. The oxidation process has an anodic branch, and the reduction reaction has a cathodic branch on the Tafel plot. Extrapolation to the free corrosion potential provides the corrosion current at the free corrosion potential for each branch. The Tafel coefficients are the slopes of the anodic and cathodic lines. Large potentials can sometimes cause undesired reactions that conceal the linear section of the anodic or cathodic branch, causing extrapolation to be erroneous and irreversibly changing the electrode surface.

The cyclic polarization test is a popular version of *potentiodynamic polarization*. This is a generally nondestructive test that is frequently used to assess corrosion rate, corrosion potential, metal pitting susceptibility, and electrolyte concentration limitation. The potential is swept once (or slightly less than once) to determine the amount of hysteresis as well as the differences between the beginning open-circuit corrosion potential and the return passivation potential. Hysteresis is commonly associated with pitting, and the size of the loop is frequently linked to the amount of pitting. The method is based on the idea that driving a metal out of its steady-state condition and watching how it reacts to the force as it is removed at a constant rate as the system is returned to its steady-state condition can predict how it would behave in an environment. The force is called applied potential, and it is raised at a constant, generally slow, rate using a potentiostat (Silverman, 1998). This is an experimental parameter known as the polarization scan rate. The applied potential is plotted

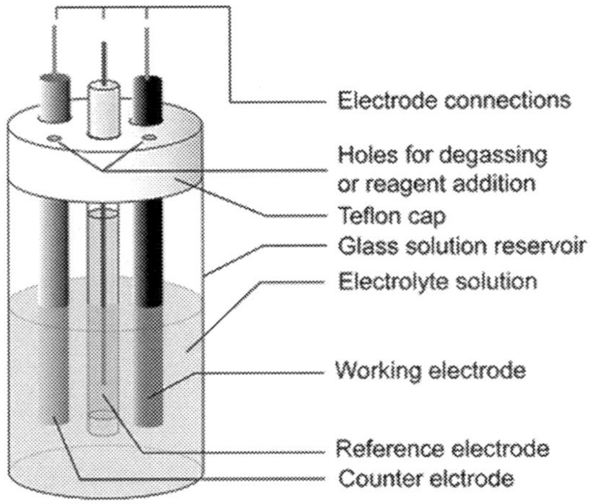

FIGURE 16–6 Schematic representation of an electrochemical cell for CV experiments.

against the logarithm of the resulting current density after the specimen's potential is altered constantly while the resulting current is observed. In all electrochemical investigations, notably in the cyclic polarization approach, the conductivity of the electrolyte (environment) is a key component to consider. The electrolyte resistance creates a potential decrease between the working and reference electrodes, which might result in mistakes. This effect has a significant impact on the interpretation and must be adjusted for.

Cyclic voltammetry is another type of potentiodynamic polarization. CV is a strong and widely used electrochemical technique for studying molecular species' reduction and oxidation processes. The study of electron transfer-initiated chemical reactions, such as catalysis, is crucial. Fig. 16–6 shows a schematic illustration of an electrochemical cell. The following sections will go through the functions of each component as well as how to put together an electrochemical cell for data collection during CV experiments.

The approach involves sweeping the potential in a positive direction until it reaches a predetermined current or potential value, then fast reversing the scan toward more negative values until the original potential value is reached. In some cases, this scan may be repeated to assess changes in the current-potential curve formed by scanning.

The potentiostaircase method is another version of potentiodynamic polarization. This is a method of polarizing an electrode in a succession of potential steps where the duration spent at each potential is constant and the current is often allowed to stabilize before moving to the next stage. The step increase can be minor or substantial, giving the procedure the appearance of a potentiodynamic curve.

Another test for determining susceptibility to intergranular or nonuniform corrosion is *electrochemical potentiodynamic reactivation (EPR)*. It is one of the most successful testing procedures for alloys and stainless steels like S30400 and S30403. EPR has the advantage of allowing for nondestructive examination on-site. This approach employs a potentiodynamic sweep from passive to active potentials (called reactivation). A cyclic polarization

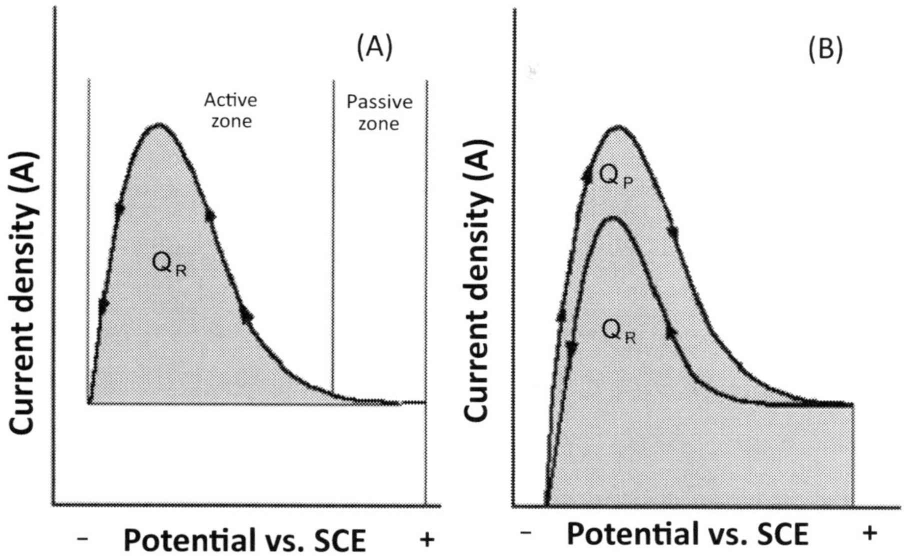

FIGURE 16–7 Schematic polarization curves in (A) single-loop and (B) double-loop potentiodynamic reactivation. QR/QP is the area of the reactivation stage and the area of the activation stage.

measurement is used in the EPR test. Based on characteristic potentials and other values found during the test, it allows for judgments on the alloy's susceptibility to intergranular corrosion, pitting and crevice corrosion, and intergranular stress corrosion cracking under specified conditions. EPR involves assessing a potentiodynamic scan of the material being examined, with the scan(s) being performed to meet conditions relating to the alloy's surface characteristics. Both the single loop scan (Fig. 16–7A) and the double loop scan (Fig. 16–7B) have become popular EPR variants, with the latter having a superior corrosion resolution potential. Although it must be understood in terms of corrosion, mechanical qualities, and alloy structure, the charge corresponding to an area under the EPR curve is the major indicator.

16.3.1 Linear polarization resistance

It is another popular polarization technique. LPR is a traditional dc (direct current) corrosion monitoring technique. Only within 10 to 20 mV of the free corrosion potential is the electrochemical corrosion test current–potential relationship linear, and the measurable current is very small in this region; the significance of this is that modern, higher-performing instrumentation has made this technique more practical.

The slope of the potential–current density (E/i) curve at the free corrosion potential (Fig. 16–8) yields the polarization resistance (R_p), which may be linked to the corrosion current using Eq. (16–8) (Macdonald, 2006)

$$R_p = \frac{B}{i_{corr}} \tag{16-8}$$

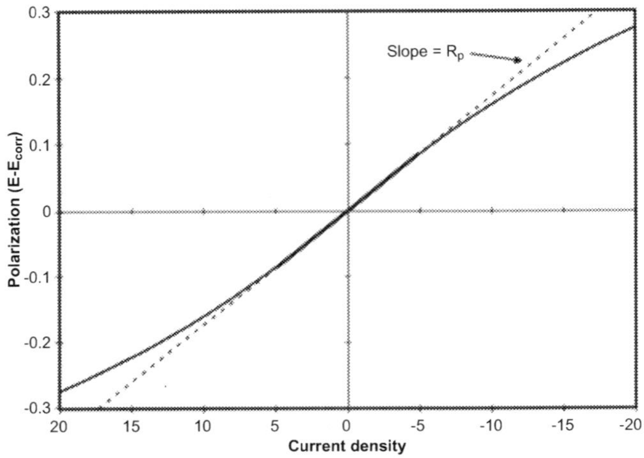

FIGURE 16–8 Hypothetical linear polarization plot

where i_{corr} is the corrosion current B is empirical polarization resistance constant that can be related to the anodic (ba) and cathodic (bc) Tafel slopes with Eq. (16–9).

$$B = \frac{\beta_a \beta_c}{2.3\beta_a + \beta_c} \tag{16-9}$$

where βa and βc are the Tafel coefficients

The properties of the Tafel slopes can also be evaluated using polarization plots. The results of these techniques can then be used to convert the currents into penetration rates using the principle of Faraday's law expressed earlier in Eq. (18–10).

$$\text{Corrosionrate} = \frac{3.3 i_{corr} E.W}{d} \tag{16-10}$$

where corrosion rate is in m/year, i_{corr} is in A/cm^2, $E.W.$ is equivalent weight of the element being oxidized, and d is the density of the element being oxidized.

Corrosion currents can also be converted using a basic conversion chart like the one shown in Table 16–2.

To be accurate, both linear polarization and Tafel extrapolation require additional precautions. The following are the primary difficulties encountered when performing polarization measurements:

16.3.1.1 Scan rate

At all potential values, the pace at which the potential is scanned may have a considerable impact on the quantity of current produced. The scan rate, if not adjusted correctly, can change the scan and lead to a misreading of the findings. Imagine the surface as a simple

Table 16–2 Conversion between current, mass loss, and penetration rates for all metals.

	mA/cm^2	mm/year	mpy	g/m^{2}day
mA/cm^2	1	3.28 M/nd	129 M/nd	8.95 M/n
mm/year	0.306 nd/M	1	39.4	2.74 d
mpy	0.00777 nd/M	0.0254	1	0.0694 d
g/m^{2}day	0.112 n/M	0.365/d	14.4/d	1

resistor in parallel with a capacitor to better understand the situation. The capacitor would represent the double-layer capacitance in this model, while the resistor would represent the polarization resistance, which is inversely proportional to the corrosion rate. The idea is to have a slow enough polarization scan rate that the capacitance remains fully charged and the current–voltage relationship at each potential simply represents the interfacial corrosion process. If this is not done, some of the current generated will be attributable to charging of the surface capacitance in addition to the corrosion process, resulting in a measured current greater than the current generated by the corrosion reactions. When this happens, the polarization measurement fails to accurately reflect the corrosion process, resulting in an incorrect prediction (Bard & Faulkner, 2001).

If the scan rate is excessively quick, the entire measurement may be thrown off, and the potentiodynamic graphs connected with the polarized electrode may be misinterpreted. If the potential scan rate is too slow, the interfacial structure of the polarized electrode will not stabilize and will alter inconstantly throughout the test. The potential scan rate is first tuned to ensure that it does not affect the interfacial structure and does not cause changes in the stable-state condition. This is very important to perform effective corrosion inhibitor research. In addition, determining the optimal scan rate for different types of corrosion inhibitors is also important because it is dependent on the type of metal and the electrolyte.

Table 16–3 shows estimated maximum scan rates for several polarization resistance, solution resistance, and capacitance values typically encountered in practice.

16.3.1.2 Effect of solution resistance

In most measurements, the distance between the Luggin probe (from the salt bridge to the reference electrode) and the working electrode is intentionally kept short to reduce the effect of solution resistance. This can have a substantial impact on solutions with exceptionally high resistance. Many materials used in corrosion measurements, such as concrete, dirt, organic solutions, and others, have high resistivity, but they can also be quite corrosive to specific metals. In these high-resistivity situations, the ability to do polarization measurements is critical. To measure the solution resistance and determine the real resistance between the reference and working electrodes, a method of interrupting the current and measuring the decay of the potential as a function of time can be used.

Table 16–3 Examples of maximum scan rates for performing valid polarization plots.

Maximum scan rate (mV/s)	Capacitance (F/cm^2)	Polarization resistance (Kcm^2)	Solution resistance (cm^2)
5.1	100	1	10
0.51	100	10	10
0.05	100	100	10
0.005	100	1000	10
6.3	100	1	100
0.51	100	10	100
0.05	100	100	100
0.005	20	1000	100
25	20	1	10
2.5	20	10	10
0.25	20	100	10
0.025	20	1000	10
50	20	1	100
2.6	20	10	100
0.25	20	100	100
0.025	20	1000	100

16.3.1.3 Changing surface conditions

Because corrosion reactions occur at the material's surface, when the surface is altered due to processing conditions, active corrosion, or other factors, the potential is usually altered as well. This has a significant impact on polarization curves.

16.3.1.4 Determination of pitting potential

The occurrence of a hysteresis (or loop) between the forward and reverse scans is commonly assumed to indicate the presence of localized corrosion when evaluating polarization curves (pitting or crevice corrosion). When the corrosion potential is greater or nobler than the pitting potential, this observation is especially true.

16.3.2 Electrical resistance probes

For any metallic equipment or building, the electrical resistance (ER) methodology is an "on-line" means of monitoring the rate of corrosion and the level of total metal loss. The electrochemical and mechanical components of corrosion, such as erosion and cavitation, are measured using the ER approach. It is the only online, instrumented method that may be used in nearly any corrosive environment.

The ER approach is particularly well suited to corrosive environments with weak or noncontinuous electrolytes, such as vapors, gases, soils, "wet" hydrocarbons, and nonaqueous liquids. (1) Oil/gas production and transmission systems, (2) refinery/petrochemical process streams, (3) external surfaces of buried pipes, (4) feedwater systems, (5) flue gas stacks, (6) architectural structures are examples of scenarios where the ER technique is effective.

ER probes measure the ER, which increases as the cross-section of the consumable coupon decreases, to assess metal consumption. Compensation is essential because temperature influences electrical resistance. The ER probe is attached to a temperature compensation reference device that has no physical contact with the corrosive media but is at the same temperature. Real-time corrosion rates can be measured with this technology, allowing for the detection of any unexpected changes in corrosion rate and prompt remedial action.

Although the ER approach can evaluate uniform corrosion, it cannot detect localized corrosion. This is because the presence of certain salts (FeS_2, PbS) and oxides (Fe_3O_4, MnO_2) on the sensor surface can lead to different findings. However, these findings can also be applied to the fouling rate.

16.3.3 Electrochemical noise

The electrochemical noise (EN) analysis method is a noteworthy way of determining the phenomena of metal corrosion at E_{OCP} without the use of any external potential disruption. The corrosion rate can be determined using time-dependent statistical treatments of random current and potential noise signals emitted by the corrosion system, and corrosion mechanisms can be anticipated using spectral fingerprints acquired from frequency-dependent noise signal analysis (Cottis, 2001; Hladky & Dawson, 1981; Kelly et al., 1996). Under certain situations, the statistically generated noise resistance, RN, has been shown to be like the traditional polarization resistance Rp (Bertocci et al., 1997; Lee & Mansfeld, 1998), but corrosion can be predicted using the so-called localization index (Eden et al., 1987). The power spectra density is important in the frequency domain for anticipating corrosion mechanisms and, in some situations, for determining noise impedance. EN consists of recording only the spontaneous fluctuations in potential (electrochemical potential noise [EPN]) and current (electrochemical current noise [ECN]) generated by corrosion at the metal–electrolyte interface of a three-electrode system.

EN was carried out according to the ASTM-G199-09 standard, which allows for the determination of noise resistance (Rn) and corrosion rate in various corrosive conditions (G199-09, 2009). Two identical specimens were utilized as working electrodes in each experiment, with a saturated calomel electrode serving as a reference electrode (Fig. 16–9) (Jáquez-Muñoz et al., 2021). A galvanic linkage between two identical working electrodes was used to assess ECN. By connecting one of the working electrodes to a reference electrode, EPN was measured simultaneously. Under open-circuit circumstances, the current and potential EN of each electrode's electrolyte mixture were measured. With a scanning rate of 1 datum/s, 1024 data points were acquired for each EN measurement set. Visual analysis of the current and potential time series was used to understand signal transients and determine the frequency and magnitude of variations as a function of time (Gaona-Tiburcio et al., 2014).

More new techniques, such as recurrence quantification analysis (G199-09, 2009) have been developed and deployed in recent years for a more accurate characterization of localized corrosion events.

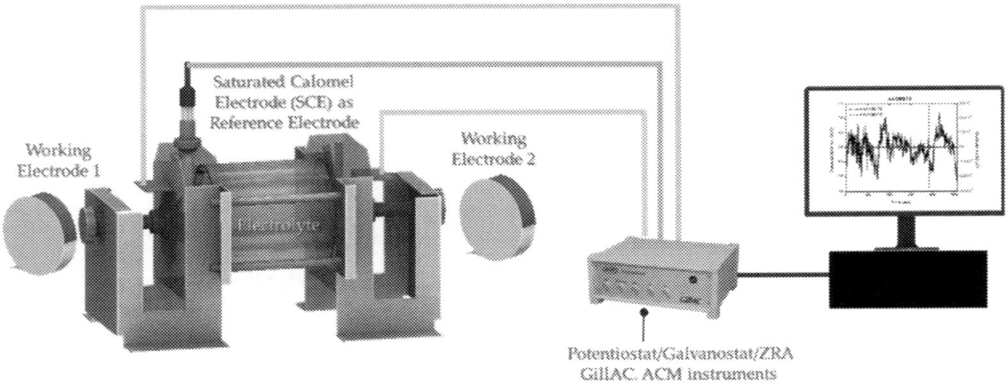

FIGURE 16–9 Experimental setup for electrochemical noise (EN) measurements. *From Jáquez-Muñoz, J. M., Gaona-Tiburcio, C., Cabral-Miramontes, J., Nieves-Mendoza, D., Maldonado-Bandala, E., Olguín-Coca, J., Estupinán-López, F., López-León, L. D., Chacón-Nava, J. & Almeraya-Calderón, F. (2021). Frequency analysis of transients in electrochemical noise of superalloys Waspaloy and Ultimet. Metals, 11(5), 702. https://www.mdpi.com/2075-4701/11/5/702.*

16.3.4 Electrochemical frequency modulation

Electrochemical frequency modulation (EFM), like EIS, is a nondestructive technology (EIS). In comparison to standard dc and ac approaches, the EFM methodology was applied. The EFM approach produced results that were consistent with those obtained by other electrochemical techniques. The EFM approach provides an instantaneous corrosion rate in a relatively short amount of time, making it perfect for online corrosion monitoring. Harmonic analysis is another name for this. It uses the same equipment as EIS but applies two sine waves of different frequencies at the same time. This approach works with alternating current (AC). It gives an uncontaminated value of the corrosion current and Tafel coefficients without necessitating the use of huge dc potentials that can damage the electrode's surface; it does not require considerable modeling and does not give mechanistic insight.

Due to the dual sine wave stimulus, the nonlinear character of corrosion processes introduces harmonics and intermodulation components into the current response of the corroding system. Although the math for analyzing harmonics is complicated, it is not necessary to adjust it to each system, as it is for EIS. The harmonic analysis results are immune to contributions from circuit parts other than Rp but obtaining a good measurement can be difficult because the excitation amplitude cannot be very big, and the magnitudes of the current harmonics are much less than the fundamental frequency. The selection of the two frequencies is also critical, as it must meet certain criteria that are dependent on the type of the corroding system. We can determine the proper frequencies and circumvent this limitation by combining EIS with harmonic analysis.

EFM was performed at two frequencies of 2 and 5 Hz. The waveform repeats after 1 second because the base frequency was 1 Hz. The upper frequency must be at least twice as high as the lower frequency. The higher frequency must also be slow enough that the double-layer charging does not affect the current response. 10 Hz is a common upper limit.

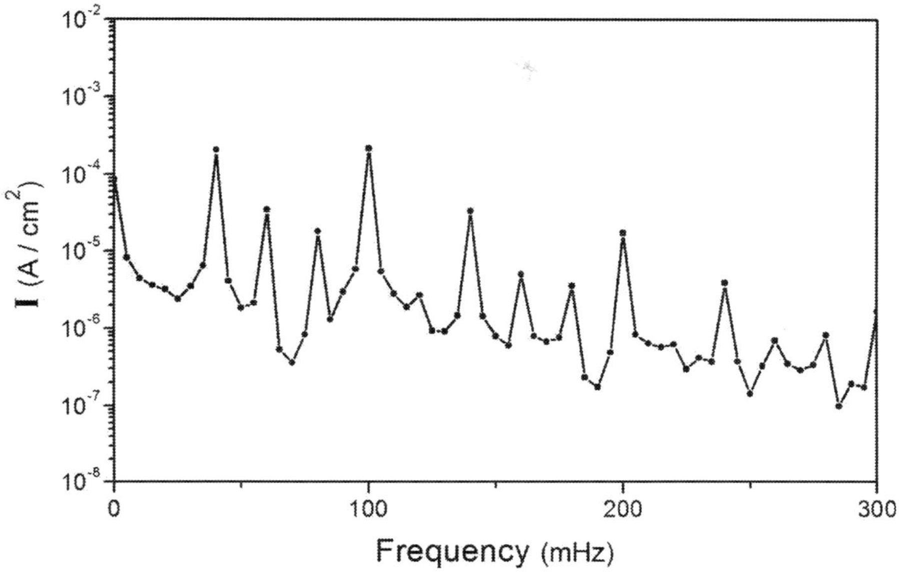

FIGURE 16–10 Intermodulation spectrum for iron in 1 M HCl. *From Abdel-Rehim, S. S., Khaled, K. F. & Abd-Elshafi, N. S. (2006). Electrochimica Acta, 51(16), 3269–3277. https://doi.org/10.1016/j.electacta.2005.09.018.*

For both perturbation frequencies of 2 and 5 Hz, we utilize a perturbation signal with an amplitude of 10 mV. The frequencies of 2 and 5 Hz were chosen based on three factors (Cesiulis et al., 2016). To begin with, the harmonics and intermodulation frequencies should not be fluent. Second, to avoid the capacitive behavior of the electrochemical double layer, the frequency should be as low as feasible. Third, to shorten the measuring time, the frequency should be as high as feasible. While these factors prevent the selection of "ideal" frequencies, the frequencies chosen were thought to be a reasonable compromise.

The spectrum is known as the "intermodulation spectrum," and an example can be seen in Fig. 16–10 (Bosch et al., 2001). The comparable current response in the intermodulation spectrum is seen in Fig. 16–10. The harmonic and intermodulation peaks can be seen clearly and are significantly larger than the background noise. The response to the 2 and 5 Hz stimulation frequencies is represented by two big peaks with amplitudes of roughly 200 A. The harmonics, summation, and differences between the two excitation frequencies are represented by the peaks between 1 and 20 A. The corrosion rate and Tafel parameters can be determined by analyzing these peaks at intermodulation frequencies. It is worth noting that the current reaction is quite tiny between the peaks. At 4.5 Hz, for example, there is hardly any response (100 nA), and the peak frequencies and amplitudes are not coincidental. They follow directly from the EFM theory.

Short questions

1. Which factors would influence the measurement of the electrochemical impedance spectroscopy (EIS)?

2. Meaning of Warburg element parameters in Zview software?
3. How to choose frequency range in electrochemical Importance spectroscopy?
4. Discuss the effects of Scan Rate on the Potentiodynamic Polarization.
5. How does polarization affect the rate of corrosion?
6. How does Tafel plot calculate corrosion rate?
7. Does polarization increase or decrease the tendency for corrosion?
8. What is cyclic potentiodynamic polarization?
9. How to prevent OCP (Open circuit potential) fluctuation due to external noise?

References

Amirudin, A., & Thieny, D. (1995). Application of electrochemical impedance spectroscopy to study the degradation of polymer-coated metals. *Progress in Organic Coatings, 26*(1), 1–28. Available from https://doi.org/10.1016/0300-9440(95)00581-1.

Bard, A. J. & Faulkner, L. R. (2001). Electrochemical methods. *Fundamentals and Applications.*

Bertocci, U., Gabrielli, C., Huet, F., Keddam, M., & Rousseau, P. (1997). Noise resistance applied to corrosion measurements: II. Experimental tests. *Journal of the Electrochemical Society, 144*(1), 37–43. Available from https://doi.org/10.1149/1.1837362, http://jes.ecsdl.org/content/by/year.

Bosch, W., Hubrecht,J., Bogaerts,W.F., & Syrett,B.C. (2001). Corrosion. 57.

Brett, C. M. A., Brett, O., & A., M. (1993). *Electrochemistry, principles, methods, and applications.* Oxford University Press.

Cesiulis, H., Tsyntsaru, N., Ramanavicius, A., & Ragoisha, G. (2016). The study of thin films by electrochemical impedance spectroscopy. *Springer Science and Business Media LLC,* 3–42. Available from https://doi.org/10.1007/978-3-319-30198-3_1.

Cole, K. S. & Cole R. H. (1941). J. Chem. Phys. 9.

Cottis, R. A. (2001). Interpretation of electrochemical noise data. *Corrosion., 57*(3), 265–285. Available from https://doi.org/10.5006/1.3290350.

Eden, D. E., John, D. G., & Dawson, J. L. (1987). *Corrosion monitoring. International Patent WO 87107022,* World Intellectual Property Organization.

Feliu. Jr S. (2020). 10

Gaona-Tiburcio, C., Aguilar, L. M. R., Zambrano, P., Estupiñán López, F., Cabral, J. A., Nieves-Mendoza, D., . . . Almeraya-Calderón, F. (2014). Electrochemical noise analysis of nickel based superalloys in acid solutions. *International Journal of Electrochemical Science, 9*(2), 523–533. Available from http://www.electrochemsci.org/papers/vol9/90200523.pdf.Mexico.

Geethanjali, R., & Subhashini, S. (2015). Investigation of corrosion inhibition efficiency of somesynthesized water soluble terpolymers on N-80 steel in HCl, NACl and simulated oil well water. *Portugaliae Electrochimica Acta, 33*(2), 85–104. Available from https://doi.org/10.4152/pea.201502085, http://www.peacta.org/pages/backIssues.

Hladky, K., & Dawson, J. L. (1981). The measurement of localized corrosion using electrochemical noise. *Corrosion Science, 21*(4), 317–322. Available from https://doi.org/10.1016/0010-938X(81)90006-8.

Huet, F. (1998). A review of impedance measurements for determination of the state-of-charge or state-of-health of secondary batteries. *Journal of Power Sources, 70*(1), 59–69. Available from https://doi.org/10.1016/S0378-7753(97)02665-7.

Jáquez-Muñoz, J. M., Gaona-Tiburcio, C., Cabral-Miramontes, J., Nieves-Mendoza, D., Maldonado-Bandala, E., Olguín-Coca, J., . . . Almeraya-Calderón, F. (2021). Frequency analysis of transients in electrochemical noise of superalloys waspaloy and ultimet. *Metals., 11*(5), 702. Available from https://www.mdpi.com/2075-4701/11/5/702.

Kelly, R. G., Inman, M. E., & Hudson, J. L. (1996). Analysis of electrochemical noise for type 410 stainless steel in chloride solutions. *ASTM Special Technical Publication, 1277*, 101–113. Available from https://doi.org/10.1520/stp37954s, https://www.astm.org/BOOKSTORE/STP_SERIES/FracMech.htm.

Kelly, R. G., Sculy, J. R., Shoesmith., & Buchheit, R. G. (2002). *Electrochemical techniques in corrosion science and engineering*. CRC Press.

Kirkland, N. T., & Birbilis, N. (2014). Magnesium biomaterials: Design. *Testing, and Best Practice*, 40–43.

Kirkland, N. T., Birbilis, N., & Staiger, M. P. (2012). Assessing the corrosion of biodegradable magnesium implants: A critical review of current methodologies and their limitations. *Acta Biomaterialia, 8*(3), 925–936. Available from https://doi.org/10.1016/j.actbio.2011.11.014, http://www.journals.elsevier.com/acta-biomaterialia.

Koutsouras, D. A., Gkoupidenis, P., Stolz, C., Subramanian, V., Malliaras, G. G., & Martin C. (2017). 4.

Lee, C. C., & Mansfeld, F. (1998). Analysis of electrochemical noise data for a passive system in the frequency domain. *Corrosion Science, 40*(6), 959–962. Available from https://doi.org/10.1016/S0010-938X(98)00033-X.

Li, X., Qin, Z., Fu, H., Li, T., Peng, R., Li, Z., ... Liu, X. (2021). Enhancing the performance of paper-based electrochemical impedance spectroscopy nanobiosensors: An experimental approach. *Biosensors and Bioelectronics, 177*. Available from https://doi.org/10.1016/j.bios.2020.112672, http://www.elsevier.com/locate/bios.

Lorenz, W. J., & Mansfeld, F. (1981). Determination of corrosion rates by electrochemical DC and AC methods. *Corrosion Science, 21*(9–10), 647–672. Available from https://doi.org/10.1016/0010-938X(81)90015-9.

Macdonald, J. R. (2011). J. Phys. Chem. A. 115.

Macdonald (2006). Electrochim. Acta. 51.

Macdonald, J. R. (2001). *Impedance spectroscopy: Emphasizing solid materials and systems*. Wiley.

Melo, R. S., Brasil, S., de. Carvalho, L. J., Filho, A. L., & Pereira, C. (2016). Assessment of the antifouling effect of exopolysaccharides incorporated into copper oxide-based organic paint. *International Journal of Electrochemical Science, 11*, 7750. Available from https://doi.org/10.20064/2016.09.37, https://materialschemistry.chemistryconferences.org/abstract/2017/assessment-of-the-antifouling-effect-of-exopolysaccharides-incorporated-into-copper-oxide-based-organic-paint.

Moreto, J. A., Marino, C. E. B., Bose Filho, W. W., Rocha, L. A., & Fernandes, J. C. S. (2014). SVET, SKP and EIS study of the corrosion behaviour of high strength Al and Al-Li alloys used in aircraft fabrication. *Corrosion Science, 84*, 30–41. Available from https://doi.org/10.1016/j.corsci.2014.03.001.

Njoku, D. I., Ukaga, I., Ikenna, O. B., Oguzie, E. E., Oguzie, K. L., & Ibisi, N. (2016). Natural products for materials protection: Corrosion protection of aluminium in hydrochloric acid by Kola nitida extract. *Journal of Molecular Liquids, 219*, 417–424. Available from https://doi.org/10.1016/j.molliq.2016.03.049.

Parveen, M., Mobin, M., & Zehra, S. (2016). Evaluation of l-tyrosine mixed with sodium dodecyl sulphate or cetyl pyridinium chloride as a corrosion inhibitor for mild steel in 1 M HCl: Experimental and theoretical studies. *RSC Advances, 6*(66), 61235–61248. Available from https://doi.org/10.1039/c6ra10010d, http://pubs.rsc.org/en/journals/journalissues.

Pourbaix, M. (1984). Electrochemical corrosion of metallic biomaterials. *Biomaterials, 5*(3), 122–134. Available from https://doi.org/10.1016/0142-9612(84)90046-2.

Raja, P. B., & Sethuraman, M. G. (2008). Natural products as corrosion inhibitor for metals in corrosive media - A review. *Materials Letters, 62*(1), 113–116. Available from https://doi.org/10.1016/j.matlet.2007.04.079.

Rechnitz, G. A. (1963). *Controlled-potential analysis*. Pergamon Press Inc.

Sastri, V. S. (1998). *Corrosion inhibitors, principles and application*.

Silverman, D. C. (2000). Uhlig's corrosion handbook, Ch. 68 In R. W. Revie (Ed.), John Wiley and Sons Inc.

Silverman, D. C. (1998). Tutorial on cyclic potentiodynamic polarization technique. *Corrosion*. 98.

Wagner, N., Schnurnberger, W., Müller, B., & Lang, M. (1998). Electrochemical impedance spectra of solid-oxide fuel cells and polymer membrane fuel cells. *Electrochimica Acta, 43*(24), 3785–3793. Available from https://doi.org/10.1016/S0013-4686(98)00138-8, http://www.journals.elsevier.com/electrochimica-acta/.

Warburg, E. (1899). Ann. Phys. Chem. 67

Zand, R. Z., Verbeken, K., & Adriaens, A. (2013). Influence of the cerium concentration on the corrosion performance of ce-doped silica hybrid coatings on hot dip galvanized steel substrates. *International Journal of Electrochemical Science, 8*(1), 548–563. Available from http://www.electrochemsci.org/papers/vol8/80100548.pdf.Belgium.

17 Surface characterization techniques

17.1 Introduction

Understanding the surface reactions that occur during the corrosion of a metallic substrate in a corrosive medium necessitates a thorough understanding of the surface composition and structure, as well as the changes that occur during corrosion. Passivity and passivity breakdown, localized corrosion (pitting), adsorption, inhibitors, atmospheric corrosion, low-temperature oxidation, early stages of high-temperature oxidation, corrosion of electronic materials and devices, surface treatments, coatings, and, more recently, biofilms have all been studied using surface analysis in corrosion science. Several methods can be used for surface chemical analysis, for example, electron spectroscopy, such as, X-ray photoelectron spectroscopy (XPS), Auger electron spectroscopy (AES), Secondary Ion Mass Spectroscopy (SIMS); Optical Spectroscopy, such as, Raman, infrared reflection absorption spectroscopy, ellipsometry; synchrotron techniques such as X-ray absorption near edge structure, extended X-ray absorption fine structure; and nuclear techniques, such as, nuclear reaction analysis, Rutherford backscattering spectrometry (RBS), radiotracer measurements, as well as for surface structure analysis, such as, glancing angle X-ray diffraction (GA-XRD), electron diffraction [low-energy electron diffraction (LEED) and reflection high energy electron diffraction (RHEED)], scanning tunneling microscopy (STM), and atomic force microscopy (AFM). All these techniques are used in combination with electrochemical measurements.

Table 17−1 shows many types of information that these analytical techniques can provide. It also includes information on usual vertical resolution (or depth probed), typical lateral resolution, typical solid specimen kinds, and popularity. In the literature, they will find basic principles and apparatus information for a wide range of analytical procedures. The analytical instrumentation area, on the other hand, moves quickly and presents spatial resolutions, sensitivities, imaging and mapping capabilities, accuracies, and instrument costs and sizes that are likely to be outdated within a year. As a result, these sources should be approached with caution. This table is only intended to be used as a quick reference tool.

17.1.1 Auger electron spectroscopy

The chemical composition of the materials that make up solid surfaces or interfaces is analyzed using the AES technique. The advantages of AES over other methods are its ability to provide a wide range of detection limits, as well as its spatial resolution. For most elements, the detection limits are as low as 0.01%−0.1% and surface sensitivity ($\sim$20 Å).

For the primary electron beam, the AES hardware can incorporate a tungsten filament or a field emission electron cannon. Secondary electron imaging is available on the devices to

Handbook of Corrosion Engineering. DOI: https://doi.org/10.1016/B978-0-323-95185-2.00017-4

Table 17–1 Some information on various analytical techniques.

Technique	Main information	Vertical resolution	Lateral resolution
Optical profiler and laser interferometry	3D and 2D imaging Morphology Profilometry Topographic mapping Film thickness Wear volume Scar and crater depth Defects	-0.1 nm	A few sub mm to a few tens of mm
Light microscopy	Imaging Morphology Damages Defects	Variable	Variable
Stylusprofilometry	Profilometry Topographic tracing Film thickness Morphology Scar and crater depthWear volume	0.5 nm	100 nm
Scanning tunneling microscopy (STM)	Topographic imaging Compositional mapping Morphology Profilometry Film thickness Spectroscopy Structure defects	$<0.03-0.05$ nm	Atomic
Atomic force microscopy (AFM) or scanning force microscopy (SFM)	Topographic imaging Friction force mapping Morphology Profilometry Film thickness Wear volume Scar and crater depth Structure Defects	$<0.03-0.05$ nm	Atomic to 1 nm

Variable-angle spectroscopic ellipsometry (VASE)	Film thickness Microstructure Optical properties	Tens of nm to mm	millimeter
X-ray fluorescence (XRF)	Film thickness to 10^4nm Elements composition Qualitative mapping	10 mm	10–150 mm
Scanning electron microscopy (SEM)	Imaging Morphology Elemental composition Damages Defects Crystallography Grain structure Magnetic domains	Few nm	50 nm in secondary electron mode
Electron probex-ray microanalysis (EPMA)	Elemental composition SEM imaging Compositional mapping	1 mm	0.5–1 mm
Energy-dispersive X-ray spectroscopy (EDS) or wavelength dispersive X-ray spectroscopy (WDS)	Elemental composition ($Z > 5$; boron touranium)Spectroscopy Imaging and mapping	0.02–1 mm	0.5–1 mm for bulk specimens as small as 1 nm for thin specimens
Auger electron spectroscopy (AES)	Elemental composition (except H and He) Chemical state Depth profiling Imaging and mapping	0.5–10 nm	A few tens of nm or less
X-ray photoelectron spectroscopy (XPS)	Elemental composition Chemical state Depth profiling Imaging and mapping	A few to several nm	5 mm
Raman spectroscopy	Identification of unknown compounds Chemical state Bonding state Structural order Phase transitions Imaging and mapping	Few mm	1 mm

(Continued)

Table 17–1 (Continued)

Technique	Main information	Vertical resolution	Lateral resolution
Secondary ion mass spectroscopy (SIMS)	Chemical state Imaging Elemental composition	0.3–2 nm	10 nm–2 mm
Rutherford backscattering spectrometry (RBS)	Elemental composition Structure Defects	2–30 nm	1–4 mm
Solid-state nuclear magnetic resonance (NMR)	Chemical state Phase identification Disordered state	–	–
X-ray diffraction (XRD)	Crystalline phases Strain Crystallite Orientation size Atomic arrangements Defect aging Concentration depth profiling Film thickness	A few mm	None; -10 mm with microfocus
Low-energy electro diffraction (LEED)	Surface crystallography Surface cleanliness Surface disorder Imaging	0.4 nm	0.1 mm (-10 mm available)
Transmission electron microscopy (TEM)	Atomic structure Microstructure Crystallographic structure Defects Imaging and mapping Morphology Chemical bonding	None	£0.2 nm
Reflection high energy electron diffraction (RHEED)	Surface crystal structure 2D and 3D defects	2–10 nm	200–4 mm

help locate specific analysis areas and obtain micrographs of the sample surface. To prevent Auger electrons from being intercepted by gas molecules between the sample and the detector, the sample chamber is kept in an ultrahigh vacuum. Special stages are available on some instruments for fracturing samples to evaluate surfaces that have been freshly exposed within the vacuum chamber. The AES data is acquired, analyzed, and displayed using a computer.

Although studies on the various components of carbon steel corrosion have been carried out, they are still in their early stages. These studies could also lead to the development of new methods for controlling the surface texture and grain boundary of the steel.

17.1.2 Low-energy electron diffraction technique

The crystallographic orientation of organic molecules on metallic surfaces can be determined using the LEED tool. This method can be used for analyzing the corrosion product properties of the materials. The data collected from the LEED pattern can then be used to calculate the porosity of the surface layers.

17.1.3 X-ray photoelectron spectroscopy analysis

XPS, also known as electron spectroscopy for chemical analysis (ESCA), is a method of obtaining chemical information from solid-state surfaces. XPS can determine the composition as well as the chemical state of surface components. Surface regions ranging from a few microns to a few millimeters across can be easily evaluated for insulators and conductors. K. Siegbahn and his team in Sweden invented XPS in the 1950s (Barlow & Raval, 2003). K. Siegbahn received the Nobel Prize for ESCA in 1960. The use of electron spectroscopy for corrosion research was pioneered during the 1970s. Fig. 17–1 shows an XPS scan exhibiting the distribution of elements of a 316 stainless steel surface (Fig. 17–2).

The sample is exposed to a high-energy X-ray source in a vacuum environment. The incoming particles cause core-level electrons to escape from the atoms. The energy of the core-level electrons is determined by their binding energy. The data collected from the tool is then used to analyze the photoemissions of the captured core electron. When the X-ray hits the core hole, an outer electron fills it. The emission of an Auger electron or a characteristic X-ray balances the energy of this transition. In addition to released photoelectrons, Auger electron analysis can be used in XPS.

An electron energy analyzer measures the energy of the photoelectrons and the Auger electrons generated by a sample. Its energy is calculated by considering the velocity of the electrons as they enter the detector. A spectrum is then obtained to show the surface composition of the sample. The energy of a peak is related to the presence of an element in the sample. The element's chemical state is shown by the peak form and its precise position. A spectral peak represents the area under which the element is present.

The energy of the electrons is greatly reduced by the presence of only electrons near the surface. Unlike other surface-sensitive techniques, the XPS method does not require the presence of photons of interest to be detected. Electrons that are more than 20–50m below

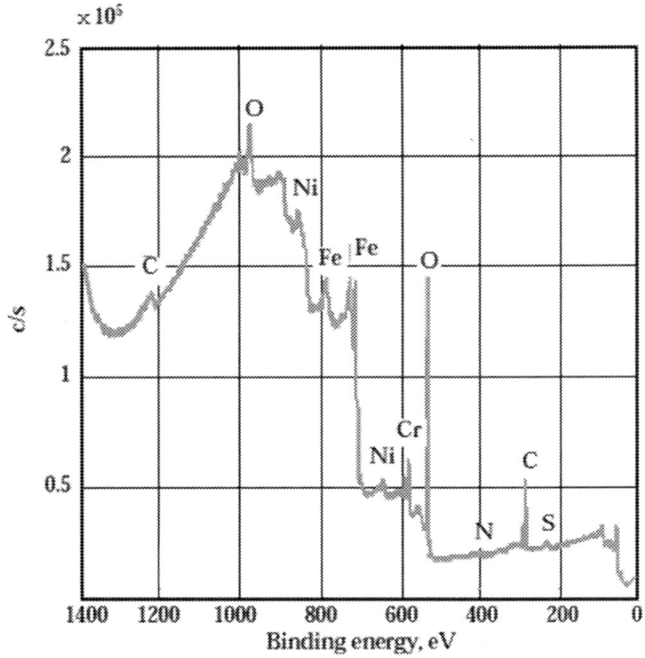

FIGURE 17–1 X-ray photoelectron spectroscopy scan showing the distribution of elements of a 316 stainless steel surface that has been exposed to the atmosphere.

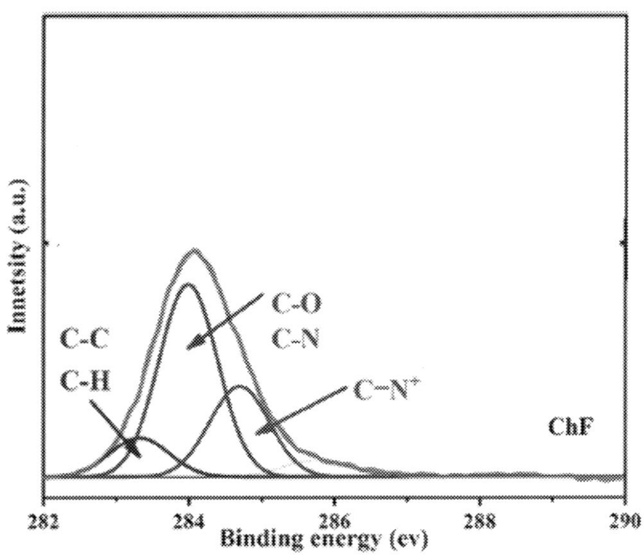

FIGURE 17–2 XPS multiplex survey for carbon.

the surface cannot escape due to the inelastic collisions in the sample's atomic structure. However, collisions within the atomic structure of the sample prevent the electrons from escaping.

Stainless steels (FeCr, FeCrNi) and Ni-based alloys (NiCr, NiCrFe) and their alloying elements (Mo, N), Al, and Al alloys (with Cu, Mg), Cu, Ti, and Ti alloys are the principal materials examined using XPS in the corrosion sector. The key results of XPS in one of the most studied topics, stainless steel passivity, will be presented in the following sections.

Passivity was and continues to be a hot topic in academia. Because the passive film exhibits a steady state that might be defined by a nonkinetic approach, XPS has provided invaluable information to this field. Furthermore, as shown below, the thickness of passive films and some atomic layers enables comprehensive characterization using surface science techniques.

Cr^{3+} enrichment was discovered in the passive coatings on stainless steels using XPS (Frankenthal & Malm, 1976; Mitchell & Graham, 1987; Olefjord, 1975). When the Cr percentage of Fe-Cr alloys was 13% Cr minimum, the surface Cr concentration in the passive film increased dramatically (Seo & Sato, 1983). This result corresponded to a well-known increase in corrosion resistance at this Cr concentration. In situ electrochemical and ellipsometric thickness determinations were compared to XPS thickness evaluations of passive films and found to be in good agreement (Schnyder & Kötz, 1992; Seo & Sato, 1983; Strehblow, 1988). Furthermore, because XPS is an ex-situ technique, the stability of the surface species after emersion of the electrode and in a vacuum is crucial (Castle, 1977). The presence of hydroxyls (OH) and H_2O molecules in ultra-high vacuum (UHV) has been noted. Surface contamination and/or oxidation in the air have been minimized in specimen transfer systems from the electrochemical cell to the spectrometer without exposure to the laboratory atmosphere (Olefjord, 1975). In the case of model surfaces, a preparation chamber coupled to the spectrometer's analysis chamber can allow for well-controlled in situ gas treatments. A lot of work was done in the 1980s to improve the capabilities of surface science approaches in corrosion science.

High-energy resolution and high counts rate, photoelectron imaging with submicrometer resolution, databases, improved data processing methods, and possibly the creation of expert systems for data collecting and evaluation are all possibilities in XPS analysis.

17.1.4 Raman spectroscopy

A laser pulse passes through a surface, which then allows a Raman spectroscopy instrument to identify the scattered photons. The molecules in the sample release energy in response to the light, which then causes a spectrum of energy losses. A spectrometer then measures these losses to identify the characteristics of the molecules that interact with the photon. Unlike surface techniques, Raman spectroscopy does not require a sampling depth of more than a meter. The advantages of Raman spectroscopy are numerous. In some studies, it can be performed in a vacuum, which would have been very challenging to do with the traditional surface techniques. Also, since the crystal environment can have unique vibrational energies, it can be used to distinguish the molecules in different crystalline structures.

Raman spectroscopy's sensitivity varies greatly, making its utility highly varied. Some substances are strong Raman scatterers that may be detected in very thin surface layers or in extremely minute particles, whereas others are invisible. Raman spectroscopy provides both qualitative and quantitative capabilities for substances that produce good signals.

The nature of the spectrum is another property of Raman spectroscopy, which makes it a useful complement to other surface techniques. Because a compound's vibrational levels are highly reliant on its molecular environment, various compounds of the same element have completely different signal patterns. Raman spectra have a "fingerprint" that can be used to distinguish distinct chemicals with high specificity.

17.1.5 X-ray diffraction

XRD is a valuable technique that can be used quantitatively. Its drawback is that to generate quantifiable signals, the substance being measured must be in crystalline form. It does, however, produce easily identifiable fingerprints and is hence extremely specific. The size of the particles can also be determined.

17.1.6 Rutherford backscattering spectroscopy

RBS (named after British physicist Ernest Rutherford) is a popular corrosion analysis method because it can determine the thickness of a protective or nonprotective corrosion product coating, as well as an inhibitor film (Chu, Nicolet, Mayer, & Evans, 1974; Sakai, Tsuji, & Naito, 1984). It can also be used to determine the composition and depth of a profile. In situ measurement of the passive layer above a substrate affected by corrosion inhibitor molecules has now been performed using RBS spectroscopy. Optical methods that can only be utilized for oxide layers smaller than 50 nm are ineffective for this measurement. The energy of backscattered projectiles is recorded with an energy-sensitive detector, commonly a solid-state detector, after an ion bombardment with energy in the MeV range (often 0.5–4 MeV). RBS provides for the quantitative analysis of a material's composition as well as the depth profiling of individual constituents. RBS is quantitative without the need of reference samples, nondestructive, and offers a strong depth resolution of several nm as well as excellent sensitivity for heavy elements in the parts-per-million range (ppm). For incident He ions, the examined depth is normally 2 m, while for incident protons, it is typically 20 m. RBS offers the ability to measure the thickness of a surface layer down to 50 nm. Ions entering and exiting the sample are subjected to an electrical "drag" that somewhat slows them down, resulting in a predictable range of backscattered energies. The peak from the dispersed ion beam is widened in proportion, and the energy spread of the signal may be precisely associated with the sample thickness. RBS can now measure absolute depth distributions, which is something no other approach can do. The ideal candidate for such measurements is a heavy-atom layer placed over a light-atom matrix. The spectrum is more complicated and difficult to interpret when a light-atom layer is placed over a heavy-atom matrix. RBS has several advantages, but it also has certain disadvantages, such as the need for an expensive accelerator. RBS is a nondestructive technology for measuring chemical depth profiles but obtaining chemical information from the sample is problematic. RBS can provide exact information on the ratio of components contained in the film, as well as its thickness and crystallinity (Marcus, Maurice, Costa, & Galtayries, 2050). It is a quantitative method that demands a smooth and thin sample but has no standard requirements. Another disadvantage of RBS is its limited sensitivity for light elements, which necessitates the use of alternative nuclear-based approaches such as NRA or in conjunction with elastic recoil detection analysis.

As a supplement to RBS, XRD was employed, and the data were combined to establish the specific phases generated on the surface. LEED could be used to examine the surface crystallography further. The need for a complex accelerator is believed to be the reason for the small number of research detailing the use of the RBS approach for corrosion of carbon steel. By using depth profiling or film thickness measurements, this technique, on the other hand, can provide insight into corrosion mechanisms.

17.1.7 Scanning probe microscopy (SPM)

An SPM is an instrument used for studying surfaces at the nanoscale level. Basically, two main types of SPM microscopes exist that are STMs and AFMs.

17.1.7.1 Scanning tunneling microscopes

The tunneling of electrons via a potential barrier between two surfaces is the principle of STM. With the width of this barrier, the tunneling current (I) decreases exponentially (the separation, s, between the two surfaces). The work function of the surfaces determines the height of the barrier. Electrons tunnel from the full electronic states of the electrode negatively biased to the empty electronic states of the electrode positively biased when the barrier is polarized at a voltage (V) well below the work function. For standard values of the work function of metals, an increase of 1 Å induces a decay of one order of magnitude of I. Because of the exponential decay, if one surface is a tip, the current is spatially restricted to the apex of that tip. In the representation of one-dimensional tunneling (Fig. 17–3), the tunneling wave of the sample electrons, ψs, and the wave of STM tip electrons, ψt, overlap in the insulating gap, allowing a current to flow.

The microscope is controlled by scanning the tip parallel to the surface along the lateral *X* and *Y* axis and measuring the tunneling current at each *X* and *Y* position. A feedback loop is employed at each location to modify the vertical position along the *Z*-axis to keep the

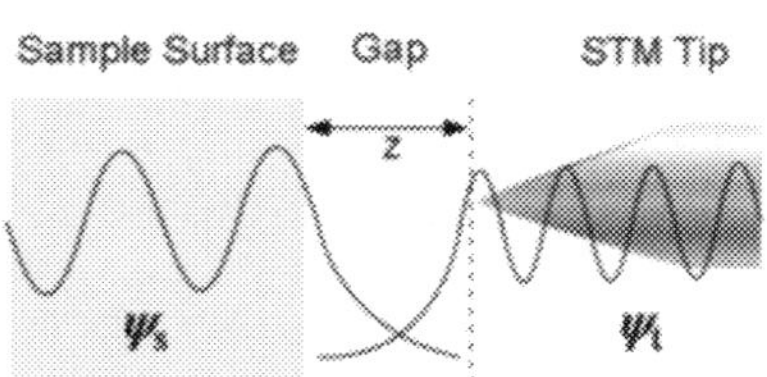

FIGURE 17–3 Schematic of STM one-dimensional tunneling configuration.

tunneling current constant. The "constant current" operating mode is the name given to this mode of operation. The Z values are then plotted against the X and Y values to create topographic images. Images can be captured at varying bias voltages V between the tip and the surface, allowing for the investigation of filled or empty electronic states of varying energies. The recording of I versus V curves at any X and Y points can also be used to do spectroscopic observations. Simultaneous imaging of I versus V curves can be used to create spectroscopic images.

17.1.7.2 Atomic force microscope

A type of surface microscope known as AFM is used to visualize the properties of a material by measuring its interactions with the surface. The data collected by the probe can range from simple topographical details to the details of the material's chemical and physical properties. After the probe has been analyzed and mapped, the generated map of the measured properties can be displayed on a graphical representation. The image is generated by the force microscope when the measured attributes change over the imaged area, such as the height or magnetic domains. Data such as surface roughness, step height, lateral spacing, and surface height can be quickly gathered. Images can be printed on two or three-dimensional or saved as digital pictures, which can then be used for electronic transfer and dissemination.

17.1.8 Scanning electron microscopy

SEM is a technique for imaging surfaces at high resolution. In the same way that a light microscope employs visible light for imaging, the SEM uses electrons. SEM has a far higher magnification ($>100,000\times$) and a much greater depth of field (up to 100 times that of light microscopy) than light microscopy. The SEM generates a stream of incoming electrons in an electron column above the sample chamber. The electrons are produced by a thermal emission source, such as a heated tungsten filament, or a field emission cathode. Depending on the evaluation objectives, the incident electrons' energy can range from 100 eV to 30 keV.

17.1.9 Energy-dispersive X-ray spectroscopy (EDS)

EDS is a technique used in conjunction with SEM to analyze the composition of a material. During the process, the X-rays from the sample are absorbed by the SEM's electron beam. The resulting electrons are then ejected from the atoms on the surface of the sample. Electrons from a higher state can fill the vacancies in the ensuing electron vacancies. An X-ray is then produced to balance the energy gap between the two states. The element from which this X-ray was produced determines the amount of energy it can emit. The EDS detector, which is a solid-state device, measures the relative quantity of the emitted X-rays compared with the energy that it can capture. When an X-ray collides with a detector, the charge pulse of the instrument will be proportional to the energy of the X-ray. A charge-sensitive preamplifier then converts this to a voltage pulse, which then passes to a multichannel analyzer. The energy from an X-ray is delivered to a computer for analysis and display.

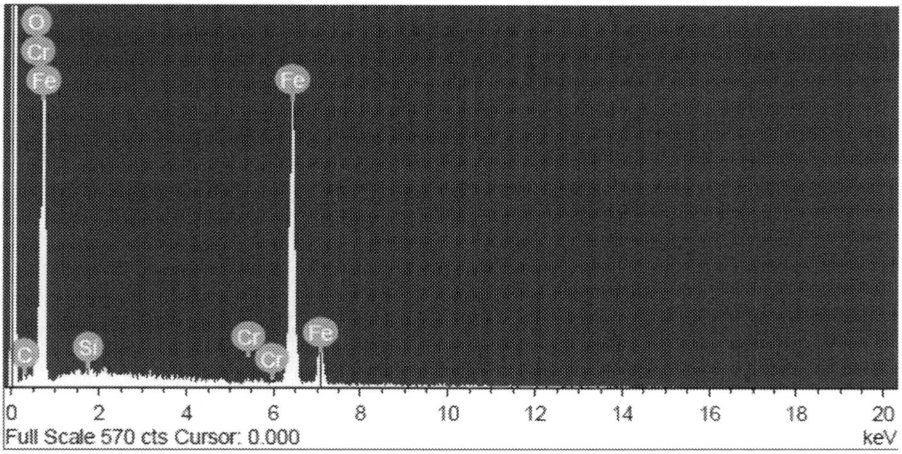

FIGURE 17–4 EDS spectrum for mild steel.

The spectrum of the energy versus its count is then analyzed to establish the composition of the measured volume. In Fig. 17–4, the EDS spectrum of mild steel is shown.

The energy values of the EDS spectrum are compared with the characteristics of known X-rays to determine if an element is present in a sample. The standard EDS analyzers can detect elements with an atomic number from 11 (sodium) upward. Depending on the sample matrix and the element's atomic number, the minimum detection limits are around 0.1 to a few atom percent.

Other applications such as coating composition analysis, quick material alloy identification, small component material analysis, phase identification, and distribution are also commonly performed with EDS.

17.1.10 Secondary ion mass spectroscopy

The SIMS approach, which allows for the detection of chemical species diffusion at grain boundaries of alloys, metals, or at the interface between the inhibitor films and substrate, could be a significant tool for carbon steel corrosion. Simultaneously, this technique aids in the collection of information about scale formation above the substrate. SIMS images show inhomogeneity very clearly. SIMS can also characterize almost all elements. The existence of hydrogen detection is critical in certain corrosion studies, particularly for hydrogen embrittlement, and is possible using SIMS. SIMS can also be used to analyze failures at grain boundaries caused by the creation of hard phases. Three-dimensional SIMS pictures with good lateral resolution can be used to detect chemical profiles. Surface properties can be investigated using SIMS, which uses a primary ion beam with an energy of 1–30 keV in both static and dynamic modes (Rothon, 2003). Static SIMS, which uses a low current density, sputters only a few monolayers of the surface film. Static SIMS's low primary ion current density is appealing for corrosion research because it allows for less exposure to corroded materials, whereas dynamic SIMS can be used to confirm the presence of trace elements. Dynamic SIMS can also be used for quantitative depth profiling because it has a high current density that can sputter the upper surface. SIMS technology has various benefits; however, it requires postanalysis data handling and ultrahigh vacuum, which makes it unsuitable for small-scale industries (BeechIB, Tapper, Gubner, & Avci, 1999; Dupont-Gillain, Mc Evoy, Henry, & Bertrand, 2010; Poleunis, Compère, & Bertrand, 2002).

The time-of-flight secondary ion mass spectrometry technique known as ToF-SIMS is used to analyze the molecular and atomic composition of surfaces. It can also detect the presence of impurities in the materials in parts per billion. This method is performed by analyzing the positive and negative mass spectra of the samples to determine the elemental and ion composition of the top 10%–20% of the studied surface. The high resolution of the ToF-SIMS allows it to identify various species with varying mass differences. It can also detect elements, such as hydrogen. The sample is then placed in the ultrahigh vacuum environment for the analysis of the materials. During the process, the primary ions bombard the surface, while the secondary ones are ejected. The ToF-SIMS then determines the mass of the ejected particles. The difference between the energy used to accelerate the particles and the energy used to analyze them is known as the velocity difference. The smaller the ion, the faster it passes through the analyzer. The mass of the secondary ions is determined by the travel time of the particles through the system. SIMS is a surface-sensitive analysis method that can detect and analyze compounds that are on the surface of a sample.

Aside from being useful for analyzing delicate samples, the use of ToF-SIMS also provides a significant advantage as it allows for the generation of minimum lasting modifications to a surface. It is also useful in the characterization of corrosion inhibitors on different types of surfaces. The ability to characterize corrosion inhibitors on different types of surfaces, such as those that are made of thin organic coatings, can be performed using ToF-SIMS or SIMS (Paul & Vickerman, 1990). The use of molecular specificity techniques for the study of corrosion inhibitors is useful because they can resolve various organic mixture structures. However, due to the complexity of the data collected by the SIMS spectra, it is not always possible to interpret them. This is why it is important to conduct studies on the active chemical species of the inhibitor. In addition, it is also possible to determine the composition of the film using the SIMS spectra.

17.1.11 Factors imposing limitations in surface analysis

Due to the complexity of the data collected by the SIMS spectra, it is not always possible to perform surface assessments. Therefore it is important to consider the various factors that affect the results.

17.1.11.1 Escape depth

Electrons and ions have finite chances of being scattered before they can escape from a solid's depths due to their kinetic energy. The depth at which an electron or ion can escape depends on its angle and its kinetic energy. When an ion or electron interacts with a molecule or atom on the surface, its energy can increase and cause it to enter the solid without being dislodged. The depth of an electron's escape tunnel increases at low energy levels. This is because the energy level of the electron can prevent it from getting stuck in regions with high electron densities.

17.1.11.2 Surface damage

In principle, an analytical measurement should not alter the substance being studied. However, if the conditions or the probe used during the analysis change, the results of the study might be erroneous. For instance, if an electron beam is used as a probe, the species on the surface might be reduced.

The beam of electrons, ions, and photons is the primary source of damage in surface spectroscopy. When a sample gets concentrated, these elements can cause it to get damaged. Photons are less harmful than X-rays, and they can also affect the materials' oxidation state.

Electrons are more harmful than photons to insulators, and they are a particular threat to organic materials. They can also cause chemical damage, which is usually caused by exposure to an electron beam. During the exposure to an electron beam, the species that are adsorbed on the surface can migrate to or from the source. The species that are adsorbed on the surface can also migrate or leave the source. Exposure to an electron beam can also cause significant damage to insulators.

Ion beams are the most dangerous materials used in the sputtering process. They are used to generate secondary ions that are used in the development of depth profiling. To minimize the damage to the sample, the current density should be maintained at a low level. To perform SIMS, the current density should be maintained at around 10^{-9} A/cm^2 or below. Dynamic measurements should only be performed if the surface damage is acceptable. Depending on the ion dosage, the effects of RBS can also be harmful.

17.1.11.3 Surface contamination

Exposure to the atmosphere can cause real-world materials to become polluted. When exposed to the air for an hour, a clean metal can absorb carbon dioxide and turn into a layer of carbon pollution. On an XPS spectrum, any metal can show a significant carbon signal. The presence of moisture in the air can also alter the surface of certain materials.

Although these layers are not particularly thick, they can still be used to sample the surface. It is important to note that these materials are still contaminated. This can affect the signal intensity of the surface.

17.1.11.4 Blooming

Blooming, an intrinsic process caused by the scattering of charged particles in materials limits the spatial resolution of numerous surface mapping techniques. Because of optics and electron scattering, when a beam of highly concentrated, high-energy charged particles, such as electrons, strikes a solid and penetrates the surface, the electrons' pathways diverge. The area impacted by the electron beam has the shape of an inkwell or an inverted light bulb.

The kinetic energy of the particles in the incident beam determines the depth and, as a result, the diameter of the impacted area. Because X-rays are generated throughout the volume and depart the sample's surface, the smallest unit area in the image will not be the diameter of the incident beam, but rather the diameter of the inkwell or light bulb's widest component. As a result, the resolution is limited.

The presence of a large amount of charged particles can also cause a significant increase in the background noise in the sample. This phenomenon can also affect the detection of light elements.

Short questions

1. Which two elements cannot be analyzed by Auger spectroscopy?
2. What is the specificity of Auger electron spectroscopy?
3. Why XPS does not detect hydrogen and helium?
4. What signal does XPS detect?
5. How does AFM work?
6. What's the difference between AFM and SPM?
7. Can EDS measure carbon content?
8. What is the difference between XPS and EDS?

References

Barlow, S. M., & Raval, R. (2003). Complex organic molecules at metal surfaces: Bonding, organisation and chirality. *Surface Science Reports., 50*(6–8), 201–341. Available from https://doi.org/10.1016/S0167-5729(03)00015-3.

BeechIB, Z. V., Tapper, R., Gubner., & Avci. (1999). Study of the interaction of sulphate reducing bacteria exopolymers with iron using X-rayphotoelectron spectroscopy and time-of-flight secondary ionization mass spectrometry. *Journal of Microbiological Methods, 36,* 3–10. Available from https://doi.org/10.1016/S0167-7012(99)00005-6.

Castle, J. (1977). *Surf. Sci., 68.*

Chu, W. K., Nicolet, M. A., Mayer, J. W., & Evans, C. A. (1974). Comparison of backscattering spectrometry and secondary ion mass spectrometry by analysis of tantalum pentoxide layers. *Analytical Chemistry, 46* (14), 2136–2141. Available from https://doi.org/10.1021/ac60350a039.

Dupont-Gillain, C. C., Mc Evoy, K. M., Henry, M., & Bertrand, P. (2010). Surface spectroscopy of adsorbed proteins: Input of data treatment by principal component analysis. *Journal of Materials Science: Materials in Medicine, 21,* 955–961. Available from https://doi.org/10.1007/s10856-009-3967-y, 09574530 3 Belgium 3 2010/03.

Frankenthal, R. P., & Malm, D. L. (1976). *Journal of the Electrochemical Society, 123.*

Marcus, P., Maurice, V., Costa, D., & Galtayries, A. (2050). The contribution of surface analysis to corrosion science: Historical background and current status. In Gerald S. Frankel (eds.) Corrosion science: *A retrospective and current status in Honor of Robert P. Electrochemical Society.*

Mitchell, D. F., & Graham, M. J. (1987). *Surface and Interface Analysis, 10.*

Olefjord, I. (1975). *Metal Science, 9.*

Paul, A. J., & Vickerman, J. C. (1990). Organics at surfaces, their detection and analysis by static secondary ion mass spectrometry. *Trace Analysis, 333,* 147–158. Available from https://shibbolethsp.jstor.org/start?entityID = https%3A%2F%2Fidp.amu.ac.in%2Fopenathens&dest = https://www.jstor.org/stable/76781&site = jstor.

Poleunis, C., Compère, C., & Bertrand, P. (2002). Time-of-flight secondary ion mass spectrometry: Characterisation of stainless steel surfaces immersed in natural seawater. *Journal of Microbiological Methods, 48*(2–3), 195–205. Available from https://doi.org/10.1016/S0167-7012(01)00323-2.

Rothon, R. N. (2003). *Particulate-filled polymer composites.* Smithers Rapra Publishing Technology Secondary ion mass spectrometry.

Sakai, H., Tsuji, T., & Naito, K. (1984). Effect of oxygen partial pressure on oxidation of iron at 573 k. *Journal of Nuclear Science and Technology, 21*(11), 844–852. Available from https://doi.org/10.1080/18811248.1984.9731123.

Schnyder, B., & Kötz, R. (1992). *Journal of Electroanalytical Chemistry, 339.*

Seo, M., & Sato, N. (1983). *Electrochimica Acta, 28.*

Strehblow, H. H. (1988). *Surface and Interface Analysis, 12.*

PART 6

Corrosion protection methods

18

Cathodic protection

18.1 Introduction

For the protection of metallic underground and subsea structures like cables, power lines, oil and gas pipelines, and building foundations, cathodic protection (CP) is a tried-and-true corrosion management technique. It is one of the few methods available for preventing corrosion on old structures (Baeckmann & Schwenk, 1999; Wyatt & Irvine, 1987; Peabody's Control of Pipeline Corrosion, 2001). Applications for CP of ordinary rebar go back well over 20 years, therefore it is well established.

CP's fundamental idea is a straightforward one. Applying a cathodic current to a shielded structure reduces anodic dissolution. The coated structures frequently receive CP, with the coating serving as the main source of corrosion defense. For uncoated systems, the present CP requirements are frequently too strict (Chess & Broomfield, 2017). CP was initially used between 1863 and 1824, several years before its theoretical underpinnings were established. The protection of oil drilling platforms, docks, jetties, ships, submarines, condenser tubes in heat exchangers, bridges, decks, civil and military aircraft, and ground transportation systems are currently common use of CP. Despite being rather complicated, the CP system design is based on straightforward electrochemical principles.

Due to the potential difference between the local action anodes and cathodes, a corrosion current runs between them. Electrons generated during an anodic reaction are used up during the cathodic reaction. If we provide more electrons to a metallic structure, more electrons would be accessible for a cathodic reaction. This would result in an increase in the rate of cathodic reaction and a decrease in the rate of anodic reaction, thereby minimizing or eliminating corrosion, and which is the main goal of the CP method. A direct electric current provides the extra electron (James, 1992; Pedeferri (Deceased), 2018) and the potential of the cathode changes to the potential of the anodic region upon application of direct current. Corrosion would finally stop happening if enough direct current was provided to eliminate the potential difference between the anode and cathode. The cathodic reaction polarizes in the direction of the local action anode potential as the cathodic current increases (more electrons are transferred). This further reduces the potential difference between the anodes and cathodes. When the metallic structure turns into a cathode, full CP is attained. Since the amount of the potential difference between the anode and the cathode directly correlates with the severity of corrosion, this difference can be reduced completely control corrosion (Bauer & Vogel, 1918; Koch, Brongers, Thompson, Virmani, & Payer, 2002; Kuhn, 1928; Scholten, 1930; Stratfull, 1959). This chapter covers the in-depth development of this method, its underlying concepts, and the standards for prevention and protection in a variety of application domains.

Handbook of Corrosion Engineering. DOI: https://doi.org/10.1016/B978-0-323-95185-2.00018-6

18.2 Principles of cathodic protection

Corrosion in aqueous solutions is caused by an electrochemical reaction that involves simultaneous cathodic and anodic electrochemical reactions, that is, (Bahadori, 2014)

$$Fe \rightarrow Fe^{2+} + 2e^{-} \text{(anodic reaction)} \tag{18-1}$$

$$2H^{+} + 2e^{-} \rightarrow H_2 \text{(cathodic reaction in acid solution)} \tag{18-2}$$

$$2\,H_2O + 2e^{-} \rightarrow H_2 + 2OH^{-} \text{(Cathodic reaction in alkaline solution)} \tag{18-3}$$

However, oxygen reduction is the most likely reaction to occur, leading to alkali at the metal's surface, unless the water is deaerated:

$$O_2 + 2H_2O + 4e^{-} \rightarrow 4OH^{-} \tag{18-4}$$

Since the anodic and cathodic processes have the same rate of progression, there is no overall net charge accumulation on the metal as a result of corrosion. By removing electrons from the metal or adding more electrons, one can increase the rates of anodic and cathodic processes. It is a well-known truth that if one of the conditions for the equilibrium of the system changes, the system will adjust to minimize the impact of the change. As a result, iron dissolution will increase and reaction (18−1) will speed up to make up for the action of removing electrons from the metal, while reaction (18−2) will slow down. Reaction (18−1) will slow down, leading to reduced corrosion, but reaction (18−2) will speed up if we add more electrons from outside the metal piece. The latter case will result in CP.

We must continue to provide the equipment with electrons from an outside source to satisfy the requirements of the cathodic reaction in order to prevent corrosion. There is a close relationship between the cathodic and anodic processes. When the anodic process speed is slowed down, the cathodic process speed will increase. These ideas can be stated quantitatively by plotting the metal's potential versus the logarithm of the anodic and cathodic reaction rates given as current densities. Where the anodic and cathodic reaction rates are equal, the corrosion current, I_{corr}, and the corrosion potential, E_{corr}, occur. If electrons are "pumped" into the metal to make it more negative, the anodic dissolution of iron is reduced to a negligible rate at a potential E_l, while the cathodic current rate is increased to I_1. As a result, in order to maintain the potential at E_1, where the rate of iron dissolution is negligible, a current I_1 must be supplied from an outside source. If the potential is reduced to E_2, the amount of external current needed will rise to I_2 (Fig. 18−1) (Bahadori, 2014).

Additional metal protection, however, is limited, and the greater current provided by the external source is wasted. The metal is therefore thought to be overprotected. In aerated neutral or alkaline solutions, cathodic corrosion typically involves the reduction of oxygen. The kinetics of this cathodic process is governed by the rate at which oxygen may diffuse to the metal's surface, which is slower than the rate at which oxygen is consumed by the cathodic reaction. As a result, unless the rate of oxygen delivery to the surface of the metal is

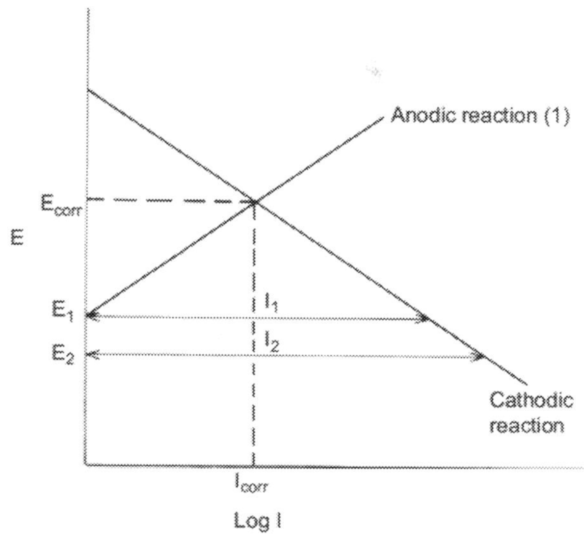

FIGURE 18–1 Kinetics of anodic and cathodic reactions.

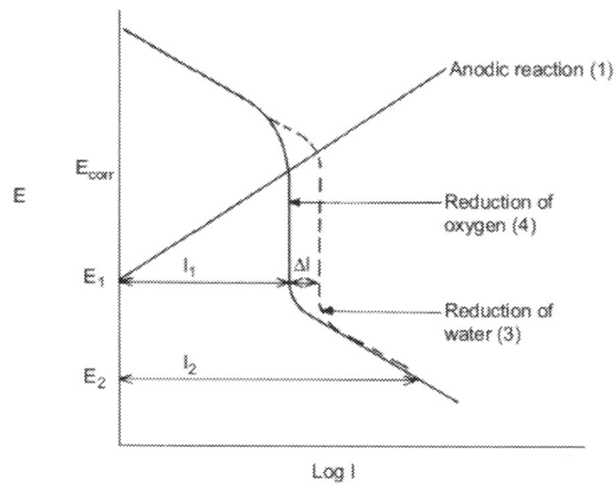

FIGURE 18–2 Diffusion control of the reduction of oxygen.

increased, such as by increasing fluid flow rate, the rate of this reaction does not increase as the potential of the metal is decreased but rather remains constant. The impact of flow velocity on the parameters of CP is shown in Fig. 18–2.

An initial current of I_1 is required to maintain the metal at the protective potential E_1. However, when the flow rate increases, both the current needed to maintain the metal at the protective potential and the limiting current for oxygen reduction increase (dotted line). As a result, depending on the service conditions, a different current density is required to maintain the correct protective potential.

It is obvious that cathodic current density cannot be used to determine if a structure is cathodically protected. Maintaining the right protective potential is necessary to prevent corrosion. More current from an external source will be needed and current will be wasted if the structure is overprotected and the potential falls to a point where water reduction (reaction 18–3) can happen. As shown in Fig. 18–2, a faster rate of water reduction results in a higher current requirement from an external source when the potential is decreased from *E*1 to *E*2. Because of the alkaline conditions thus formed at the cathode, high negative potentials can induce quicker corrosion of lead and aluminum. These alkaline circumstances may damage some paint systems, which could lead to the loss of the paint layer. High-strength steels may experience hydrogen embrittlement, which weakens them, as a result of hydrogen evolution at the cathode surface. Additionally, it might degrade any insulating coatings, which would prevent them from shielding cathodic-protection currents.

18.3 Electrical basis of cathodic protection

Consider an underground pipe to better understand the CP system's electrical foundation. There is a potential difference between the anode and the cathode, that is, $E = (E_a - E_c)$, where E denotes the potential difference. If $R_a + R_c$ is controlling resistance and $E_a - E_c$ has a limited degree of circuit resistance, the corrosion current (I_c) will flow. I_c must be set to zero if complete corrosion resistance is needed (Ahmad, 2006).

$$I_c = \frac{E_a - E_c}{R_a + R_c} = 0\left(I = \frac{E}{R}\right)(\text{Ohm's Law}) \qquad (18-5)$$

where I_c = corrosion current (A), E_a = anodic potential, E_c = cathodic potential, R_a = anode resistance (ohms), R_c = cathode resistance (ohms).

The corrosion current can be made zero by bringing $R_a + R_c$ to infinity, for example, by pointing. You can also do this by balancing the potential difference between E_c and E_a, or by setting $(E_c - E_a)$ to zero. By running an electric current through the structure from the outside until its potential equals that of the anode, the difference between E_c and E_a vanishes completely.

$$E_A - E_C = I_1 R_A + R_C(I_1 - I_2) \qquad (18-6)$$

$$I_1 = \frac{(E_A - E_C + R_C I_2)}{R_A + R_C} \qquad (18-7)$$

where E_A = open circuit anode potential, E_C = open circuit cathode potential, R_A = effective anode resistance, R_C = effective cathode resistance, I_1 = current from anodic area, I_2 = current from the external anode.

I_1 must be zero if corrosion is to be avoided, hence $(E_C + R_C.I_2) = E_A$. It means that enough current must flow through R_C for the anode's potential to match the cathode's potential (open circuit potential). According to the aforementioned connection, an external anode must generate enough current to pass through the cathode resistance R_Q in order for the cathode potential (E_C) to match the anode potential (E_A). There will be no corrosion and the structure will be cathodically protected if $E_A = (E_C + R_C.I_2)$ is attained. The complete and incomplete protection is depicted in Fig. 18–3A. Fig. 18–3B shows that the current is not flowing to all anodic areas, whereas Fig. 18–3C shows that the current is flowing to all anodic areas (Spencer, 1960).

CP may be achieved in either of two ways, either by employing:

1. Galvanic anode system or sacrificial anode system, or
2. Impressed current anode system.

18.3.1 Galvanic anode system or sacrificial anode system

Connecting sacrificial anodes to a structure can provide CP. The basic idea is to make a galvanic cell, with the anode symbolizing the less noble material that is consumed during the galvanic reaction. In an ideal world, the galvanic current flow will protect the structure. In practical applications, a number of anodes are usually linked to a structure to ensure overall protection levels. Examples of typical uses include buried tanks, underground pipes, buried communication and power cables, water and gas distribution systems, internal heat exchangers and hot water tank protection, ships, and maritime structures. Fig. 18–4 schematically depicts the sacrificial anode CP system for a structure buried in soil and made of reinforced concrete. In both instances, a porous material surrounds the metal (soil or concrete) (Angst, 2019).

18.3.1.1 Components

18.3.1.1.1 Anode

In this type of system, magnesium and zinc anodes are commonly employed. Because of their high current production, magnesium anodes are the most popular ones. They offer the following benefits:

a) favorable position in the emf series (very active),
b) nonpolarizing nature, and
c) high current output.

With 65%, an average current production of 500–600 A-h/lb can be obtained. Magnesium anodes are often used in soils with a resistance of less than 3000 ohms-cm. For example, a conventional 17-pound magnesium anode will provide 1 A of current for a year, 0.1 A for 10 years, and so on. The current output, on the other hand, is determined by the soil resistivity. The current produced would be 0.1 A with a resistance of 1000 ohm-cm. The produced current would be smaller if the resistance is very high. Backfill is a mixture of gypsum and bentonite that surrounds the anodes to ensure optimal effectiveness. A typical

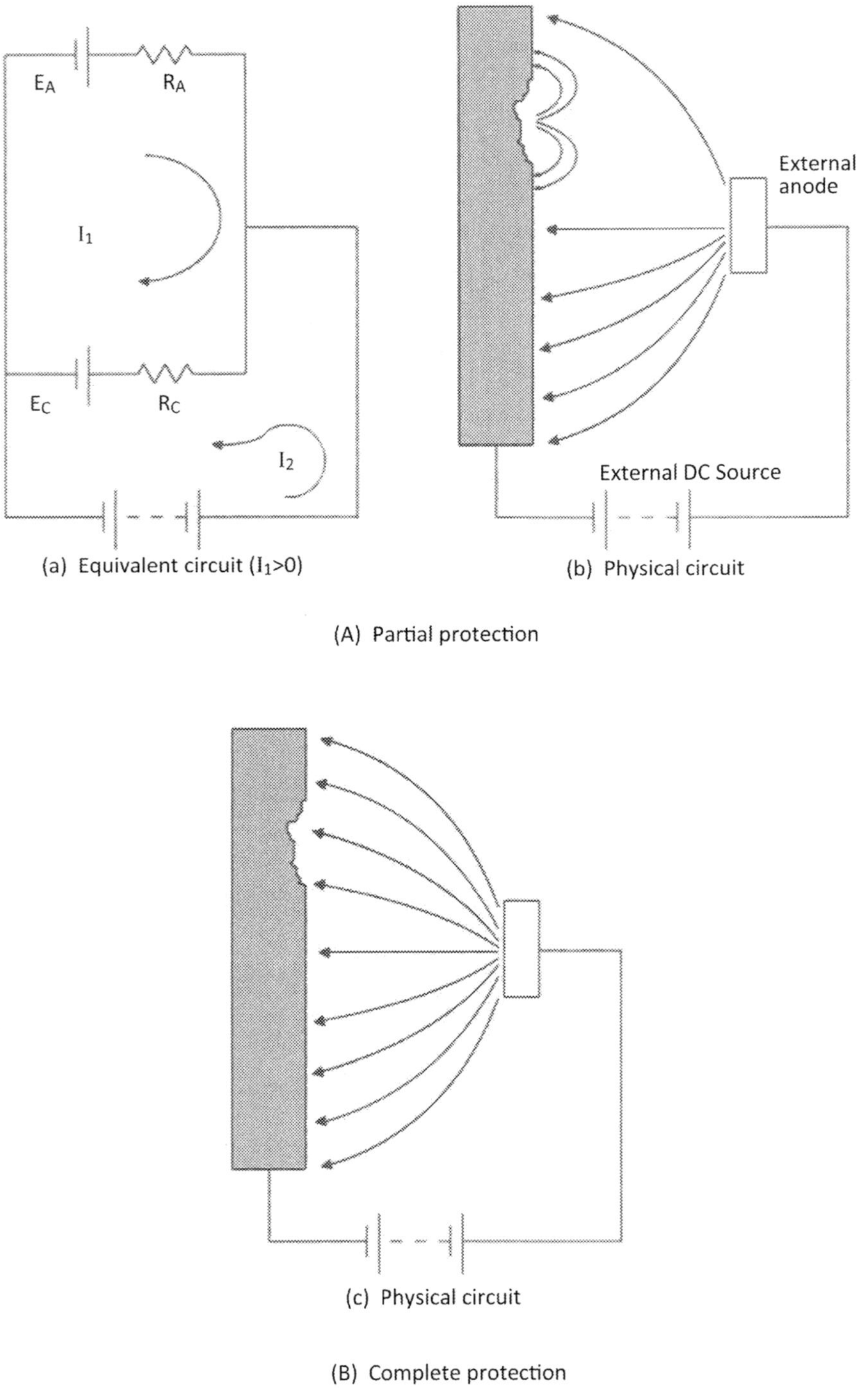

FIGURE 18–3 Equivalent circuit of cathodically protected metals.

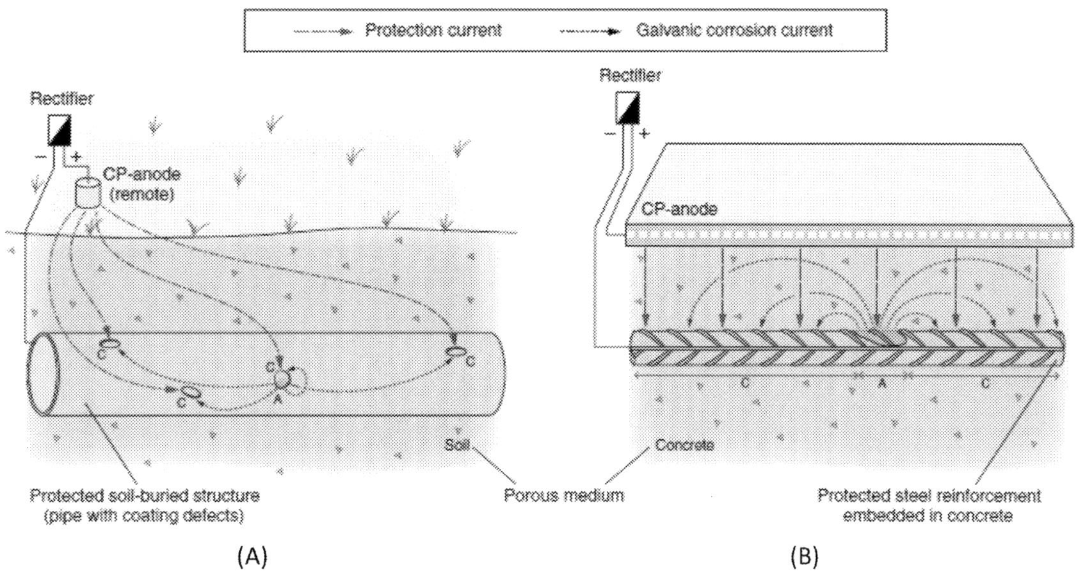

FIGURE 18–4 The sacrificial anode cathodic protection system for a soil-buried structure and for reinforced concrete.

backfill consists of ground hydrated gypsum (75%), powdered Wyoming bentonite (local brand) (20%), and anhydrous sodium sulfate (5%).

Magnesium anode: Fig. 18–5 depicts the details of a magnesium anode. In comparison to the CU-Q1SO4 reference electrode, the open circuit potential (no current draw) of a normal alloy magnesium anode is − 1.55 V. In comparison to a copper–copper sulfate half-cell, the open circuit voltage of a high manganese magnesium anode is − 1.75 V. When iron is polarized to − 0.85 V, ordinary alloy magnesium has a driving potential of (1.55–0.85 = 0.70 V), while a high manganese magnesium anode has a driving potential of (1.75–0.85 = 0.90 V). Table 18–1 indicates the composition of standard and high manganese magnesium anodes.

Local corrosion cells limit the output to 100% efficiency. Magnesium is alloyed with Al and Zn to avoid this. In practice, only 60%–70% efficiency may be achieved. For seawater service, galvomag anodes are employed. Cu (0.02%), Al (0.01 maximum), Fe (0.03%), Ni (0.001%), Sn (0.01%), Pb (0.01%-max), and Mg (0.01%-max) are the specifications for Galvomag (Balance). With a potential of − 1.70 V Ag/AgCl and a capacity (A-h) of 1230 per kg 560 per lb, this anode has a 50% efficiency. The annual consumption rate is 17.52 lb.

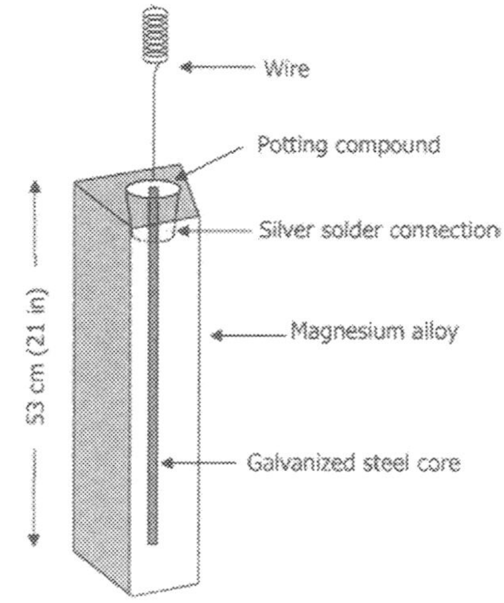

FIGURE 18–5 Typical magnesium anode.

Table 18–1 Composition of the magnesium anode.

S. no.	Anode	Composition
1.	Standard magnesium anode	Aluminum (5.30%–6.7%), manganese (0.15-min%), zinc (2.50%–3.5%), silicon (0.30-max %), copper (0.05-max %), nickel (0.003-max %), iron (0.003-max %), other impurities (0.3-max %), and magnesium (balance %)
2.	High manganese magnesium anode	Aluminum (0.010-max %), manganese (0.5%–1.31%), zinc (0.02-max %), copper (0.02-max %), nickel (0.001-max %), iron (0.003-max %), other impurities (0.3-max %), and magnesium (Balance %)

Zinc anode: Submarine pipelines are usually protected with zinc anodes. Weights ranging from 5 to 60lb are commercially available. If anodes are to be installed in the ground, prepared backfill should be used. When compared to a Q1-CUSO4 reference electrode, they have a driving potential of -1.10 V. Fig. 18–6 shows the details of zinc anodes.

Some of the characteristics are:

- Because corrosion products shield the anodes, they must be positioned below the water table in soils devoid of free carbonate or phosphate to avoid passivity.
- Zinc is not commonly polarized anodically, and the theoretical current production for 99.99% pure zinc can be as high as 372 A-h/lb.
- Zinc may operate at up to 95% efficiency, resulting in a current output of 335 A-h/lb (0.90 372 = 335 A-h/lb).
- Zinc has an open circuit potential of − 1.10 V when compared to a copper–copper sulfate reference electrode.
- The driving potential of zinc is 0.25 V, based on a polarization potential of − 0.85 V for steel.

Aluminum anode: The majority of them are used in seawater applications. Aluminum makes about 98%–99% of the basic metal. The following are some of the characteristics:

- They are inexpensive and light in weight.
- The corrosion products do not contaminate the water in any way.
- The annual consumption rate fluctuates between 7 and 9 pounds. Between 87% and 95% efficiency is achieved.
- The anodes are readily passivated and must be reactivated by rinsing with NaCl. With aluminum anodes, backfill is required.

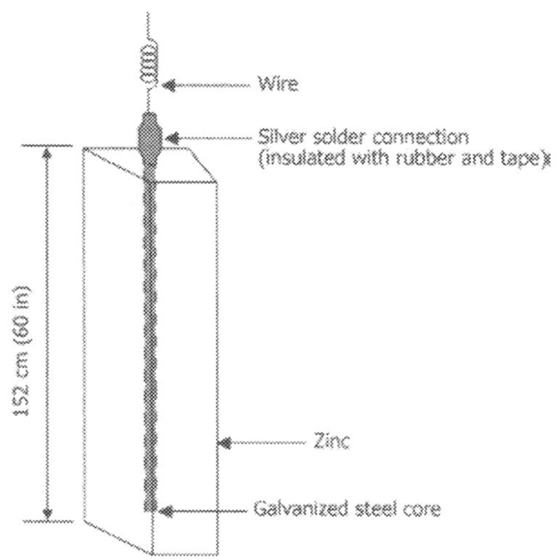

FIGURE 18–6 Typical zinc anode.

- The consumption of Galvalum-Aluminum anodes (Fe = 0.08, Si = 0.11–0.21, Zn = 0.35–0.50, and Hg = 0.035–0.40) is 1285 A-h/lb. It is a very popular anode.

18.3.1.1.2 Installation

In comparison to current installations, galvanic anode installations are simple. For a packaged installation, 17-pound packaged magnesium anodes are usually used. The anodes are connected to a header wire that is directly attached to the pipeline when numerous magnesium or zinc anodes are to be installed at a single spot. This shows how packaged anodes can be put in multiples. The anode line might run parallel to the pipeline or perpendicular to it. If the pipelines are in parallel lines, zinc anodes should be kept at a distance of 5 feet. Magnesium anodes are typically placed 15–20 feet away from pipelines. Galvanic anodes are positioned below the pipe if space is limited and soil resistivity is extremely low. To insert several anodes, very deep holes are required. Deep soil anodes are safer since they are not impacted by changes in moisture content or soil conditions. Nonpackaged anodes are regularly utilized due to seasonal variations. In this arrangement, the anode and backfill are put separately rather than as a single unit. A horizontal galvanic anode can also be used.

18.3.1.1.3 Advantages

The following advantages are associated with sacrificial anode CP systems:

- There is no need for external power sources.
- Installation is simple (and has relatively low installation costs).
- Cathodic interference in other structures is unlikely to occur.
- Low-maintenance systems are those that do not require a lot of attention (assuming low current demand).
- The system is basically self-regulatory.
- Overprotection is a rare occurrence.
- Potential distributions that are rather uniform.

18.3.1.1.4 Disadvantages

- Current and power output is limited.
- In high-resistivity conditions or with massive structures, an excessive number of electrodes may be required. Depending on coating quality, a maximum resistivity of 6000 to 10,000 Ω.cm is commonly considered the limit.
- Under high current demand, anodes may need to be replaced often.
- If anodes are physically attached to a building, they can add weight to it.

18.3.1.2 Impressed current method

Fig. 18–7 shows how to safeguard a pipeline that is buried. The underground pipe receives current from a direct current (DC) power source via an additional inert electrode. The anode is the secondary electrode, while the cathode is the pipe. An additional electrode can occasionally be made out of scrap iron. The anode in this instance is referred to as a consumable anode because reaction (18–1) dissolves the iron from it. If the anode is a noble metal or

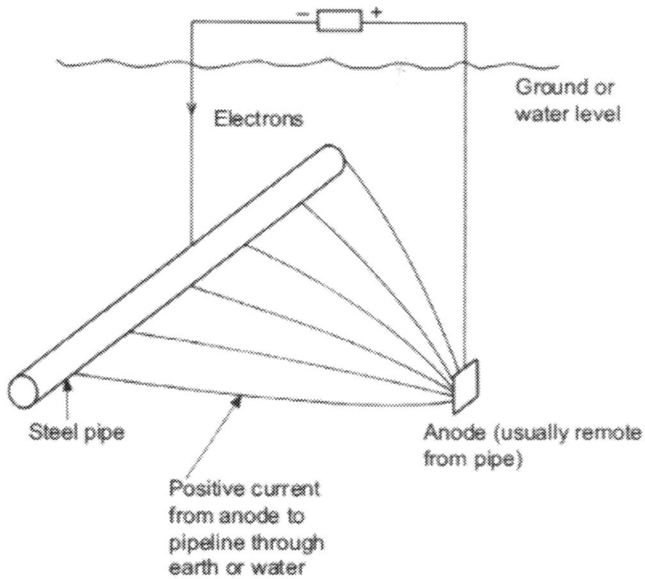

FIGURE 18–7 Application of cathodic protection by an impressed current.

an electrochemically inert substance, the surrounding environment will oxidize, and we will have:

However, chlorine can be generated at the anode in saline situations. In confined locations, this could cause issues. For impressed-current systems, a variety of materials have been employed as nonconsumable anodes. These anodes require the following characteristics:

- excellent electrical conductivity,
- low corrosion rate, strong mechanical characteristics (ability to withstand pressures encountered during installation and service),
- capacity to be quickly and easily formed into a range of shapes
- economical or of low cost,
- ability to endure large current densities without producing resistive barrier oxide layers at their surfaces, etc.

Magnetite, carbonaceous materials (graphite), high silicon iron (14%–18% Si), lead/lead oxide, lead alloys, and platinized materials have all been utilized as anodes (e.g., tantalum, niobium, and titanium). Platinum, with its excellent corrosion resistance, would be an ideal anode material, but it has the big problem of being extremely expensive. On impressed-current anodes, voltages up to 100 V and significant current densities are attainable in practice (Table 18–2). As a result, a single anode may protect extensive portions of a structure, and the anode can be located far away from the structure due to the high driving voltage.

Table 18–2 Properties of impressed-current anodes.

S. no.	Anode material	Max volts (dependent on the environment)	Typical anode current density (A/m^2)
1.	Platinum (Pt)/niobium (Nb)	100	250–1500
2.	Lead (Pb)/silver (Ag)/antimony (Sb)	100	250–1000
3.	High silicon iron	100	10–100
4.	Graphite	-	200

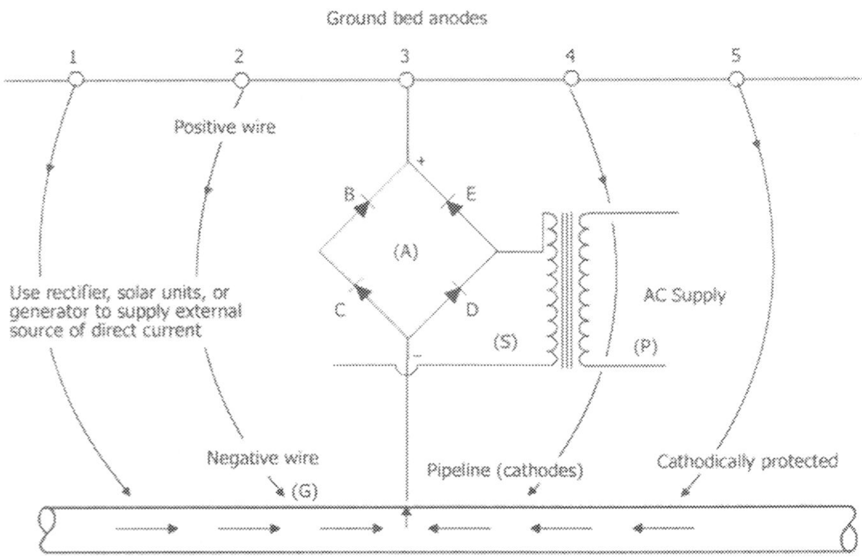

FIGURE 18–8 Schematic diagram of impressed current cathodic protection system.

18.3.1.2.1 Components

In the impressed current system, unlike the galvanic anode system, the current flow from the anode to the cathode is pushed from a DC source. In the impressed system, the electrode works as a conductor and corrodes very little, whereas in the anodic galvanic system, the current is given by the electrode corrosion, and the AC input is translated and rectified to a changing DC voltage. The most crucial component of the system is the transformer rectifier. A schematic representation of an impressed current CP system is shown in Fig. 18–8.

The direct terminal supply is provided by a transformer rectifier (T/R unit), which steps down the typical alternating mains voltage before rectifying it to direct current. To suit the demands, a wide range of adjustments can be made to this output. An AC power supply with an ammeter powers the rectifier and to protect the rectifier from overloads in the AC circuit, a switch box with a magnetic or thermal circuit breaker is installed. The AC is delivered to the rectifier device. The resulting magnetic field oscillates as it passes through the primary coil, growing, contracting, and reversing its direction (P). Due to the shifting field, the secondary coil produces AC proportionate to the ratio of turns between the two coils. The electrons from the rectifier enter the pipe or any other structure that requires protection. The anodes are placed in a ground bed, which is a trench. The soil with the lowest resistance is often where the anodes are placed. The specifics of a typical impressed current installation are shown in Fig. 18–9.

The AC power unit is connected to the rectifier (R). The ground is connected to the positive terminal, whereas the structure (S) is connected to the negative terminal. The silicon

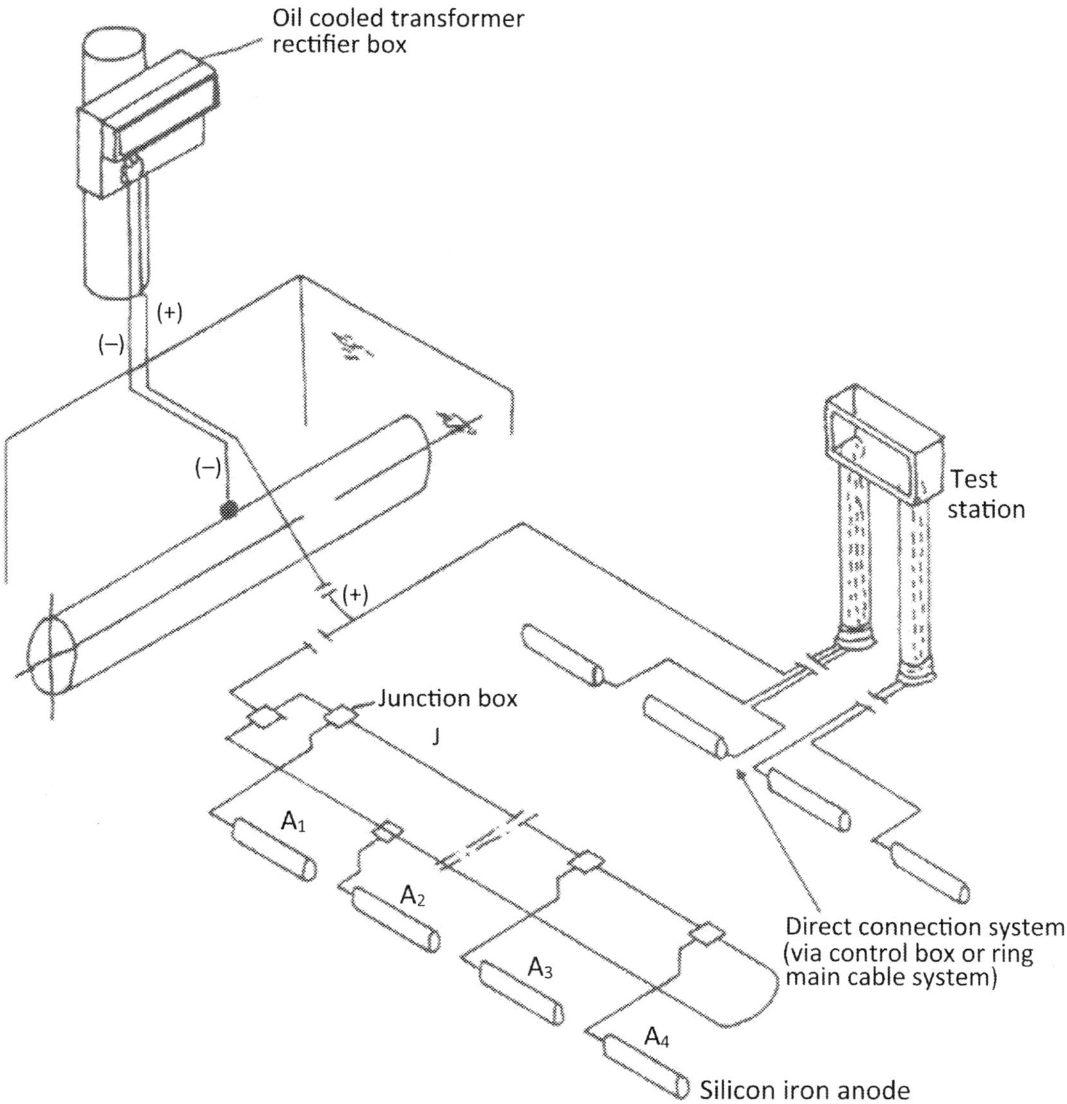

FIGURE 18–9 Details of typical power impressed ground-bed installation.

cast iron anodes are identified in as A1 through A4. A shunt is installed into each anode lead after it is placed in a junction box (J). The junction box's IR drop can be utilized to measure the current output of certain anodes. The anodes can be connected directly to the positive terminal of the rectifier using a ring main cable system, as shown in, or a control box (direct connection system) can be used to track the current output above the ground.

18.3.1.2.2 Advantages

The major advantages of the impressed current CP system are discussed below:

- A single installation can safeguard a considerable amount of metal.
- The system can handle a wide range of voltage and current demands.
- If necessary, schemes can be developed to last for more than 20 years.
- Current requirements and potentials are easily adaptable to changing security needs.
- It can be used on a variety of buildings.
- Requires a limited total number of anodes in most cases.
- Simple controls that can be automated are required.
- Soil resistivity is less of a constraint.
- A single installation can protect a huge region.

18.3.1.2.3 Disadvantages

- Interference effects on other subterranean constructions are a possibility.
- Maintenance must be done on a regular basis.
- The cost of electricity is really expensive.
- Power outages can be disastrous, and defects might go unchecked for lengthy periods of time.

18.4 Examples of cathodic protection

In 1987, the Rio Veggio bridge deck (3700 m^2) in the Apennine section of the Al highway between Bologna and Florence received a CP system after suffering from significant chloride-induced corrosion that necessitated extensive steel rehabilitation. Concrete overlay spraying; application and fixing of the mesh using appropriate plastic pins inserted into pre-drilled holes. The monitoring system is based on fixed reference electrodes (silver/silver chloride and graphite) implanted in concrete in each of the eight anodic zones that make up the bridge's deck. These electrodes are placed in the most crucial regions for each anodic zone. Every 7 days, protection conditions are automatically managed by the "100 mV" decay criterion: if depolarization exceeds 250 mV, the current density is decreased; if it is less than 100 mV, the current density is increased. Absolute potential readings are used twice a day to check for overprotection conditions. It was decided to set an extremely low limit of -850 mV. The feeding system for the concerned zone is automatically turned off by the control unit if a reference electrode exceeds the overprotection limit.

The repair method involved removing the damaged concrete and replacing it with new concrete, installing new reinforcement bars and welding them to the old ones, checking the electrical continuity of the new reinforcement bars with the old ones, and adding welded links where necessary prior to pouring, installing electrical connections and monitoring equipment, pouring concrete, and curing it.

The bridge's two lanes were renovated using the same concrete restoration method and two distinct anodic systems on activated titanium mesh. The anode mesh was either fastened to the concrete surface and sprayed with a concrete overlay substance, or it was directly implanted into concrete, 1.8–2 cm away from reinforcement (south lane) (north lane and curbs of both lanes). After 7 years of service, there is no discernible variation in the CP of the two lanes.

With a feeding voltage of 1.8–2.5 V, the start-up protection current was 10 mA/mm^2 of concrete surface (i.e., 10 mA/mm^2 of rebars of the first-row surface of approximately 5 mA/m^2 of the total rebar surface). After 5 months, the protection criteria were fulfilled. The protective current density dropped by 30–50 times the next year, and it has remained essentially constant ever since in the range of 5–8 mA/mm^2 of concrete surface. According to changes in anodic overvoltage and, to a lesser extent, ohmic drop, regular changes in the feeding voltage over time have been seen on the shielded span in relation to ambient temperature. More than 7 years later, the system is still functioning well.

18.5 Cathodic protection market

Currently valued at US$ 5.3 billion in 2022, the global CP market is anticipated to grow to US$ 7.3 billion by the end of 2028. From 2022 to 2028, the demand for CP is expected to grow at a compound annual growth rate of 5.6%. Currently, CP accounts for 23% of the market for corrosion protection worldwide. The last few years have seen a rise in the importance of corrosion prevention as a means of cutting down on industrial losses and maintenance expenses. Governments all over the world are putting more emphasis on industrial facility safety and maintenance in order to prevent unfavorable results in the event of unexpected disasters.

18.6 Conclusion

This chapter assessed the state-of-the-art CP from a scientific and engineering standpoint. Without making any changes to the structure's electrified CP system, a variety of essential corrosion characteristics can be simply monitored under controlled conditions. The measurement principles are straightforward. In practice, a decision must be made on the amount of CP current that will be used. Excessive current (or "overprotection") can cause coating disbonding and hydrogen embrittlement, whereas too little current causes excessive corrosion damage. Furthermore, corrosion potentials and protection requirements are not uniform across the full surface of corroding structures. Such variances must be taken into account in practical protection standards. An analysis of the most recent CP evaluation standards reveals the unreliability of these technical methods. They may either lead to overprotection and the related harmful side effects if they are too conservative, or they may suggest corrosion protection when this is not the case. This is due to the empirical nature of the assessment criteria, which are incapable of appropriately accounting for the complexity of the underlying processes.

Short questions

Q1 What Is Cathodic Protection? Explain with example.
Q2 How does cathodic protection works to prevent corrosion?
Q3 What is the basic principle involved the cathodic protection method?
Q4 What is galvanic or sacrificial anode protection system?
Q5 What materials are used for Sacrificial Anodes?
Q6 What materials are used for Sacrificial Anodes?

References

Ahmad, Z. (2006). *Principles of corrosion engineering and corrosion control*. Saudi Arabia: Elsevier Ltd. Available from http://www.sciencedirect.com/science/book/9780750659246.10.1016/B978-0-7506-5924-6.X5000-4.

Angst, U. M. (2019). A critical review of the science and engineering of cathodic protection of steel in soil and concrete. *Corrosion, 75*(12), 1420–1433. Available from https://doi.org/10.5006/3355.

Bahadori, A. (2014). 3-Design considerations on cathodic protection for buried pipelines and marine structures. *Cathodic Corrosion Protection Systems*, 91–131. Available from https://doi.org/10.1016/B978-0-12-800274-2.00003-X.

Baeckmann, W. V., & Schwenk, W. (1999). *Handbuch des kathodischen Korrosionsschutzes*. Wiley-VCH.

Bauer, O., & Vogel, O. (1918). Mitteilungen aus der Materialprüfanstalt zu. 36.

Chess, P. M., & Broomfield, J. P. (2017). *Cathodic protection of steel in concrete and masonry*. CRC Press.

James, F. A. J. L. (1992). *Physis, 29*, 205–225.

Koch, G. H., Brongers, M. P. H., Thompson, N. G., Virmani, Y. P., & Payer, J. H. (2002). *MP, 42*.

Kuhn, R. J. (1928). In *Proceedings of soil corrosion conference bureau of standards galvanic current on cast iron pipes. Their causes and effects. methods of measuring and means of prevention*.

Peabody's Control of Pipeline Corrosion. NACE International (2001).

Pedeferri (Deceased), P. (2018). *Cathodic and anodic protection engineering materials* (pp. 383–422). Italy: Springer Science and Business Media B. V. Available from http://springer.com/series/4288.10.1007/978-3-319-97625-9_19.

Spencer, K. A. (1960). The protection of gas plant and equipment from corrosion. In *Joint Symp. Corrosion Group of the Society of Chemical Industry*.

Stratfull, R. F. (1959). Progress report on inhibiting the corrosion of steel in a reinforced concrete bridge. *Corrosion, 15*(6), 65–68. Available from https://doi.org/10.5006/0010-9312-15.6.65.

Scholten, W. üllen Wvan (1930). *Das Gas- und Wasserfach, 73*, 403–404.

Wyatt, B. S., & Irvine, D. J. (1987). Review of cathodic protection of reinforced concrete. *Materials Performance, 26*(12), 12–21.

19

Anodic protection

19.1 Introduction

Anodic protection (AP) is a corrosion-reduction method that involves polarizing a metal into its passive area and keeping it there. When coating and cathodic protection measures are insufficient to preserve a structure from corrosion, AP may be explored. In aggressive media such as sulfuric and phosphoric acids, sulfite digesters, and aqueous ammonia, it is typically employed on carbon steels, stainless steels, nickel alloys, and titanium. In extreme conditions, such as sulfuric acid and mild solutions, the AP method is typically quite successful (Wagner, 1965; Faraday, n.d.).

The most important criterion for AP is that the material is to be protected as an active–passive behavior, but the passive potential range must be large enough, and the passive current must be low enough in comparison to the corrosive current. As a result, a potentiostat can provide a potential in the passive zone of a polarization diagram, but potential monitoring is required; otherwise, a potential variation can cause the material to become active, resulting in the loss of AP. This method involves attaching the anode to the positive terminal of an electrochemical circuit to safeguard it. The negative cathode, on the other hand, is formed of steel or graphite. In cathodic protection, on the other hand, the structure that has to be protected becomes the cathode (positive).

Only material–environment combinations with relatively large passive zones are suitable for AP. Type 304 stainless steel in phosphoric acid, titanium in sulfuric acid, steel, and stainless steel are only a few examples. The anodic polarization has the effect of putting the steel into a state of passivity. This encourages the creation of a protective coating on the steel surface, which prevents corrosion in general. Corrosion and hydrogen evolution is caused by insufficient or excessive polarization. Only wetted surfaces can be protected, similar to cathodic protection; this eliminates condensed films or droplets on vessel roofs and walls above the liquid line. Corrosion attack in these regions is common.

Edeleanu (1960) proved the viability of AP on small-scale stainless steel boilers used for sulfuric acid solutions for the first time in 1954. This method was created utilizing electrode kinetics principles and is difficult to explain without introducing sophisticated electrochemical theory topics. AP is based on externally delivered anodic currents forming a protective film on metals. AP has a number of advantages. The applied current, for example, is normally equal to the corrosion rate of the protected system. As a result, AP not only shields but also provides a direct way of monitoring the corrosion rate of the system. AP can significantly reduce the rate of corrosion. AP's main benefits are its ability to work in severely corrosive situations and its low current needs.

Handbook of Corrosion Engineering. DOI: https://doi.org/10.1016/B978-0-323-95185-2.00019-8

The most prevalent use of AP is the protection of steel in H_2SO_4 ($>78\%$ concentration) storage containers. One of the most profitable initiatives for this technology has been the sale of anodically shielded heat exchangers used to cool H_2SO_4 production facilities. However, there is minimal effort being made to develop applications to protect metals against corrosion by other chemicals. AP is employed to a lower extent than the other corrosion control strategies. This is due to the limitations of AP's ability to decrease corrosion in metal–chemical systems. Furthermore, if suitable controls are not established, equipment corrosion might be accelerated.

19.2 Passivity of metals

If the rate of anodic dissolution in a corrosive environment reduces as the applied potential approaches more noble levels or as the concentration of an oxidizing agent in the solution increases, the metal is in a passive condition (Wagner, 1965). Faraday was the first to notice that following a rapid dissolution in concentrated nitric acid, iron had significant resistance to corrosion. The passive state of iron was characterized by the metal's observed stability in concentrated nitric acid (Faraday, n.d.). By interacting with oxidizing substances like nitric acid, an oxidizing redox pair like $Fe^{3+}|Fe^{2+}$, or anodic polarization, the metal becomes passivated. Fig. 19–1 shows a typical anodic polarization curve of an active–passive metal, for electrochemical parameters; corrosion potential (E_{corr}) = 0.5 V versus Standard Calomel Electrode (SCE); corrosion current (I_{corr}) = 1×10^{-4} A/cm^2; $b_a = 0.05$ V/decade; passivation potential (E_{pp}) = 1×10^{-5} A/cm^2; and potential in the transpassive region (E_{tr}) = 1.0 V versus SCE.

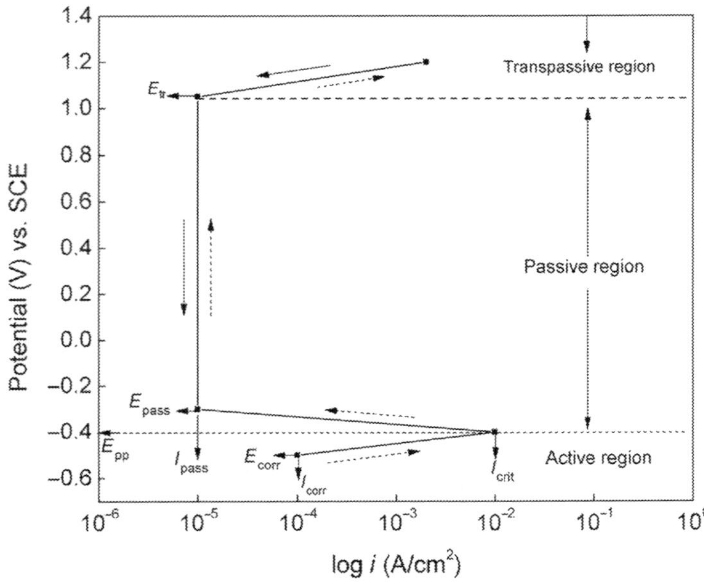

FIGURE 19–1 Potentiostatic polarization curves (forward and reverse scan).

The polarization curve of an active–passive metal has active, passive, and transpassive portions. In the active region, the corrosion potential and corrosion current are controlled by the Tafel kinetics of the individual redox reactions. An increase in the anodic potential or increase of solution oxidizing power moves the system into the passive region, in which the corrosion current is low and independent of the potential. The onset of the passivation region is defined by the passivation potential, where a sufficient critical current density is provided for the formation of the passive film. The thickness and the composition of the passive film are determined by the potential, time, temperature, and composition of the electrolyte. In the passivation region, the corrosion rate is equal to the passivation current. The corrosion rate starts to increase again in the transpassive region due to the oxygen evolution reaction or the breakdown of the passive film. The corrosion rate of the active–passive metal significantly decreases by shifting the corrosion potential to anodic potentials in the passive region by applying the necessary current using an external power source. The impressed anodic current must be greater than the critical current density for passivation. Impressed anodic potential also maintains the passivation current above the passivation potential. Once the system is anodically protected, a small current density corresponding to the passivation current is necessary to maintain the passivity. The polarization behavior of the substrate is used to optimize the AP parameters. The surface of the active–passive alloy is easier to passivate by an impressed anodic current when the active–passive metal has a lower critical current density and a higher cathodic passivation potential. A desirable active–passive metal should have a passive region with a large voltage range and a low corrosion rate in the corrosive solution. The optimum applied anodic potential should be midway in the passive region. Otherwise, in the case of an alloy with a small voltage range of the passive region, a slight variation of the applied protection potential may cause a transition from passive to active corrosion and result in a drastic increase in corrosion rate.

19.2.1 Basic principle of the passivation

Passivation can be achieved by developing an oxide layer (metallic coating) on the anode surface as reacting ions accumulate. As a result of the increased electric resistance of the metallic coating, the current flowing from the anode to the electrolyte decreases.

From Fig. 19–2, it is clear that the anodic electrochemical process in the active zone is metal oxidation. The Tafel kinetics of individual redox reactions describes the corrosion potential and current in this area. The standard electrode potentials and exchange current densities of the different redox processes govern the corrosion rate (I_{corr}), and the corrosion potential (E_{corr}). The reduction of the oxidizer and hydrogen or oxygen evolution processes are cathodic polarization reactions. The corrosion current increases in the active state as the corrosion potential is polarized more. The current grows exponentially with the applied potential in this area, and the semilogarithmic plot shows a straight line (Fig. 19–2). Further increases in potential result in a sharp drop in corrosion current in the passive zone. A sufficient critical current density (I_{crit}) is delivered to the system at the passivation potential (E_{pp}) to produce a passive film on the surface. The current stops growing at this point and begins

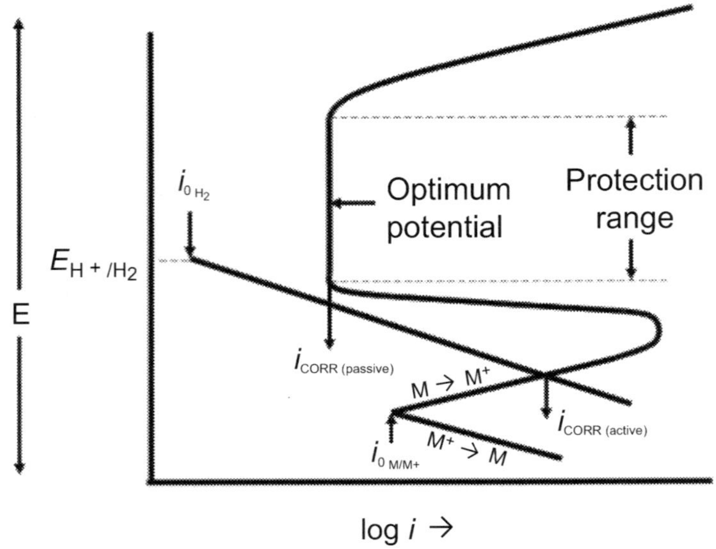

FIGURE 19–2 Schematic diagram showing the protection range and optimal potential for anodically protecting an active–passive metal.

to decrease toward the passivation current (I_{pass}), which is many orders of magnitude less than the active surface current. The thickness and composition of the passive film are determined by the electrolyte composition, voltage, time, and temperature. The prepassivating state is the portion of the potentiostat curve in Fig. 19–2 that connects the active and passive states. In this state, a thick insulating coating forms on the stainless steel surface. This layer begins to disintegrate over time, leaving a much thinner, more corrosion-resistant passive film on the surface. The passivation qualities of the active–passive metal surface are determined by the critical current density and the passivation potential. When the active–passive metal has a lower critical current density and a larger negative passivation potential in the electrolyte, the surface is easier to passivate. The current increases in the transpassive area as a result of the oxygen evolution process or the breakdown of the passive film.

19.2.2 Criterion of passivation

Increases in the oxygen diffusion limiting current result in a rise in the passivation potential of the active–passive film, as seen in Fig. 19–3. At this potential, the system receives a sufficient anodic critical current density of 10^2 A/cm^2, causing the corrosion rate to drop to the passivation current density of 10^4 A/cm^2, leading to the creation of a passive oxide film on the surface. The corrosion potential and corrosion current of the passive portion of the active–passive metal are unaffected by rotational velocity when the cathodic limiting current is larger than the anodic critical current density. The passive film of an active–passive alloy is stable only when the cathodic current density or oxygen diffusion limiting current density is larger than the

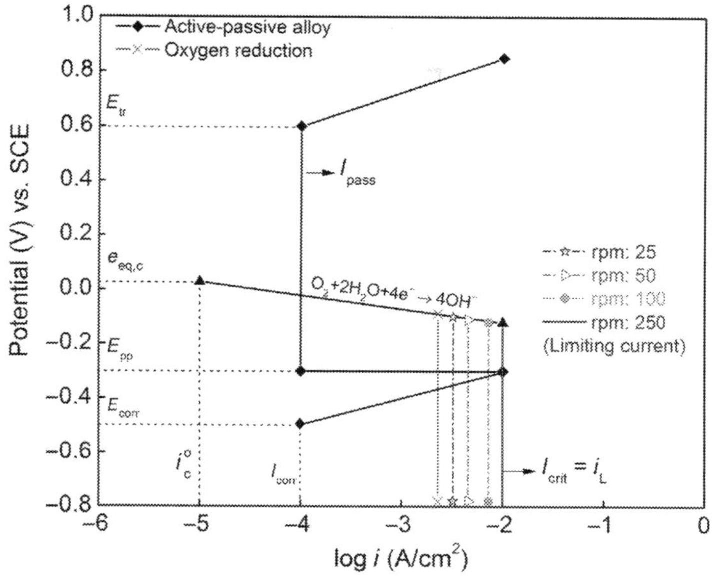

FIGURE 19–3 Effect of solution velocity on corrosion of active–passive alloy in deaerated and aerated alkaline solution (pH = 11).

anodic critical current. One of the goals of corrosion engineering is to figure out how to make an active–passive alloy that is easier to passivate, which should have a very low critical current density and a highly negative (cathodic) passivation potential.

19.2.3 Theories of passivity

In order to explain the nature of the passivity, two theories, that is, film-formation and adsorption theories have been put forward (Walker & Ward, 1969).

19.2.4 Film-formation theory

The film theory assumes the formation of a metal oxide thin film that acts as a barrier layer protecting the underlying metal from interacting with the corrosion medium. The thickness of the oxide layer is estimated to be in the range of several monolayers, proving the molecular nature of the passive films (Sato & Okamoto, 1981; Sato, 1990). Oxide passive films are amorphous when thin, but partially crystalline when the film grows (Leach & Pearson, 1988). According to their structure and composition, Fehlner and Mott (1970) and Barr (1978) divided the passive oxide films into two groups: network glass oxides consisting of Al, Ti, Zr, Mo, or Si, and network modifiers Pt, Ni, Cu, Co, or Fe. The network glass oxides are characterized by the formation of passive oxide films in one oxidation state, such as Al/Al_2O_3 or Ti/Ti_2O_3. The passive oxide films form compacts and completely dehydrated oxide layers with no impurities from the electrolyte. The oxides grow by inward diffusion, forming passive layers with excellent barrier properties and very low conductivities. In the network modifiers, the metals at high anodic potentials form mixed stable oxide films that are combinations of low- and high-oxidation state oxides. The lower valence oxides usually have poor barrier properties. Examples of such mixed oxides are $Cu/Cu_2O/CuO$ and $Co/CoO/Co_2O_3$. The poor corrosion-resistance properties of these mixed oxides result from the outward metal ion diffusion growth, which incorporates anions and impurities from the electrolyte (Leach & Pearson, 1988).

Lead, copper, and platinum metal oxides act as electrodes of the second kind. The passivation and activation potentials of these metals are equal to the reversible potentials of their corresponding metal oxide electrodes, proving the validity of film theory. When immersed in the solution, they have the same electrolyte anion in the surface oxide of the electrode. However, the reversible Flade potentials calculated using the Gibbs free energy of formation of barrier layers of FeO and Fe_2O_3 are 0.051 and 0.085 V, respectively, which are not consistent with the experimentally measured value of the Flade potential of 0.6 V versus Standard Hydrogen Electrode (SHE) (Uhlig, 1958). A Flade potential of 0.57 V versus SHE for iron oxides, close to the experimentally measured value of 0.6 V versus SHE, is calculated by substituting enthalpy for O_2 (H = -75 kcal/mol) and entropy values for the chemisorption of oxygen on Fe (S = -46.2 cal/mol), as shown in following equations:

$$2FeO + 4H^+ \rightarrow 2Fe + 2H_2O \tag{19-1}$$

$$2Fe + O_2 \rightarrow 2FeO \tag{19-2}$$

$$G(2FeO) = H(2FeO) - TS(2FeO) \tag{19-3}$$

19.2.5 Adsorption theory

The adsorption theory is based on the decrease in the metal dissolution rate by adsorbed oxygen atoms (Uhlig, 1958; Evans, 1927). The monomolecular passive layer formed by adsorbed oxygen atoms inhibits the hydration of metal ions by increasing the overvoltage of the anodic reaction, thus electrochemically retarding the metal dissolution. Initial partial coverage of the metal surface with oxygen causes an abrupt change in metal dissolution resulting in an increase in polarization and a shift in potential in the anodic direction. As a result, as the positive surface charge rises, circumstances emerge that allow a continuous monomolecular layer of oxygen to cover a larger area. This selective adsorption method constantly builds the passive layer while reducing metal dissolution. The deposited oxygen layer does not operate as a barrier film; rather, it increases polarization resistance, lowering the exchange current density of the metal dissolution reaction. The uncoupled electrons in their "d" shell orbital generate the high adsorption of oxygen on transition metals (Uhlig, 1976). In addition to ionic bonding, the oxygen with unpaired electrons forms a covalent link with the d orbital of the transition metals. Passivation of stainless steel and chromium happens through chemisorption of oxygen from solution or directly from the air, due to the high energies required for oxygen adsorption on transition metals.

19.2.6 Methods for spontaneous passivation

Many technologies and industries rely on the usage of high-corrosion-resistant metals and alloys, and passivation is critical to their success. During the first part of the 20th century, the development of stainless steel hastened the development of contemporary technologies in the transportation, building, chemical, and pharmaceutical industries. The improved passive films produced on silicon were crucial in the development of semiconductor devices. Engineers can create new materials with enhanced corrosion characteristics thanks to a greater knowledge of the nature of passivity and the development of novel procedures for creating passivity in metallic materials. The composition of an active–passive alloy that is more easily passivated is a primary goal of corrosion engineering. This alloy should have a lower critical current density and a greater negative (cathodic) passivation potential. An increase in the oxygen diffusion limiting current, for example, results in a change in the corrosion potential to the passivation potential, resulting in spontaneous passivation. At this potential, the system receives enough anodic critical current density to reduce the corrosion rate to the passivation current density, resulting in the formation of a passive oxide film on the surface. The limiting diffusion current density of oxygen in air-saturated solutions is 100 $\mu A/cm^2$. Some active–passive metals have a lower critical current density than 100 $\mu A/cm^2$, and the oxygen dissolved in the solution causes spontaneous passivation on their surfaces. Surface pretreatments can be used to reduce critical current density and passivate stainless steel, nickel, and titanium spontaneously. The corrosion potential of an active–passive metal is shifted from active to passive by impressed anodic potential, which keeps the passivation current above the passivation potential. Iron in 1 N sulfuric acid has a threshold current density of 105 A/cm^2 and cannot be passivated in acidic solutions, as shown in Fig. 19–4 (Steigerwald & Greene, 1962).

In 1 N sulfuric acid, adding 10.5% chromium to iron reduces the critical current density to 104 A/cm^2, which is still quite high. In the case of 304L stainless steel (18Cr–8Ni), the addition of chromium and nickel to iron reduces the passivation critical current density to 100 A/cm^2 (Steigerwald & Greene, 1962; Fontana & Greene, 1968). The anodic polarization curves obtained for steels with chromium content ranging from 3.54% to 19.20% (Fig. 19–5) demonstrate that increasing chromium content improves the Fe-Cr-Ni alloy passivation performance by lowering the critical current density from 10^6 to 10^2 A/cm^2 (Osozawa & Engell, 1966). In sulfuric acid, the passive current drops from 50 A/cm^2 for a 3.54% Cr alloy to 10^2 A/cm^2 for a 19.2% Cr alloy.

Fig. 19–6 shows a schematic description of the alloying metals that alter the anodic polarization curve of stainless steel. The addition of 8% nickel to an alloy with 18% chromium results in the austenitic structure of SS Type 304. The addition of Mn and N to austenitic steel improves its stability. The anodic polarization curves are affected by the chromium

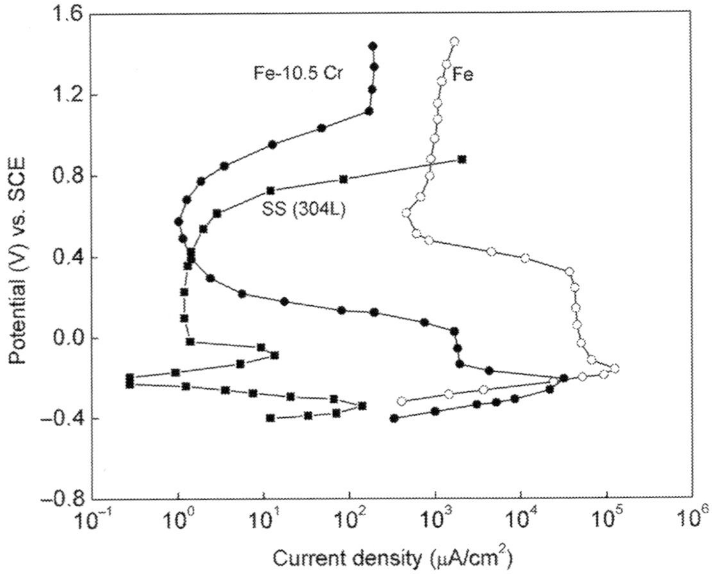

FIGURE 19–4 Anodic polarization curves for Fe-10.5%C and SS (304L) in 1 N sulfuric acid.

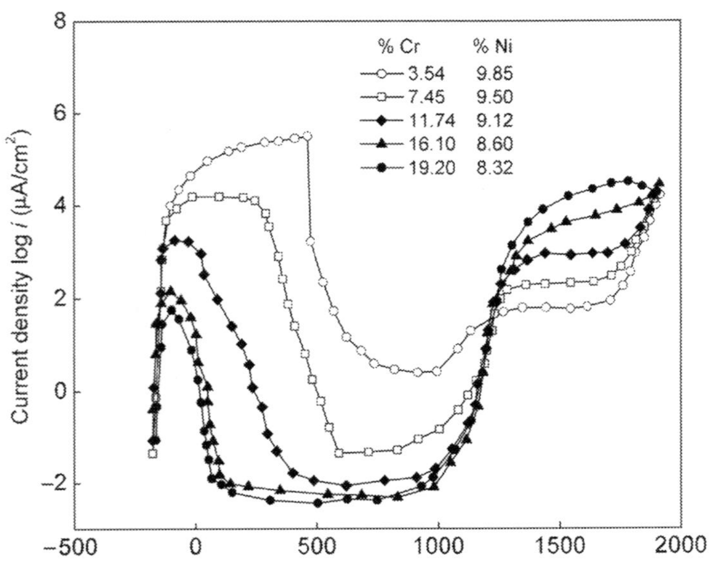

FIGURE 19–5 Anodic polarization curves for Fe-Cr-Ni alloys with different contents of Cr and Ni.

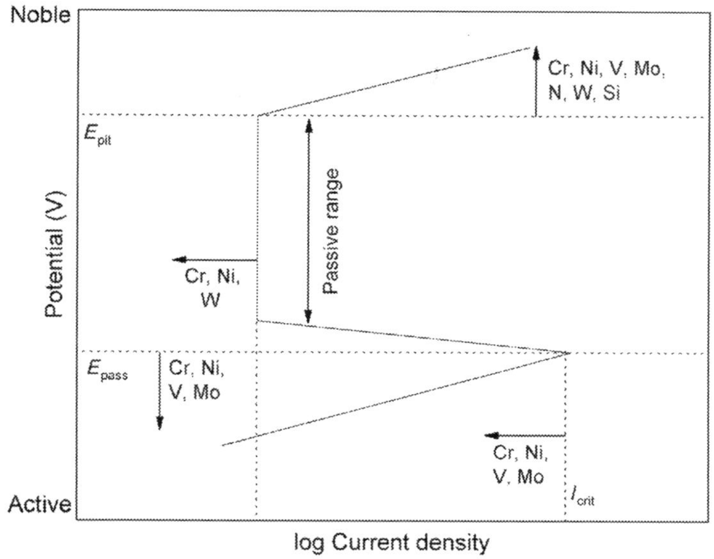

FIGURE 19–6 Effect of alloying elements on stainless steel anodic polarization curve.

concentration of stainless steel, as illustrated in Fig. 19–6. In a corrosive environment, nickel promotes re-passivation; however, concentrations of more than 30% diminish the passivation current and the critical current density and raise the critical pitting potential. The addition of nitrogen to austenitic stainless steel, up to 0.2 wt.%, stabilizes and promotes the strength of the austenitic phase (Sedriks, 1979; Craig & Anderson, 1995; Gunn, 1997). In chloride electrolytes, molybdenum in conjunction with chromium improves the corrosion resistance of ferritic stainless steel and is useful in reducing pitting and crevice corrosion. Cr-Ni-Mo-Cu alloys improve passivity in sulfuric acid concentrations ranging from 20% to 70%. The corrosion resistance of an alloy with a nickel concentration of more than 30% increases in 75% sulfuric acid (Pierre & Roberge, 2000). Pitting, chloride stress corrosion cracking, crevice corrosion, intergranular corrosion, and overall corrosion resistance are all excellent in acidic conditions for alloys containing 29Cr–4Mo–2Ni. These alloys are very ductile, weldable, and strong (Pilling & Ackerman, 1929).

19.2.7 Factors affecting passivity

Solution pH, velocity, and oxidizer concentration change the properties of the anodic curve of the active–passive metal. For example, the equilibrium potential of the cathodic reaction shifts according to the Nernst equation in the noble direction by increasing the oxidizer concentration. Mixed potential theory, in this case, may predict the intersection of the cathodic and anodic Tafel lines and corrosion rate or extent of passivation of the metal.

19.3 Equipment required for anodic protection

Fig. 19–1 depicts the impressed current AP system of a steel storage tank storing sulfuric acid. Experimental polarization measurements are used to determine the design parameters and operating conditions for AP. The storage tank is protected by the power supply, which maintains a constant voltage between the tank and the $Hg/HgSO_4$ reference electrode. The applied voltage is the total of the cathodic overvoltage, anodic overvoltage, and ohmic losses of the structure.

As AP is a potential-control technique in which the power supply is a potentiostat capable of supplying a constant potential, the potential E is assumed to be constant; otherwise, a significant change in potential causes a change in the current, and the structure may become unprotected due to film breakdown. Not only does a good anodic protection design necessitate a regulated potential/current, but it also necessitates a high-quality passive film that is insoluble in the aggressive solution. A schematic representation of the anodic protection system for the sulfuric acid storage tank is represented in Fig. 19–7.

The basic components of the anodic protection system are discussed below:

19.3.1 Cathode

The cathode should be a permanent-type electrode that is not dissolved by the solution or the currents impressed between the vessel wall and electrode. The cathodes used in most of the first applications of anodic protection were made of platinum-clad brass. These electrodes were excellent electrochemically but were costly, and the active area contacting the solution was limited by this cost. Because the overall resistance is a direct function of current density, it is advantageous to use large surface area electrodes. Many other, less costly metals have been used for cathodes instead of these costly materials, and some of these metals are listed in Table 19–1, with the chemical environments in which they were used. The electrode size is chosen to conform to the geometry of the vessel and to provide as large a

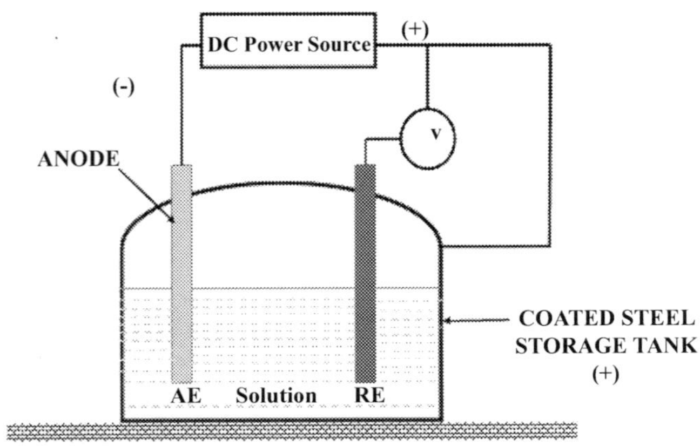

FIGURE 19–7 Schematic representation for the anodic protection system for sulfuric acid storage tank.

Table 19–1 Cathode materials used in the field installations.

S. no.	Environment	Cathode metal
1.	Sulfuric acid of various concentration	Platinum-clad brass
2.	Kraft pulping liquor	Steel
3.	H_2SO_4 (78%–105%)	Chromium nickel steel
4.	H_2SO_4 (89%–105%)	Silicon cast iron
5.	Hydoxylamine sulfate	Copper
6.	Liquid fertilizers (nitrate solution) Sulfuric acid	S30400
7.	Chemical nickel-plating solution	Nickel-plated steel
8.	Liquid fertilizers (nitrate solution) Sulfuric acid of various concentrations Kraft digester liquid	Hastelloy C

Table 19–2 Reference electrodes used for anodic protection installations.

S. no.	Environment	Electrode
1.	Sulfuric acid Kraft digester	Calomel
2.	Sulfuric acid Kraft solution Fertilizer solution Sulfonation plant	Ag/AgCl
3.	Sulfuric acid solution Hydroxylamine sulfate	$Hg/HgSO_4$
4.	Sulfuric acid solution	Pt/PtO
5.	Alcohol solution	Au/AuO
6.	Sodium carbonate solution Kraft digester Green or black liquors	Mo/MoO_3
7.	Sulfuric acid solution	Platinum
8.	NH_4OH	Bismuth
9.	Fertilizer solution Sulfuric acid solution	S31600 Steel
10.	Fertilizer solution Nickel plating solution	Nickel
11.	Fertilizer solution	Silicon

surface area as possible. The location of the cathode is not a critical factor in simple geometry, such as storage vessels, but in heat exchangers, it is necessary to extend the electrode around the surface to be protected. Multiple cathodes can be used in parallel to distribute the current and to decrease circuit resistance.

19.3.2 Reference electrode

The voltage of the vessel to be protected must be precisely regulated, reference electrodes must be employed in anodic protection systems. The reference electrode must have an electrochemical potential that remains constant throughout time and is unaffected by temperature or solution composition changes. For anodic protection, a variety of reference electrodes have been utilized, including those indicated in Table 19–2. Because of its brittle nature, the reference electrode has been a source of numerous issues in anodic protection systems.

19.3.3 Potential control and power supply

With one exception, the DC power supply used in anodic protection systems has the same design and criteria as the rectifiers used in cathodic protection systems. The currents necessary to keep the potential of the vessel wall in the passive range might become very minimal due to the nature of the vessel's active–passive behavior. Some DC power supply designs must be adjusted to lower the minimum amount of current that the power supply can output. In anodic protection installations, the potential control serves two purposes. The potential must first be measured and compared to a target value. The power supply must then receive a control signal to drive the DC between the cathode and the vessel wall. Because of the high cost of electronic circuitry, this control function was done in an ON–OFF manner in early systems. Because of recent advances in power electronics, all systems now have continuous proportional-type control. Because most installations are built-in chemical facilities, packaging these electrical components occasionally necessitates unique considerations. Some facilities need explosion-proof enclosures, while others necessitate chemically resistant enclosures.

19.4 Design concerns

An understanding of the system's fundamental electrochemical behavior and the geometry of the equipment to be protected, as well as any particular operational conditions, is required when designing an anodic protection system. The voltage at which the vessel must be maintained for corrosion protection, the current necessary to establish passivity, and the current required to maintain passivity are the electrochemical parameters of concern. Polarization curves may be used to establish the electrode potential, and the needed currents can be approximated using the polarization data. As the current is so heavily time-dependent, its fluctuations over time must be carefully evaluated. The greatest source for this sort of information is empirical data from field installations. Estimating a system's solution resistivity requires special care and attention since it is crucial in calculating the total circuit resistance.

To cut running expenses, the supply of the DC power needs should be as low as feasible. The solution resistivity should generally be low enough that the cathode surface area controls the circuit resistance. A throwing power of the system or the capacity of the applied current to achieve the desired value over complicated geometries and varying distances, is critical. Anodic protection can easily provide a uniform distribution of potential over a regular-shaped passivated surface.

Surface imperfections, such as the recessions surrounding sharp slots, grooves, or fissures, are far more difficult to protect since the appropriate current density cannot be generated in these locations. This imperfect passivation has the potential to be disastrous. This problem may be solved by either designing the surface to prevent these imperfections or selecting a metal or alloy that is easy to passivate and has a low critical current density. A current of 5000 A, for example, may be required for a storage tank with a surface area of

1000 m^2. As a result, one of a few strategies (Walker, 1994) must be used to prevent these extremely high currents. It may be conceivable and practicable, for example, to reduce the critical current density by lowering the temperature of the electrolyte before passivating the metal. If a vessel has a modest floor area, it can be treated in stages by passivating the base, then the lower portions of the walls, and finally the top regions of the walls; however, this method is not practicable for big storage tanks with a significant floor area. The following are some of the prerequisites for designing anodic protection systems:

1. The potentiodynamic polarization curve should be used to calculate the polarization potential. The protective potential's fluctuation over time must be correctly measured.
2. Avoid polarizing the system at a high rate through the passivation potential (E_{pass}) because of a drastic increase in critical current density (I_{crit}) at a high polarization rate.
3. In the corrosive solution, the cathode should be inert and electrochemically stable.
4. To prevent hydrogen permeability and embrittlement, the cathode materials must have a low hydrogen overvoltage and a small surface area.
5. The resistivity of the solution should be very low. The entire system resistivity should be controlled by the cathode surface area.
6. Surface imperfections in the structure should be avoided in the system design to avoid a non-uniform current distribution.
7. The critical passivation current density of the active–passive metal or alloy should be low.

19.5 Applications

Stainless steel in phosphoric acid, steel in sulfuric acid, nickel in nitric acid, nickel alloys in nitrate solutions, and titanium in ammonia solution are some of the alloys that are protected by anodic protection. Anodic protection has recently been employed to protect stainless steel heat exchangers used in sulfuric acid manufacturing (Riggs, 2012). Sudbury and Locke (1965) reviewed the current requirements for some of the steels that have had anodic protection added to them. Anodic protection has also been employed in the production, storage, and transportation of sulfuric acid in plants. Iron dissolution in sulfuric acid is reduced to 1 ppm every day of storage. During the transportation of sulfuric acid in trucks, portable anodic protection devices are employed to prevent iron dissolution (Riggs, 2012).

Storage containers, process reactors, heat exchangers, and transportation vessels that hold corrosive solutions have all been treated with anodic protection. The bulk of anodic protection applications involves the production, storage, and transportation of sulfuric acid, which is generated in more significant quantities than any other chemical on the planet. In terms of vessel life, storage of 93% H_2SO_4 and higher in low-carbon steel vessels has shown to be success. Portable anodic protection has been employed for such vessels to preserve acid purity and increase storage time (Riggs, 2012). Anodically protected heat exchangers have developed a sizable market as a replacement for cast iron coolers. Shell and tube, spiral, and plate-type heat exchangers have been supplied with anodic protection as a standard feature. Anodic protection was used to handle sulfuric acid with a concentration of 96% to

98% at temperatures up to 110°C in S31600 stainless steel heat exchangers. Corrosion rates have been lowered from more than 5 mmy^{-1} unprotected to less than 0.025 mmy^{-1} protected, resulting in significant cost savings due to increased equipment life and higher-purity acid generated by employing these protected heat exchangers. Anodically shielded heat exchangers have also been used to handle a variety of other corrosive systems (Riggs, 2012). When the rate of corrosion of steel in acid is reduced, the formation of hydrogen is reduced as well, which has been cited as a cause of explosions in phosphoric acid systems (Walker, 1994). The presence of hydrogen can cause blisters to develop at metal inclusions and grooving on vertical surfaces. Anodic protection has been proven to reduce hydrogen production, hence reducing such issues considerably.

Short questions

1. What is anodic protection method of corrosion prevention and control?
2. What are the applications of the anodic protection method?
3. What are the limitations associated with the anodic protection method?
4. Explain the phenomenon of passivation and the principle involved.

References

Barr, T. L. (1978). An ESCA study of the termination of the passivation of elemental metals. *Journal of Physical Chemistry, 82*(16), 1801–1810. Available from https://doi.org/10.1021/j100505a006.

Craig, B. D., & Anderson, D. S. (1995). *Handbook of corrosion data.* ASM International.

Edeleanu, C. (1960). Corrosion control by anodic protection. *Platinum Metals Review* (3), 86–91.

Evans, U. R. (1927). CXL.—The passivity of metals. Part I. The isolation of the protective film. *J. Chem. Soc,* 1020–1040. Available from https://doi.org/10.1039/JR9270001020.

Faraday, M. *Faraday, experimental researches in electricity.*

Fehlner, F. P., & Mott, N. F. (1970). Low-temperature oxidation. *Oxidation of Metals, 2*(1), 59–99. Available from https://doi.org/10.1007/BF00603582.

Fontana, M. G., Greene, N. D., & Klerer, J. (1968). , Corrosion engineering. *Journal of the Electrochemical Society, 115*(5), 142C. Available from https://doi.org/10.1149/1.2411256.

Gunn, R. N. (1997). *Duplex stainless steel.* Abigton Publishing.

Leach, J. S. L., & Pearson, B. R. (1988). Crystallization in anodic oxide films. *Corrosion Science, 28*(1), 43–56. Available from https://doi.org/10.1016/0010-938X(88)90005-4.

Osozawa, K., & Engell, H. J. (1966). The anodic polarization curves of iron-nickel-chromium alloys. *Corrosion Science, 6*(8), 389–393. Available from https://doi.org/10.1016/S0010-938X(66)80022-7.

Pierre, R., & Roberge, P. R. (2000). *Handbook of corrosion engineering.* McGraw-Hill.

Pilling, N. B., & Ackerman, D. E. (1929). Resistance of iron-nickel-chromium alloys to corrosion by acids. *Transactions of the AIME., 83,* 248–279.

Sato, N.; & Okamoto, G. (1981). *Electrochemical passivation of metals* (pp. 193–245). Springer Science and Business Media LLC. Available from https://doi.org/10.1007/978-1-4757-4825-3_4.

Riggs, O. (2012). *Anodic protection: Theory and practice in the prevention of corrosion*. Springer Science & Business Media.

Sato, N. (1990). An overview on the passivity of metals. *Corrosion Science, 31*(C), 1–19. Available from https://doi.org/10.1016/0010-938X(90)90086-K.

Sedriks, A. J. (1979). *Corrosion of stainless steels*. John Wiley.

Steigerwald, R. F., & Greene, N. D. (1962). The anodic dissolution of binary alloys. *Journal of the Electrochemical Society, 109*(11), 1026. Available from https://doi.org/10.1149/1.2425230.

Sudbury, J. D., & Locke, C. E. (1965). Anodic control of corrosion. *Proceedings, American Petroleum Institute - Section III, 43,* 115–119.

Uhlig, H. (1958). The adsorption theory of passivity and the Flade potential. *Z. Electrochem, 62,* 626–632.

Uhlig, H.H. (1976). Passivity and its breakdown on iron and iron based alloys.

Wagner, C. (1965). Passivity and inhibition during the oxidation of metals at elevated temperatures. *Corrosion Science, 5*(11), 751–764. Available from https://doi.org/10.1016/S0010-938X(65)80003-8.

Walker, R. (1994). *Anodic protection* (10, pp. 155–170). Elsevier BV. Available from https://doi.org/10.1016/b978-0-08-052351-4.50088-1.

Walker, R., & Ward, A. (1969). The theory and practice of anodic protection. *Metallurgical Reviews, 14*(1), 143–151. Available from https://doi.org/10.1179/mtlr.1969.14.1.143.

20

Corrosion inhibitors

20.1 Introduction

A variety of corrosion control technologies are used to protect metals from corrosion. The objective of corrosion control techniques is to use the most effective and economical approaches to reduce corrosion to an acceptable level. Each has better complexity and goals. Corrosion control techniques require a broad understanding of the underlying corrosion mechanism. Every technique for managing corrosion seeks to bring corrosion down to a tolerable level so that the material can last the desired or anticipated amount of time. The employment of several chemical compounds as corrosion inhibitors is one of the most widely used methods for preventing internal corrosion (Bedair, Soliman, & Metwally, 2016).

In a corrosive environment, a compound known as a corrosion inhibitor efficiently slows down or stops the corrosion of exposed metal. To keep an inhibiting surface film in place, inhibitors are introduced in minute amounts to cooling water, acid, and steam. Some inhibitors obstruct the cathode reaction, some obstruct the anode reaction, and some obstruct both cathodic and anodic reactions. However, the majority of them are ineffective at retarding or preventing localized corrosion, attacks including crevice corrosion, pitting, or SCC, even though they are frequently employed to control general corrosion. A different mechanism is followed by corrosion inhibitors to prevent corrosion. It is common for inhibitors to form one of three interactions with metal surfaces: a passive film, a barrier in the form of a thin film of unreacted reaction products or inhibitors, or an adsorbed inhibitor film, which may only be a monolayer in thickness.

The corrosion inhibitors can be employed in all three forms and can be added to solid, liquid, or gaseous media. Concrete, organic coatings and coal slurries are examples of solid media. Liquids include water, aqueous solutions, and organic solvents. The atmosphere or water vapor are considered examples of gaseous media. They are selected based on their solubility or dispersibility in the solvent that corresponds to the media in which corrosion must be controlled or inhibited (Hoar & Holliday, 1953).

In the oil extraction, refining, and chemical sectors, corrosion inhibitors are the first line of defense. A corrosion inhibitor can also be found in antifreeze used in cars, as well as in a variety of other processes, including recirculation systems for water, chemical processing, water treatment, and additives (Morad, 2008; Revie & Uhlig, 2011).

The inhibitor molecules form a thin, protective film on the metal's surface, as was already indicated, which needs to be protected from corrosion. The resulting barrier lessens the interaction between the metal surface and the corrosive environment (McCafferty, Bernett, &

Handbook of Corrosion Engineering. DOI: https://doi.org/10.1016/B978-0-323-95185-2.00020-4

Murday, 1988). The efficiency of corrosion inhibitors is influenced by a number of variables. Amongst the several requirements, the principal prerequisites are:

- the capability to oxidize metal, thus resulting in the formation of a surface passive metal layer;
- large molecular structure, bond, and presence of the functional groups in order to donate the lone pair of electrons;
- ability to cover a large surface area of the metals;
- economic and cost-effectiveness;
- water solubility;
- nontoxic to man and nature, greenness, etc.

For inhibitors to work and, in some cases, to prevent corrosion from getting worse, a critical concentration must be reached or exceeded. As a result, it is vital to frequently verify the inhibitor additions and assess the corrosivity of the environment using techniques like linear polarization or coupon tests etc. The growth in publications in the corrosion literature reflects an increasing interest in the research of corrosion inhibitors. The major goal of this chapter is to lay the groundwork for a fundamental comprehension of corrosion inhibitors to develop a basic understanding of the field.

20.2 Types of inhibitors

Various types of inhibitors can be classified according to their mechanisms of inhibition, their applications (pickling, acid cleaning, descaling, etc.), and their chemical nature (organic or inorganic). Major classes are discussed comprehensively below (Samuel et al., n.d.):

20.2.1 Anodic inhibitors or passivating inhibitors

Anodic inhibitors raise anode polarization to the critical passivation potential of metal or alloys. The corrosion current is significantly reduced by these "passivating inhibitors," which is why they are called passivators. Passivators retards corrosion using a number of mechanisms:

- They slow down corrosion by stabilizing the passive film;
- They repassivate the metal if the protective film thus formed by the inhibitors got disfigured;
- Several passivators aid in the repairment of the protecting film, by developing compounds that get plugged into the pores formed in the film.
- They retard the adsorption of aggressive anions, such as CI^-, by following competitive adsorption of inhibitive anions.

The polarization and passivation effects of anodic inhibitors are depicted in Fig. 20–1A and B, respectively. These inhibitors are strong oxidizers that alter the corrosion potential of metals in a noble manner by forming a passive barrier film.

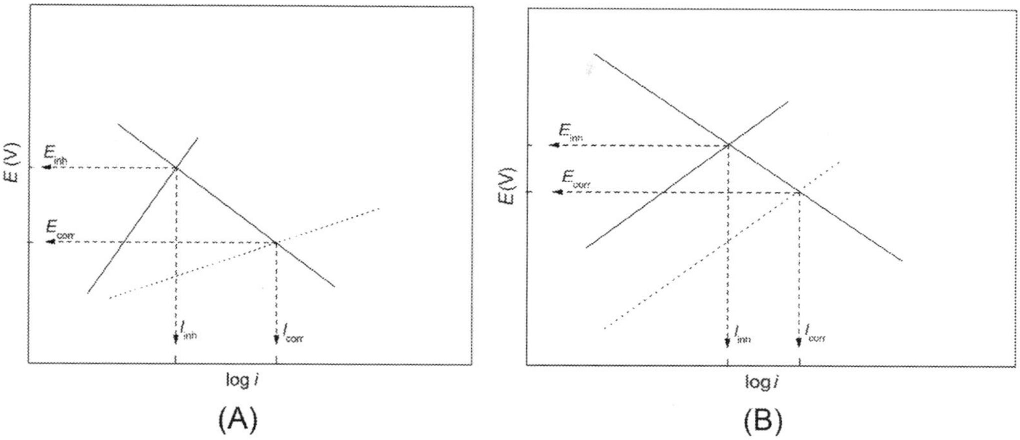

FIGURE 20–1 Polarization diagram depicting (A) the polarization effect of an anodic inhibitor and (B) the effect of anodic passivation.

Due to the possibility of pitting corrosion when the inhibitor concentration is insufficient, anodic inhibitors are sometimes known as "hazardous" inhibitors (below the critical value for oxidation). Nonoxidizing and oxidizing anions are two primary types of passivating inhibitors used to stop corrosion. Oxygen must be present for nonoxidizing anions such as tungstate, molybdate, and phosphate to passivate the metal. Polyphosphate, sodium benzoate, and sodium cinnamate are organic nonoxidizing inhibitors (Kuznetsov & Thomas, 1996). Through oxygen adsorption, they polarize the corrosion system in the passive zone. Passivation of steel without oxygen is achieved by oxidizing anions. Among the examples are chromate, nitrite, and nitrate. They frequently become reduced because they absorb electrons from metals. By depolarizing the cathodic curve in the anodic direction in the passive zone, sodium chromate intersects the anodic curve. Additionally, it formed an ionic connection and adsorbed in the anodic area. As a result of the addition of these inhibitors, the anodic process becomes polarized because it reduces the critical current density required to change the potential in the passive zone. The mixed inhibitor chromate changes the polarization curves of both the cathodic and anodic processes. The inhibitor's effects on the cathodic or anodic polarization process dictate how the corrosion potential shifts.

20.2.2 Direct passivators or oxidizers

Direct passivators, also called "oxidizers," have anions that interact with metal surfaces and integrate into passive layers, repairing, fortifying, and completing them. The best oxidizers are chromates (CrO_4^{2-}) and nitrites (NO^{2-}), which passivate steel even in deaerated solutions. To surpass the critical current density, molybdates (MoO_4^{2-}) and tungstate (WO_4^{2-}), which are less efficient, require more oxygen in the solution. Due to their potential to damage the environment and human being, these oxidizers cannot be employed. Nitrite decomposes to hazardous nitrosoamines, while chromium is carcinogenic. Microbes gradually break down these materials into harmless compounds. The following are some practical examples of where oxidizers have either been highly effective or ineffective:

Reinforced concrete: The only inhibitor that has been proven to work well when added to cement to preserve steel reinforcing is calcium nitrite.
Automotive antifreeze: All the metals in the cooling system—iron, aluminum, copper, brass, and solder—are protected by the addition of sodium molybdate when combined with other inhibitors.
Sour gas wells: Oxidizers got depleted and were entirely inefficient at reducing hydrogen sulfide to elemental sulfur.

20.2.3 Indirect passivators

Indirect passivators do not actually act as oxidizers; instead, they help dissolved oxygen better adhere to metal surfaces, where it can then passivate the metal. Only the presence of dissolved oxygen or an oxidizing inhibitor, such as nitrite or molybdate, in the environment, will cause these passivators to function. Alkaline substances known as indirect passivators produce a high concentration of OH^-, which combines with H^+ ions deposited on cathode sites to clean them so oxygen may get adsorbed. Indirect passivators made of both organic and inorganic materials work well. Inorganics most typically utilized are alkaline compounds, such as NaOH, Na_3PO_4, Na_2HPO_4, Na_2SiO_3, and $Na_2B_4O_7$ (borax). For instance, steel is routinely protected in water by the use of Calgon, a mixture of vitreous phosphates. Organic indirect passivators such as sodium benzoate and sodium cinnamate do not cause pitting even at high chloride ion concentrations. There is a natural rise in corrosion rates, but the corrosion occurs uniformly.

Although the inhibitor businesses have long benefited from it, the synergistic impact of a mixture of inhibitors has only recently come to common awareness.

20.2.4 Cathodic or precipitation inhibitors

Cathodic or precipitation inhibitors increase the cathodic depolarization overvoltage to slow the rate of corrosion (hydrogen evolution and oxygen reduction). These inhibitors cause thick deposits to accumulate and reduce the depolarizer's ability to diffuse to the metal surface, preventing corrosion and the cathodic reaction. Through the inhibition of cathodic polarization, cathodic inhibitors also lower the corrosion current. Cathodic polarization and passivation are depicted in polarization diagrams in Fig. 20–2A and B in the presence of cathodic or precipitation inhibitors. They are categorized based on the inhibitory mechanism as oxygen scavengers, cathodic poisons, and cathodic precipitates.

20.2.4.1 Cathodic poisons

Sulfides, selenides, arsenates, bismuth, and antimony are examples of cathodic poisons that adhere to or deposit on metal surfaces to slow down the rate of the hydrogen evolution

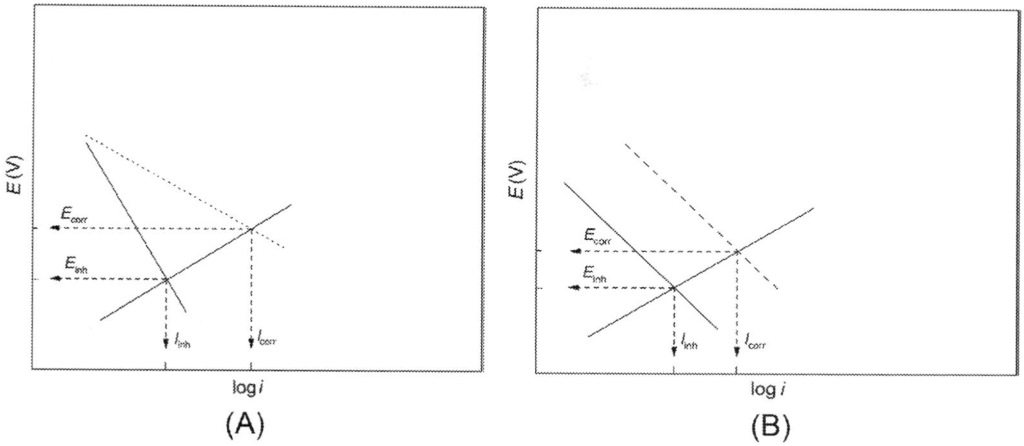

FIGURE 20–2 (A) Impact of cathodic inhibitors on the cathodic polarization behavior, and (B) the depiction of cathodic passivation in the presence of cathodic inhibitors.

reaction. These inhibitors prevent hydrogen from recombining, which may result in hydrogen blistering and make metals more vulnerable to hydrogen embrittlement.

20.2.4.2 Cathodic precipitators

With the precipitation of insoluble substances, cathodic precipitators promote cathodic site passivity. Cathodic precipitation inhibitors such as $CaCO_3$, $MgCO_3$, or zinc sulfates that precipitate as $Zn(OH)_2$ are frequently utilized. The only way to regulate the effectiveness of these inhibitors is to change the pH. The calcium carbonate of limestone dissolves as calcium bicarbonate in water $(HCO_3)_2$. Calcium carbonate barrier films made with care are smooth and durable. The pH must be carefully regulated after the precipitate has formed to prevent film dissolution at lower pH levels:

$$Ca^{2+} + 2HCO_3^- \rightarrow Ca(HCO_3)_2 \rightarrow CaCO_3 + CO_2 + H_2O \tag{20-1}$$

In the presence of oxygen,

$$O_2 + 2H_2O + 4e^- \rightarrow 4OH^- \tag{20-2}$$

Through a process in Eq. (20–3), the hydroxide ions formed by Eq. (20–2) speed up the precipitation of calcium carbonate:

$$Ca(HCO_3)_2 + OH^- \rightarrow CaCO_3 + HCO_3 + H_2O \tag{20-3}$$

For aerated cooling systems, polyphosphates work well as cathodic inhibitors (Boffardi, 1987). They are also referred to as polymer or condensed phosphates. By forming a barrier film with calcium ions, they form positively charged colloidal particles on the cathode

(Uhlig, Triadis, & Stern, 1955). This inhibitor has the disadvantage of hydrolyzing the phosphorus oxygen bond, thereby generating orthophosphate, which makes it a less desirable inhibitor.

20.2.5 Oxygen scavengers

Sulfite and hydrazine prevent the cathodic corrosion of steel, copper, and cast iron in oil production and steam generation by preventing oxygen cathodic reactions. These inhibitors work by interacting with oxygen to create salts and water, which lowers the concentration of dissolved oxygen. Both sulfur dioxide and sodium sulfite are frequently used oxygen scavengers in water. At low temperatures, the reaction between oxygen and sulfites is catalyzed by cobalt or manganese. In boiler steam water, sodium sulfite (Na_2SO_3), ammonium sulfite ($(NH_4)_2SO_3$), and hydrazine (N_2H_4) have been found to be effective oxygen scavengers (Baker & Marcy, 1956; Ongman, 1929).

High steam temperatures cause hydrazine to break down, releasing ammonia or nitrogen. New scavengers, such as ammonium isoascorbate and hydroquinone carbohydrazide, have been suggested since hydrazine promotes cancer. Examples of precipitation inhibitors that function by film formation include silicates and phosphates. They lessen the rate of metallic corrosion by combining the effects of cathodic and anodic polarization. The outcome is an increase in overvoltage for both oxygen reduction and hydrogen evolution. When oxygen is present in neutral solutions, these inhibitors act as anodic inhibitors and cause steel to passivate:

$$N_2H_4 + O_2 \rightarrow N_2 + 2H_2O \tag{20-4}$$

$$(NH_4)_2SO_3 + \frac{1}{2}O_2 \rightarrow (NH_4)_2SO_4 \tag{20-5}$$

$$Na_2SO_3 + \frac{1}{2}O_2 \rightarrow Na_2SO_4 \tag{20-6}$$

$$NH_4HSO_3 + \frac{1}{2}O_2 \rightarrow NH_4HSO_4 \tag{20-7}$$

20.2.6 Ohmic inhibitors

The formation of a thick, nonconducting film on the metal surface or a potential drop in the solution are two ways that ohmic inhibitors can increase resistance. Depending on where the film is deposited, they increase the overvoltage of anodic or cathodic reactions (Fig. 20–3).

20.2.7 Vapor phase inhibitors/volatile corrosion inhibitors

Some volatile substances called vapor phase corrosion inhibitors are added to a closed system to prevent corrosion. These inhibitors shield the metal by developing a barrier layer on its surface and have a high vapor pressure at room temperature. In general, the most

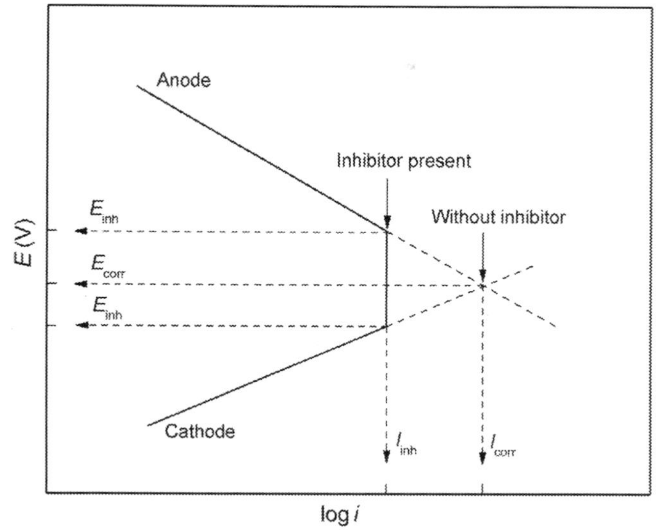

FIGURE 20–3 Illustration of the impact of the ohmic inhibitors with the help of polarization curve.

Table 20–1 Vapor phase inhibitors.

S. no.	Substance	Vapor pressure (mmHg)	Temperature (°C)	Melting point (°C)
1.	Benzylamine	1.0	29	–
2.	Cyclohexylamine carbonate	0.397	25.3	–
3.	Diisopropylamine nitrite	4.84×10^{-3}	21	139
4.	Dicyclohexylamine nitrite	1.3×10^{-4}	21	179
5.	Dicyclohexylamine caprylate	5.5×10^{-4}	21	–
6.	Guanadine chromate	1×10^{-5}	21	–
7.	Morpholine nitrite	3×10^{-3}	21	–
8.	Morpholine	8.0	20	–

efficient inhibition is provided by a weak volatile acid or base that hydrolyzes quickly. In boilers, which employ steam, volatile alkaline substances like octadecylamine and morphine are used to prevent corrosion in condenser tubes by neutralizing the acidic carbon dioxide. The vapor pressure of organic structures that have inhibitory anions like nitrite (NO^{2-}) and amine (NH^{2-}) substituted into them is higher, which increases their resistance to hostile ions in atmospheric corrosion (Estevão & Nascimento, 2001). As an adsorbed monolayer, volatile corrosion inhibitor (VCI) is thought to produce a weak link that shields the surface from water inclusion and harmful ions like sulfates and chlorides. The adsorbed monolayer modifies the kinetics of the anodic (metal dissolution) and cathodic reactions (Bastidas, Mora, & Feliu, 1990). Table 20–1 provides a summary of the vapor phase inhibitors utilized in corrosion protection.

20.3 Inorganic inhibitors

For ferrous alloys, inorganic inhibitors, including chromate, nitrite, nitrates, and arsenate, are highly effective. They create monolayer or multilayer passivating layers to stop the corrosion of stainless steel pits. For chromate, iron oxidation (anodic inhibitor) creates the film (Zehra, Mobin, & Aslam, 2022). Iron and chromium oxide make up the passive film.

Environmentally beneficial anodic inhibitors are nitrates. In alkaline settings (pH levels 9–10), they combine to create a passive coating that prevents the corrosion of copper, nickel, and tin alloys. However, aggressive ions like chloride and sulfate ions attack and destroy the barrier film. As seen in (Jones, 1984), they slow the rate of anodic dissolution on steel. Because they oxidize to nitrates when exposed to oxygen, nitrites are only employed in closed systems. They are not as effective as chromates as inhibitors.

The majority of inorganic inhibitors are hazardous to the environment. Molybdate inhibitors have been utilized in place of chromate-based inhibitors in open-recirculating cooling systems in neutral environments (Pryor & Cohen, 1953). Aluminum, galvanized steel, and zinc have all been protected against corrosion by nitrates (Cabral et al., 2006; Wang, Tang, Qiu, Gan, & Chen, 2005). The toxicity of tungstate is also extremely low, just like that of molybdates (Fujioka, Nishihara, & Aramaki, 1996).

20.3.1 Organic inhibitors

Numerous inorganic substances, including chromate, dichromate, nitrite, and nitrate, have been studied in-depth as inhibitors. These toxicity of several chemicals restricts their application in corrosion prevention. Organic heterocyclic compounds with polar groups and organic corrosion inhibitors with heteroatoms like sulfur, nitrogen, or oxygen are substitutes (Aslam, Serdaroglu et al., 2022). On the surface of the metal, these substances adsorb and forms a protective film (Trabanelli, 1970). In order to slow the rate of depolarization and completely cover the corroding metal's surface area with a thick film consisting of several monolayers, organic inhibitors modify the structure of the double layer at the metal-to-metal contact. By obstructing anodic and cathodic active sites or reducing the pace at which electroactive species are transported to or from the metal surface, they may also function as a barrier layer. Only for strong adsorption is a lower inhibitor concentration effective. The film is created as a result of the adsorption of positively or negatively charged soluble organic molecules. Organic substances that produce films may have cathodic, anodic, or mixed characteristics. Among the adsorption-based film-forming inhibitors are amines, imidazolines, quaternary, and acetylenic alcohols. Primary amines (RNH_2), secondary amines (R_2NH), tertiary amines ($R\text{-}N(CH_3)_2$), diamines ($R\text{-}NHCH_2CH_2CH_2NH_2$), amides ($R\text{-}CONH_2$), and polyethoxylated substances are examples of positively charged anodic inhibitors. Sulfonates are chemicals that have a negative charge.

The polar groups hydroxyl, nitrogen, sulfur, phosphorous, and selenium are a few examples of important polar groups. The charged ionic species initially form a physisorbed film due to van der Waals forces. Chemisorption is then used to stabilize the film. The position of the corrosion potential in relation to the metal's potential at zero charge (PZC) determines the surface charge of the metal. The term "overpotential" refers to the difference between the corrosion potential and the PZC. Cations are adsorbed at negative overpotentials, whereas anions are at positive overpotentials. Inhibiting organic ions or dipoles are attracted to the charged surface of the metal through electrostatic adsorption. The ions are not in direct touch with the metal during electrostatic adsorption. A layer of water molecules separates them. The electrostatic adsorption technique has very low activation energy and is virtually completely independent of temperature. During chemisorption, unshared electron pairs from organic molecules, also known as "p" electrons, engage with the metal orbitals to form a coordinate-type link. The contact can go on as long as the adsorbed molecules contain heteroatoms (P, Se, S, N, and O) with a single pair of electrons and/or aromatic rings (Aslam et al., n.d). Chelate is created when an electron is transferred from an organic chemical to metal, creating a coordinate covalent connection. The activation energy of chemisorption is larger than that of electrostatic adsorption. It is a slow-moving, temperature-dependent phenomenon that cannot be reversed. Because of the unsaturated bonds on the heteroatoms, organic molecules are chemisorbed onto the metal's surface (Zehra, Mobin, Aslam, & Parveen, 2017). They favor electronic binding to the metal substrate. The charge of the metal surface, the molecular mass, and the dipole moment all contribute to an increase in the adsorption of surface-active organic compounds. By raising the concentration of the inhibitor in the solution, the amount of inhibitor adsorbed on the surface rises. Adsorption bond strength, increased molecular weight, asymmetry, and electron density all increase inhibitor potency (Boulhaoua et al., 2021; Parveen, Mobin, & Zehra, 2016). The adsorption mechanism of anodically polarized surfaces and cathodically shielded surfaces are comparable. Imidazolines, amines, diamines, pyrimidines, and the fatty acid, naphthenic acid, and sulfonate derivatives of these chemicals are examples of inhibitors used in refinery operations (Aoufir & Zehra, 2020; Lgaz, Zehra, Toumiat, et al., 2019). The polar group of these molecules, which is electron-rich, adsorbs on the metal surface, while the nonpolar hydrocarbon component creates a hydrophobic coating that repels water. Inhibitors and cathodic protection of pipelines have been found to work in concert. When the pipeline is cathodically polarized when an organic inhibitor is present, an increase in efficiency is seen. Nitrogen-containing organic inhibitors, such as aryl and alkyl amines, aniline, imidazoline derivatives, and ketones, are used in industrial acid cleaning, acid-descaling, and acid-pickling (Lgaz, Zehra, Albayati, et al., 2019).

The inhibitory effect on steel polarization behavior while corroding in pickling acid is depicted in Fig. 20–4. Inhibitors were subsequently added to the pickling acid solution, which significantly reduced the corrosion current. For the preservation of steel, nitrite, and glycerophosphate work together effectively.

20.4 Corrosion inhibition in various environments

The majority of corrosion inhibitors application in aqueous, or partially aqueous systems are concerned with four key categories of environments:

1. In acid solution
2. In near-neutral water systems

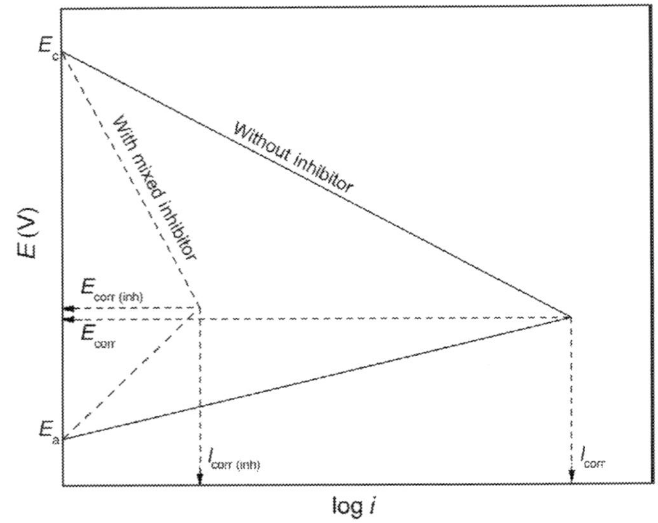

FIGURE 20–4 Effect of inhibitor on polarization diagram for steel corroding in pickling acid.

3. In oil and gas industries
4. In atmospheric and gaseous corrosion.

The following environment can be used to explain the corrosion inhibition mechanism:

20.4.1 Corrosion inhibition in acid solution

The most efficient inhibitors of metal corrosion in the acid solution are halide ions, CO, and organic compounds containing heteroatoms of elements from Groups V and VI of the Periodic Table (N, P, As, O, S, and Se). Additionally, organic molecules with numerous bonds, particularly triple bonds, can function well as inhibitors. The first stage of the inhibition in the acid solution is frequently thought of as adsorption onto the surface of the metals, which are usually free of any oxides in the acidic solutions that can impede the cathodic and/or anodic electrochemical corrosion reactions. Metals can be affected by acid solution corrosion inhibitors in a variety of ways, some of which may occur simultaneously and lead to a corrosive reaction. Additionally, because the mechanism may change under test conditions, applying an inhibitor to a single general method of action is not always necessary. How the corrosion inhibitors interact with one another in the acid solution depends on the nature of the metal to be protected from corrosion, the pH of the solution, the concentration, the type of anions used in the acid, and the other species present in the solution. It also depends on how much reaction takes place to form secondary corrosion inhibitors. The mechanism also depends on the size of the hydrocarbon component of the molecule, the functional group's electron density, and the molecular structure of inhibitors with identical functional groups (Zehra et al., 2022).

20.4.2 Mechanism of inhibition action of organic inhibitors

When a metal initially encounters an inhibitor, it often causes physical adsorption on the metal's surface, which may or may not be followed by chemisorption. Adsorption kinetics and heat of adsorption are two variables that can be used to distinguish between pure physical adsorption and chemisorption. Physical solubility is the result of electrostatic attraction between inhibited organic ions or dipoles and electrically charged metal surfaces. In contrast, chemisorption necessitates the exchange of the charge or the movement of the charge from the organic molecule to the metal surface in order to create a bond of the coordinate form. The metal that needs to be safeguarded, the inhibitor, and the hostile medium or electrolyte all play a role in the inhibition process. The characteristics of all three of the aforementioned substances have an impact on the inhibition process, regardless of whether it happens by physical adsorption or chemisorption. Because water exhibits a weak bipolarity, it is assumed that the electrolyte's water molecules are adsorbed on the metal surface prior to the injection of the inhibitor. It is possible that some aggressive ions, such as the chloride of the electrolytes, will adsorb. However, according to the following equation, organic molecules replace water or other adsorbent molecules with the presence of an inhibitor:

$$\mathrm{Org}_{(\mathrm{sol})} + n\mathrm{H_2O}_{(\mathrm{ads})} \rightarrow \mathrm{Org}_{(\mathrm{ads})} + n\mathrm{H_2O}_{(\mathrm{sol})} \qquad (20-8)$$

where each molecule of the organic inhibitor that has been adsorbed has replaced n water molecules, that "n" is dependent on the proportionate cross-sectional area of organic and water molecules given. Organic molecules are preferentially adsorbed to the metal surface because the interaction energy between an organic molecule and a metal surface is higher than the interaction energy between a water molecule and a metal surface.

20.4.3 Adsorption of the inhibitor on metal surface

The extensive consensus is that the primary stage in the process of organic molecules inhibiting the corrosion of metals is their adsorption onto the surface of the metal (substrate) via their functional group (s). As a result, it is necessary to understand the adsorption behavior of organic molecules (inhibitors) on the metal surface to comprehend the corrosion inhibition mechanisms (Zehra et al., 2022). The adsorption process depends upon (1) the topography, uniformity, and physicochemical properties (charge, conformation, and chemistry) of the metal surface; (2) the propagation of the charge in the molecule; (3) the aggressive medium (electrolyte); and (4) the chemical composition of the organic molecule. The adsorption of ions or neutral molecules on bare metal surfaces immersed in solution is determined by the mutual interactions of all species present at the phase boundary.

Depending on the existence of the forces at play, there are two forms of adsorption: (1) **Chemisorption**: This kind of adsorption involves the largely irreversible attachment of a single layer of molecules, atoms, or ions to the surface through chemical bonds. (2) **Physical adsorption**: In this type of adsorption, bonds are formed using weaker van der Waals forces, whose energy levels are similar to those of condensation. Adsorption utilizing electrostatic/van der

Waals forces has a lower heat of adsorption than chemisorption. Coulombic interactions as well as other interactions exist in the case of physisorption between the ions or dipoles of the inhibitor and the charged metal surface (van der Waals forces and hydrogen bonding). Although the adsorption process in this instance is quick, the species that are adsorbing are easily washable off the surface. To donate electrons to the unoccupied d-orbital of transition metals, species with loosely bonded electrons, such as electrons in aromatic rings, multiple bonds, or unpaired electrons in functional groups containing elements such as oxygen, nitrogen, sulfur, or phosphorus, are frequently involved (substrate). The performance of corrosion inhibition of the sequence homology set of organic substances differing only in the heteroatom is usually as follows: phosphorus > sulfur > nitrogen > oxygen. Factors influencing the absorption of ion inhibitors on metal surfaces are:

20.4.3.1 Metal surface charge

Electrostatic attraction may be present during adsorption between the electrical charge on the metal at the metal–solution interface and the ions or dipole loads on the adsorbed inhibitor.

20.4.3.2 The functional groups and structure of inhibitors

Inhibitors can connect to the metal surface by electron transfer and create a coordinated sort of bond that aids in strong binding and effective inhibition. Anions, neutral compounds, lone pairs of electrons, triple bond or organic ring system-related electron systems, functional groups with periodic table elements in groups V or VI, and species with relatively loosely bound electrons all to favor quick electron transfers and better bond formation, which leads to effective inhibition. Decreasing electronegativity increases the likelihood of stronger coordinate bonds forming, and this tendency follows the order: oxygen < nitrogen < sulfur < phosphorus.

20.4.3.3 Synergism and antagonism, the interaction between adsorbed inhibitor species

Adsorbed species may enter into various interactions on the surface of an electrode that may significantly influence their inhibitive properties and the mechanism of their action.

20.4.3.4 The reaction of adsorbed inhibitors

Adsorbed inhibitor species can typically react electrochemically to create a product that is also inhibitive. Primary and secondary inhibition is used to describe the effects of an additional drug and reaction products, respectively.

Isothermal adsorption is a term used to describe the adhesion of organic inhibitors on metal surfaces. It is generally acknowledged that adsorption isotherms offer priceless insights into the process of corrosion inhibition and provide crucial information on the nature of interactions between metals and inhibitors. The adsorption isotherm, which displays the contrast between the inhibitor surface concentration and bulk solution, can be used to infer the strength of the adsorption link. Numerous isotherms of adsorption (Langmuir, Temkin,

Frumkin, Flory–Huggins, Dhar–Flory–Huggins, Bockris–Swinkels, and Freundlich) have been published in order to demonstrate the method of adsorption of inhibitors onto the metal surface.

20.4.4 Corrosion inhibition in near-neutral solutions

For two reasons, the corrosion of metals in nearly neutral solutions (and in the open air) is different from that in acidic solutions. First off, the metal surface is free of oxide in acidic solutions but covered with oxide in neutral solutions. Second, the primary cathodic reaction in acid solutions is the hydrogen evolution reaction, whereas the cathodic reaction in air-saturated neutral solutions is oxygen reduction.

While metal surfaces in acid solutions are coated with films of oxides, hydroxides, or salts, preventing the solubilization of these species, corroding metal surfaces in acid solutions are oxide-free. These variations mean that compounds that prevent corrosion in acid solution by adhering to oxide-free surfaces usually do not prevent corrosion in neutral solution. The anions of weak acids are frequently used as inhibitors for solutions that are close to neutral, with chromate, nitrite, benzoate, silicate, phosphate, and borate among the most significant examples in daily life. Metals with passivating oxide layers strongly resist metal ion diffusion, and the anodic reaction of metal dissolution is suppressed. In near-neutral solutions, these inhibitory anions often referred to as anodic inhibitors are more frequently utilized than cathodic inhibitors to prevent the corrosion of iron, zinc, aluminum, copper, and their alloys. The precipitation of substances on a metallic surface that can build or solidify protective coatings can also cause inhibition in neutral solutions. By precipitation or reactivity, the inhibitor may produce a surface coating of an insoluble salt.

20.4.5 Corrosion inhibition in oil and petroleum industries

Oil corporations initially introduced many chemical compounds, often successfully, to reduce corrosion damage to oil wells and surface handling equipment. The use of amines and imidazolines significantly improved the effectiveness of corrosion prevention in oil fields. Inhibitors are currently deployed in the field at doses of 15–50 ppm, continuously or in irregular batches, based on total liquid output. There is now a far wider variety of inhibitor chemistry available to combat oil-field corrosion than there was even ten years ago. Organic compounds with diverse combinations of sulfur, phosphorus, and nitrogen have been created recently. These inhibitors have improved oil field inhibitor effectiveness, particularly in the areas of oxygen contamination tolerance and corrosion control under high CO_2 and low H_2S environments (Aslam, Aslam, Zehra, & Rizvi, 2022).

The majority of the inhibitors now utilized in well development are organic nitrogenous chemicals. Long-chain hydrocarbons are included in the fundamental forms as part of the structure (usually C18). Today's most successful inhibitors frequently target either long-chain aliphatic diamines or imidazolines with lengthy carbon chains. To change the material's physical characteristics, a number of alterations to these structures are being used, such as the conventional reaction of ethylene oxide with these compounds at various molecular ratios to produce polyoxyethylene derivatives with different levels of salt miscibility. Many carboxylic acids are employed to turn these amines or imidazolines into salts. There is now a far wider variety of inhibitor chemistry available to combat oil-field corrosion than there was even 10 years ago. One of the traditional theories is known as the "sandwich principle," where the link between the molecule's polar end and the metal's surface makes up the bottom layer of the sandwich. This link is crucial to the defensive action's success.

The nonpolar end of the molecule, which forms the middle of the sandwich, contributes to defense to the extent to which it may cover or wet the surface. The top layer of the protective sandwich is the hydrophobic layer of oil connected to the long carbon tail of the inhibitor. By protecting the inhibitor's surface and acting as a barrier to both ferrous ion outward diffusion and corrosive species inward diffusion, this oil coating serves as the external protective surface. Corrosion will not be detrimental if it only contains water or water and salt solutions, unless corrosion agents, such as CO_2, H_2S, and their breakdown products, are confined. The wells for oil and gas are either sweet or sour.

20.4.5.1 Sweet corrosion

In CO_2 gas wells, corrosion can be categorized into three ambient temperatures. The corrosion product is nonprotective under 60°C, and there will be high corrosion rates. Magnetite is formed above about 150°C, and the wells are not corrosive even in the presence of high brine levels. The iron carbonate corrosion product layer is protective in the middle-temperature regime, in which most gas well conditions lie, but is adversely affected by chlorides and fluid velocity. The dispersibility feature of the oil and the brine being produced is among the significant physical properties of oil-field inhibitors. A properly selected inhibitor based on the corrosion mechanism will not be successful unless it has access to the corrosive metal. There are also some major variations when it comes to handling oil and gas wells. There is no clear cut in the difference between an oil well and a gas well. The fact that a large volume of gas is generated by many oil wells and a large volume of liquid is produced by many gas wells, plus the fact that wells frequently undergo a shift in production over their life makes it difficult to create a technical distinction. There are more substantial distinctions, however. Typical gas wells are much hotter than oil wells, and they are much lighter in terms of hydrocarbon liquids. Normally, gas wells are much deeper and generally produce brines of lower total dissolved solids. In gas well corrosion, oxygen is not a consideration to consider but may cause major issues in artificial lift oil wells. Corrosion mechanisms may change due to the wide temperature gradient in many gas wells, resulting in various forms of corrosion in the same well, whereas oil wells do not exhibit this behavior. Oil wells usually contain more liquid than gas wells, resulting in a shorter batch treatment existence. Being corrosion in oil wells is electrochemical in nature, an electrolyte must be present for corrosion to occur. The source of water in oil wells is almost always the formation source, and at concentrations ranging from trace to saturation, the water may contain dissolved salts. Corrosion-related water can be in a thin layer, in droplets, or even in a major process. Results of the corrosion control analysis by inhibitors in the development of oil wells in fields flooded with carbon dioxide showed that imidazolines are effective in protecting CO_2 brines.

It was found that the inhibitor was integrated into the layer of the carbonate corrosion product, but if the surface film contained sulfide, it was even more successful. Inhibitors, such as nitrogen–phosphorus compounds or sulfur compounds in organic molecules, have also had better performance.

20.4.5.2 Sour corrosion

Aldehydes, cyanamide thiourea, and urea derivatives are examples of corrosion inhibitors that have been employed in the past to treat the corrosion issue that sour wells experience. The sour wells' most often used inhibitors are by far organic amines. The addition of hydrogen sulfide greatly speeds up amine inhibition, even though organic amines are thought to be less potent inhibitors of acid solutions. The principal corrosive substance in sour wells is hydrogen sulfide, however, carbon dioxide is also frequently found there. The impact of various hydrogen sulfide concentrations and iron sulfides on corrosion products has also been identified. Oil-field inhibitors work by forming a thin layer by incorporating corrosion products on the metal surface. This surface film may be sulfide or carbonate and may be anaerobic or partially oxidized.

20.4.5.3 Acidizing

Acidizing is a useful strategy for increasing oil and gas well productivity. Due to the extremely low permeability of such hydrocarbon-containing rocks, they cannot flow easily into the well. These limestone/dolomite formations can be treated with HCl or a solution including HF if the rock is sandstone. During the acidifying process, the acid (e.g., HCl at a concentration of 7%–28%) is forced down the tube into the well where it reaches the perforations and hits the structure. The acid etches channels that create a conduit for oil and gas to enter the pipe. For good acidizing procedures, acetylenic alcohols like 1-octyn-3-ol and high molecular weight nitrogenous chemicals, alkyl or alkylaryl nitrogen compounds, are used as inhibitors. However, these materials provide serious handling challenges because of their extremely hazardous nature; as a result, an operator's choice of product will ultimately depend on this. Their effectiveness is also constrained in terms of time and output. The average length of an acid soak is 12–24 hours, after which the inhibitor's effectiveness may start to alarmingly decline. Oxygen-containing inhibitors that work well in concentrated HCl include cinnamonaldehyde and alkynols with unsaturated groups together with the oxygen role known as α-alkenyl phenones (Aslam, Aslam et al., 2022; Jones, 1988). They offer defense comparable to that of acetylenic alcohols, especially when mixed with minor quantities of surfactants.

20.4.6 Corrosion inhibition in atmospheric and gaseous corrosion

The use of Volatile Corrosion Inhibitors (VCIs) can significantly lessen the harm that atmospheric or gas corrosion does to metals. The environment and a very modest concentration of inhibitors mostly determine how much VC is inhibited. In addition to volatility, several other electrochemical phenomena are required to operate as a VCI in order to provide the protective effect, such as changes in the potential of the diffusion part of the double layer that controls the migration of the electrode reaction components. The ability of a vapor phase inhibitor to reach the metallic surface that has to be protected is the first prerequisite for good efficiency. The second is that the molecule's rate of transfer should not be too sluggish to avoid the aggressive environment first attacking the metal surface before the inhibitor can start working. The distance between the source(s) of the inhibitor and the metal surfaces, the accessibility of the surfaces, and the vapor pressure of the inhibitor are all factors in these two circumstances. The crystal lattice structure and the type of atomic bonds in a molecule will determine a chemical compound's vapor pressure. In this regard, the molecule's organic components will typically guarantee its volatility. Because enclosures are sometimes not airtight, a handy volatile inhibitor shouldn't have a very high vapor pressure because this will cause protection to decline. For effective chemicals, a practical partial vapor pressure will range between 10^{-5} and 10^{-1} mm Hg (i.e., 10^{-3} to 10 Pa). The chemicals that exhibit noticeable vapor pressure behave as inhibitors of the electrolyte layer under specific atmospheric circumstances. They acted in this way by changing the kinetics of the electrode reaction, and they are essentially considered VCIs. However, neutralizing amines are not categorized as VCIs even if they have a sizable vapor pressure. The protection of ferrous metals is frequently effective with them. The modification of the pH values of the electrolyte determines the mechanism of its inhibition. This results in a number of circumstances that make rust formation difficult. The volatile component typically quickly achieves the protective vapor concentration, but if the enclosure in which it is used is not airtight, their protective efficiency may only last for a very short period and their consumption may be very high. However, volatile substances with very low vapor pressure show prolonged protection; the issue with these substances is that more time is needed for them to reach the protective vapor pressure. The other issue is the potential for corrosion, which could happen during the initial saturation period and prevent even the effective inhibitor concentration from being reached if the enclosure is not hermetically sealed. Therefore, neither an excessively high nor an excessively low vapor pressure is acceptable for volatile chemicals used as corrosion inhibitors. They must have the best possible vapor pressure.

20.5 Mechanistic aspects of corrosion inhibition

According to the previous sections, the adsorption process on a metal surface is a trustworthy mechanism for the corrosion inhibition process. The chemical composition of the inhibitors and the surface charge of the metal, respectively, have an impact on the adsorption phenomena. The surface charge of the metal is a result of the electrical charge that forms at the point where it is submerged in the electrolyte. At the electrolyte/metal interface, the probable contact of the corrosion inhibitor with the metal atom led to the surface covering of adsorbed inhibitor molecules, which slowed the electrochemical response of the corrosion process. Usually, one or more of the following mechanisms contribute to the blockage of corrosion inhibitors into the metal surface.

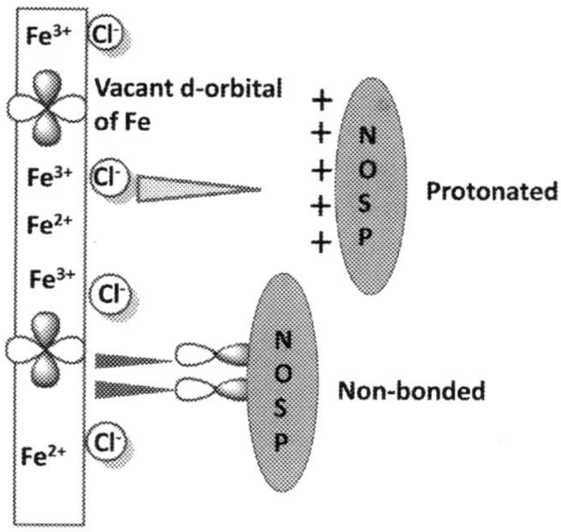

FIGURE 20–5 Illustration of various adsorptive interactions between the inhibitor molecules and the vacant d-orbitals of the metal atoms.

20.5.1 Adsorptive interaction of corrosion inhibitors on the metal surface

Many of the chemicals used as corrosion inhibitors have a propensity to adsorb on metal surfaces. The adsorbed corrosion inhibitor will typically react chemically or electrochemically to generate a product that will potentially have an inhibiting effect. In most cases, organic inhibitors were utilized to effectively prevent corrosion (as described in the previous section). They feature strong polar properties, such as electrons in double bonds that are conjugated, electronegative groups, and heteroatoms (N, S, or O) in their molecular structures, which enable the adsorption process to occur on the metal surface. The inhibiting effect of the corrosion inhibitors is explained by the adsorptive interactions between adsorbates and the metal surface. Additionally, when metal has low energy unoccupied electron orbitals, the inhibitors are linked to the metal surfaces by electron transfer, creating a coordinate bond. Because the electrons are rather loosely bound, electron transfer from the adsorbed species is encouraged. Additionally, Fig. 20–5 depicts the plausible elementary processes that served as the foundation for the proposal of an adsorptive interaction between the inhibitor molecules and the unoccupied d-orbitals of the metal atoms. These processes include:

- the charged inhibitor molecules and the charged metal are drawn together by electrostatic forces (physical adsorption),
- Metal atoms having unoccupied d-orbitals, like Fe, interacting with heteroatoms' lone pair electrons or protonated compounds like amine and phosphate,
- when possible, vacant d-orbital of Fe atoms in the metal interacts with p-electrons in the conjugated bond of inhibitor, and,
- a combination of the above three methods.

20.5.2 Inhibitive performance of protective layer formation

Due to the migration of ions or water molecules, which subsequently react with inhibitor molecules, the adsorbed inhibitor molecules may create a surface layer known as a

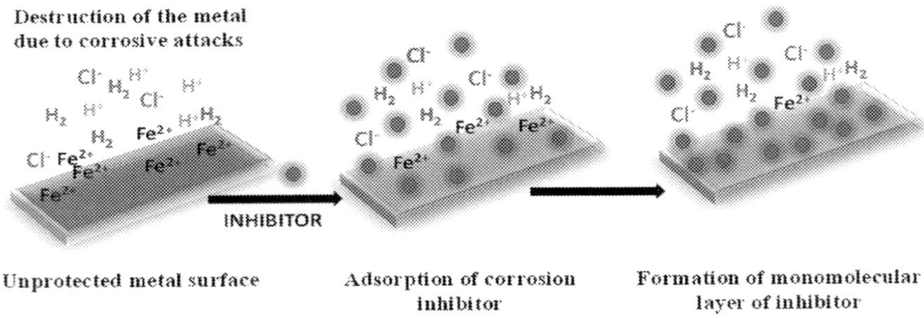

FIGURE 20–6 Illustration of the formation of protective film by the corrosion inhibitors on the metal surface.

protective layer that may operate as a physical barrier to the metal surface, as shown in Fig. 20–6. These shielding layers, also referred to as monomolecular layers, can influence both cathodic and anodic processes, thereby preventing the electrochemical corrosion process (Aslam, Mobin, Aslam, Lgaz, & Chung, 2019). Additionally, because of the lateral interaction between the dipoles of the adsorbed molecules and the ionic charges on the metal surface, the inhibitive performance of corrosion inhibitor is typically dependent on the surface coverage of adsorbed inhibitor, which is also influenced by the presence of the surface charge on the metal surface. However, the substitution of inhibitor adsorption for water molecules could potentially be a contributing factor in the rising inhibitive efficiency. In other words, the displacement reaction of the adsorbed inhibitor molecules causes the adsorbed water molecules to be removed from the metal surface. Additionally, the number of adsorption sites (surface coverage), the coordination bond (orbital character and electrical structure of the molecule), and the density of the electron at the donor atom (surface charge) can all be used to establish the inhibitive performance of a corrosion inhibitor against the corrosion of metal in an aggressive solution. Because of these interactions, the rate of corrosion reactions will be slowed down. This is because the interactions between adsorbed inhibitor molecules may significantly increase the surface coverage and the adsorbed species due to the physicochemical properties and the electrostatic force of attraction between the electric charge on the metal at the metal/solution interface.

20.5.3 Inhibition of electrochemical process

The electrochemical process is hindered by delaying any anodic and/or cathodic reaction sites because corrosion reactions involve the interaction of adsorbed inhibitor molecules with metal atoms in order to build a barrier layer that prevents further corrosion reaction(s). As shown in Fig. 20–6, when the required concentration of OH- is created, an insoluble oxide/hydroxide of corrosion inhibitor is formed as a result of the interaction with hydroxide ions on the cathodic sites. Depending on the type of corrosion inhibitors used, oxidation or reduction reactions may release the inhibitor's active site and cause it to further react with

H_2O_2 or OH-, which was created at the cathodic site as an intermediate of oxygen reduction. As a result, the aggressive electrochemical corrosion process that takes place at the cathodic site is inhibited by this process. It is interesting to note that measurements might also be made of corrosion parameters including corrosion potential, current density, and charge transfer resistance. These inhibitory reactions are identified by a change in impedance, gradient of polarization slope of the metal's anodic dissolution, and cathodic rate of hydrogen evolution on the metals that impact the mechanism of the reaction sites (cathodic, anodic, or both).

The capacitive surface layer in which the double-layer capacitance might be measured to determine the thickness of the surface layer can also change as a result of ions adhering to the metal surface. At this speed, the barrier film might act like a form of pseudocapacitor that uses charge transfer procedures. The working electrode's surface inhomogeneity and roughness, which will be covered in the following section, are to blame for the departure from the capacitive behavior. Utilizing a corrosion inhibitor that lowers the corrosion rate by stopping the anodic metal dissolution and obstructing the cathodic reaction rate, and/or both, based on the inhibitor molecule's adsorption on the metal surface, is a straightforward way to explain corrosion inhibition. At least one of the accompanying methods could be involved with the inhibition surface:

20.5.3.1 Chemisorption

Chemically adsorbed, the inhibitor's molecular surface charge combines with the metal's surface charge to create a shield-like coating.

20.5.3.2 Physisorption

A protective layer is created by the interaction and combination of adsorbate and adsorbent.

The potential corrosive ion in the electrolyte causes the corrosion inhibitor to react, and a complex byproduct is formed at the electrolyte/metal interface.

20.6 Examples of corrosion inhibitors

There are several applications of corrosion inhibitors known so far (as given in Fig. 20–7). Some of them are discussed below (Table 20–2):

20.6.1 In acid

Acid picking, acid well acidizing operations, cleaning of oil–gas industrial equipment, heat exchangers, and other key applications of acids in which metals or their alloys are in direct or indirect contact. Most often, multiple types of corrosion inhibitors are used to protect them from corrosion (Table 20–2).

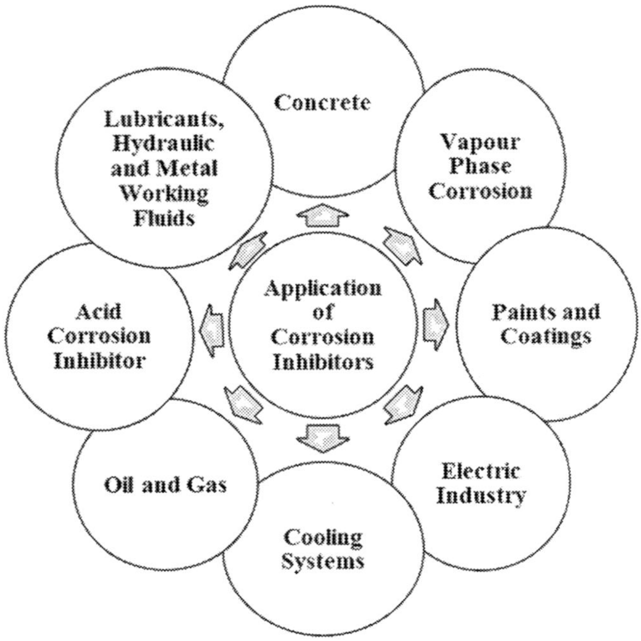

FIGURE 20–7 Various applications of corrosion inhibitors.

Table 20–2 Corrosion inhibitors in various environments.

S. no.	Atmosphere	Environment inhibitors	Conditions	Metals to be protected
1.	Water, potable	i. $Ca(HCO_3)_2$	10 mg/L	Steel, cast iron, etc.
		ii. Polyphosphate	5–10 mg/L	Fe, Zn, Cu, Al
		iii. $Ca(OH)_2$	To pH 8.0	Fe, Zn, Cu
		iv. Na_2SiO_3	10–20 mgL	Fe, Zn, Cu
2.	Water, cooling	i. $Ca(HCO_3)_2$	10 mglL	Steel, cast iron, etc.
		ii. Na_2CrO_4	0.1%	Fe, Zn, Cu, Al
		iii. Na polyphosphates	pH6.5–7.5	Fe, Cu alloys
		iv. Mercaptobenzotriazole (MBT)	If no Cl_2	Cu, brass
		v. Tolyltriazole	–	Cu
		vi. Zn phosphonates	pH6.5–9	Fe
		vii. $NaNO_2$	0.05%	Fe
		viii. NaH_2PO_4	1%	Fe
		ix. Morpholine	0.2%	Fe
3.	Water, boiler	NaH_2PO_4	10 mglL	Fe, Zn, Cu
		Polyphosphate	10 mglL	Fe, Zn, Cu
		Morpholine	variable conc	Fe
		Hydrazine	02 scavenger	Fe
		Ammonia	Neutralizer	Fe
		Octadecylamine	Variable conc	Fe

(Continued)

Table 20–2 (Continued)

S. no.	Atmosphere	Environment inhibitors	Conditions	Metals to be protected
4.	Brines	$Ca(HCO_3)_2$	10 mglL	Fe, Zn, Cu
		Na_2CrO_4	0.1%	Fe, Zn, Cu
		Na benzoate	0.5%	Fe
		$NaNO_2$	(NaCl5%)	Fe
5.	Brines, dil field	Na_2SiO_3	0.01%	Fe
		Na_2SO_3 or SO_2	02 scavenger	Fe
		Quaternary amines	10–25 mglL	Fe
		Imidazoline	10–25 mglL	Fe
		Rosin amine acetate	5–25 mglL	Fe
		Coco amine acetate	5–15 mglL	Fe
		Formaldehyde	50–100 mglL	Fe
6.	Engine coolants	Na_2CrO_4	0.1%–1%	Fe, Pb, Cu, Zn
		$NaNO_2$	0.1%–1%	Fe
		Borax	1%	Fe
7.	Glycol-water coolant	Borax + MBT	1% + 0.1%	All
		Na benzoate + $NaNO_2$	5% + 0.3%	All
8.	Seawater	Na_2SiO_3	10 mglL	Zn
		$NaNO_2$	0.5%	Fe
		$Ca(HCO_3)_2$	pH-dependent	All
		$NaHlPO_4$ + $NaNO_2$	10 mglL + 0.5%	Fe
9.	HC1Acid	Ethylaniline	0.5%	Fe
		MBT	1%	Fe
		Pyridine + phenylhydrazine	0.5% + 0.5%	Fe
		Rosin amine + ethylene oxide	0.2%	Fe
10.	H2SO4	Phenylaeridine	0.5%	Fe
		Diethylaniline	Pickling	Fe
11.	HNO3	$K_2Cr_2O_7$	–	Fe
		KMnO4	–	Fe
		KI	–	Fe
12.	H3PO4,conc	NaI	200 mglL	Fe
13.	Most acids	Thiourea	1%	Fe
		Sulfonated eastor oil	0.5%–1%	Fe
		As_2O_3	0.5%	Fe
		Benzotriazole	–	Cu
14.	Vapor condensate	Morpholine	Variable conc.	Fe
		Ammonia	Variable conc.	Fe
		Ethylenediamine	Variable conc.	Fe
		Cyclohexylamine	Variable conc.	Fe
15.	Cement	$Ca(NO_3)_2$	Variable conc.	Fe
16.	Drilling muds	Na_2CrO_4	pH > 8	Fe
17.	Vapor-phase inhibitors	Cyclohexylamine carbonate	30 g/m^3	Fe, Al, Sn, Zn, solder
		Dicyclohexylamine nitrite	10 g/m^2	Fe, Al, Sn
		Amylamine benzoate	Variable conc.	Fe
		Diisopropylamine nitrite	Variable conc.	Fe
		Methylcyclohexylamine carbonate	Variable conc.	Fe
18.	Coating inhibitors	$ZnCrO_4$ (yellow)	Variable conc.	Fe, Zn, Cu
		$CaCrO_4$ (white)	Variable conc.	Fe, Zn, Cu
		$Pb_3Ü_4$ (red lead)	Variable cone.	Fe

20.6.2 In water

20.6.2.1 Potable water

Potable water is corrosive unless a protective film or deposit is formed on the surface of metal or alloys. Potable water is generally saturated with dissolved oxygen. Some of the commonly utilized corrosion inhibitors for potable water are mentioned in Table 20–2.

20.6.3 In cooling water system

Evaporation is the main source of the recirculating systems. The number of dissolved mineral salts rises as evaporation continues. Different metals and nonmetals may be used in cooling systems. Galvanic corrosion is caused when metals that are picked up from one area of the system are deposited in another. Nitrate and chromate are examples of anodic (passivating) inhibitors of corrosion, as are cathodic (e.g., zinc salt) inhibitors. Corrosion inhibitors mainly employed are discussed in Table 20–2. Additionally, organic corrosion inhibitors are used.

20.6.4 In automobiles

To reduce internal corrosion (fluid system corrosivity), which is typically brought on by coolant aeration temperature, pressure, etc., inhibitors are utilized. Some of the commonly used antifreeze corrosion inhibitors include nitrates, phosphates, silicates, arsenates, and chromates (anodic inhibitors), amines, benzoates, mercaptans, and organic phosphates (mixed inhibitors), as well as polar or emulsifiable oils (film formers). Protecting against exterior corrosion is another application for corrosion inhibitors in automobiles (corrosion of metal surfaces that are exposed to the environment). Automobiles are exposed to moist air, and wet SO_2 gas, which reacts with moisture in the air to produce sulfuric acid, and deicing salt (NaCl and $CaCl_2$). Grease, wax resin, and resin emulsion are employed as corrosion-proofing formulations, as well as metalloorganic and asphaltic compounds, to control exterior corrosion. Fatty acids, phosphonates, sulfonates, and carboxylates are common inhibitors used in rust-proofing applications.

20.6.5 In organic coatings and paints

The primers contain polar compounds that are inhibitory pigments that are finely powdered. They usually displace the water and orient themselves in a way that towards the environment is their hydrophobic tails. Red lead (Pb_3O_4) is widely utilized in paints on iron. It helps to preserve the physical properties of the paints by preventing the formation of local cells. Lead azelate, calcium plumbate, and lead suboxide are some other inhibitors that are employed.

20.6.6 Internal corrosion of steel pipelines

Similar to refineries and petrochemical plants, assembly pipes that transport oil and gas between wells and processing facilities contend with corrosion issues. Multiphase fluid flow has an impact on the pace of pipeline corrosion. Corrosion occurs at rapid rates due to flow- and erosion-induced processes, and the pipes' sediments are flushed out by the strong flow tendency. Pitting is the typical form of corrosion at the low flow rates as it is associated with the settlement of the subsurface sediments. The corrosion of the pipelines or pigging is controlled by cleaning them by adding the continuous or batch of the corrosion inhibitors.

20.6.7 Petroleum production

The petrochemical and refinery sectors use a variety of film-forming inhibitors to retard wet corrosion. Many of the inhibitors are long-chain amines and amides, as well as other nitrogenous organic molecules. It is common practice to utilize both oil-soluble and water-soluble dispersible oil inhibitors (batch inhibitors) and water-soluble and oil-soluble dispersible oil inhibitors (water-soluble and oil-soluble). Due to their interfacial nature, which makes them active at liquid–liquid and/or liquid–gas interfaces, inhibitors can cause emulsification. As a result, foaming is rarely noticed in the presence of inhibitors. The polar group of the film-forming inhibitors serves as an anchor for the metal.

20.6.8 Others

Inhibitors have been employed to reduce corrosion in steel reinforcement in concrete, hot chloride dye baths, cooling brines, and fuel oil reservoirs. Several corrosion inhibitors are also used to shield artifacts against corrosion.

20.7 Selection of the efficient corrosion inhibitors and challenges related

The technique of application and the characteristics of the system in which they are used must be taken into account when choosing corrosion inhibitors. Numerous aspects should be considered in order to choose the right inhibitors, but one of them is matching the right inhibitor chemistry with the corrosion conditions. Choosing the physical characteristics, such as whether the inhibitor should be in a solid or liquid condition, is the first step in the selection of inhibitors. It is important to take into account both the melting and freezing points. Is their deterioration with time and temperature critical or not? is a further issue that needs to be taken into account. It ought to work well with the other system additives. Are particular solubility properties necessary? Although this list may be lengthy, it is significant since it clears the way for potential corrosion inhibitors. It is regarded as the initial step in the evaluation of any new system. Physical quantifications are carried out as part of least-quality affirmation testing as is customary. The simplest corrosion test should be carried out to evaluate the effectiveness of the inhibitors and weed out any inappropriate candidates. The goal of the initial screening is to determine which inhibitors should be tested further and which ones should not. Even if some good inhibitors may also be excluded, the inhibitor user must apply test methods that carefully exclude inferior inhibitors. The design of the experimental settings (real-world scenarios) required to determine the performance of the inhibitor system presents the biggest hurdle. Temperature, pressure, and velocity must all be taken into account, as well as the characteristics of the metal and the corrosive ambient chemistry.

In order to effectively inhibit corrosion, the inhibitive species must have simple access to the metal surface. Therefore surfaces should ideally be free of dirt, debris, corrosion products, water hardness scales, and other contaminants. And additional steps ought to be taken to prevent the buildup of solid particles. This conditioning is frequently challenging to accomplish, and there are several instances when systems have not been adequately prepared for inhibitive treatment. The inhibitor must also make contact with all areas of the metal surfaces. It is important to ensure that the inhibited fluid contacts all dead ends, pockets, and crevice regions, especially when first filling a system. The movement of the fluid in service will favor this in many systems, but in ostensibly static systems, it will be beneficial to establish a flow regime periodically to offer a fresh supply of the inhibitor.

The type of metals present, their combinations, the environment's propensity for corrosion, and the flow, temperature, and heat transfer requirements must all be considered while selecting inhibitors. Regular inhibitor concentration checks should be performed, and losses should be corrected by adding the required amount of inhibitor or, if recommended, by completely replacing the fluid with engine coolants. While it must be kept in mind that the results from monitoring devices, probes, coupons, and other similar tools apply to the behavior of that specific component at that specific section of the system, continuous monitoring should be used wherever practical. Nevertheless, it must be acknowledged that corrosion monitoring in an inhibited system is well-established and frequently utilized despite this warning.

References

Aoufir, Y. E. & Zehra, S. (2020). Evaluation of inhibitive and adsorption behavior of thiazole-4-carboxylates on mild steel corrosion in HCl. *Colloids and Surfaces A: Physicochemical and Engineering Aspects*. Available from https://doi.org/10.1016/j.colsurfa.2020.125351.

Aslam, R., Mobin, M., Aslam, J., Lgaz, H., & Chung, I.-M. (2019). Inhibitory effect of sodium carboxymethylcellulose and synergistic biodegradable gemini surfactants as effective inhibitors for MS corrosion in 1 M HCl. *Journal of Materials Research and Technology, 8*.

Aslam, J., Aslam, R., Zehra, S., & Rizvi, M. (2022). Corrosion inhibitors for sweet (CO2 corrosion) and sour (H2S corrosion) oilfield environments. Elsevier BV. Available from https://doi.org/10.1016/b978-0-323-85405-4.00021-5.

Aslam, R., Serdaroglu, G., Zehra, S., Kumar Verma, D., Aslam, J., Guo, L., . . . Quraishi, M. A. (2022). Corrosion inhibition of steel using different families of organic compounds: Past and present progress. *Journal of Molecular Liquids, 348*, 118373. Available from https://doi.org/10.1016/j.molliq.2021.118373.

Aslam, R., Mobin, M., Zehra, S., & Aslam, J., A comprehensive review on the application of inhibitors for corrosion mitigation of stainless steel in different environments. *Journal of Molecular Liquids*, Accepted. (n.d).

Baker, M. D. & Marcy, V. M. (1956). 2 1 1956/02/01 Journal of Fluids Engineering, Transactions of the ASME https://doi.org/10.1115/1.40136511528901X2299-308. American Society of Mechanical Engineers (ASME) United States Hydrazine as an Oxygen Scavenger-A Progress Report on Tests at Springdale Station. https://asmedigitalcollection.asme.org/fluidsengineering/issue/browse-by-year78.

Bastidas, J. M., Mora, E. M., & Feliu, S. (1990). The protective action of two vapour phase inhibitors on the corrosion of mild steel. *Materials and Corrosion/Werkstoffe und Korrosion, 41*(6), 343–347. Available from https://doi.org/10.1002/maco.19900410605.

Bedair, M. A., Soliman, S. A., & Metwally, M. S. (2016). *Journal of Industrial and Engineering Chemistry, 41*.

Boffardi B. P. (1987). Metals handbook, In Corrosion 487–497 ASM International Control of environmental variables in water-recirculating systems XIII.

Boulhaoua, M., El Hafi, M., Zehra, S., Eddaif, L., Alrashdi, A. A., Lahmidi, S., . . . Lgaz, H. (2021). Synthesis, structural analysis and corrosion inhibition application of a new indazole derivative on mild steel surface in acidic media complemented with DFT and MD studies. *Colloids and Surfaces A: Physicochemical and Engineering Aspects, 617*, 126373. Available from https://doi.org/10.1016/j.colsurfa.2021.126373.

Cabral, A. M., Trabelsi, W., Serra, R., Montemor, M. F., Zheludkevich, M. L., & Ferreira, M. G. S. (2006). The corrosion resistance of hot dip galvanised steel and AA2024-T3 pre-treated with bis-[triethoxysilylpropyl] tetrasulfide solutions doped with Ce(NO3)3. *Corrosion Science, 48*(11), 3740–3758. Available from https://doi.org/10.1016/j.corsci.2006.01.010.

Estevão, L. R. M., & Nascimento, R. S. V. (2001). Modifications in the volatilization rate of volatile corrosion inhibitors by means of host-guest systems. *Corrosion Science, 43*(6), 1133–1153. Available from https://doi.org/10.1016/S0010-938X(00)00106-2.

Fujioka, E., Nishihara, H., & Aramaki, K. (1996). The inhibition of pit nucleation and growth on the passive surface of iron in a borate buffer solution containing Cl- by oxidizing inhibitors. *Corrosion Science, 38*(11), 1915–1933. Available from https://doi.org/10.1016/S0010-938X(96)00075-3.

Hoar, T. P., & Holliday, R. D. (1953). The inhibition by quinolines and thioureas of the acid dissolution of mild steel. *Journal of Applied Chemistry, 3*(11), 502–513. Available from https://doi.org/10.1002/jctb.5010031105.

Jones, D. A. (1984). Polarization studies of brass-steel galvanic couples. *Corrosion, 40*(4), 181–185. Available from https://doi.org/10.5006/1.3581936.

Jones, L. W. (1988). *Corrosion and water technology for petroleum producers*, Tulsa, Okla, Oil and Gas Consultants International.

Kuznetsov, Y. I., & Thomas, J. G. N. (1996). Organic inhibitors of corrosion of metals. Plenum Press.

Lgaz, H., Zehra, S., Toumiat, K., Chaouiki, A., El Aoufir, Y., Ali, I. H., . . . Chung, I. M. (2019). Mild steel corrosion inhibition by furocoumarin derivatives in acidic media. *International Journal of Electrochemical Science, 14*(7), 6699–6721. Available from http://www.electrochemsci.org/papers/vol14/140706699.pdf, https://doi.org/10.20964/2019.07.11.

Lgaz, H., Zehra, S., Albayati, M. R., Toumiat, K., El Aoufir, Y., Chaouiki, A., . . . Mohamed, S. K. (2019). Corrosion inhibition of mild steel in 1.0 M HCl by two hydrazone derivatives. *International Journal of Electrochemical Science, 14*(7), 6667–6681. Available from http://www.electrochemsci.org/papers/vol14/140706667.pdf, https://doi.org/10.20964/2019.07.08.

McCafferty, E., Bernett, M. K., & Murday, J. S. (1988). *Corrosion Science, 28*.

Morad, M. S. (2008). *Corrosion Science, 50*(2).

Ongman, H. D. (1929). Reducing solutions at steam-boiler temperatures. *Combustion, 24*, 40–44.

Parveen, M., Mobin, M., & Zehra, S. (2016). Evaluation of l-tyrosine mixed with sodium dodecyl sulphate or cetyl pyridinium chloride as a corrosion inhibitor for mild steel in 1 M HCl: Experimental and theoretical studies. *RSC Advances*, *6*(66), 61235–61248. Available from http://pubs.rsc.org/en/journals/journalissues, https://doi.org/10.1039/c6ra10010d.

Pryor, M. J., & Cohen, M. (1953). The inhibition of the corrosion of iron by some anodic inhibitors. *Journal of the Electrochemical Society*, *100*(5), 203–215. Available from https://doi.org/10.1149/1.2781106.

Revie, R. W., & Uhlig, H. H. (2011). *Uhlig's corrosion handbook*. Wiley.

Samuel, C. Bradford, C., Van Nostrand. Reinhold, Available from https://doi.org/10.1007/978-1-4684-8845-6.

Trabanelli, G. (1970). Chemical industries: Corrosion mechanism. Marcel Dekker.

Uhlig, H. H., Triadis, D. N., & Stern, M. (1955). Effect of oxygen, chlorides, and calcium ion on corrosion inhibition of iron by polyphosphates. *Journal of the Electrochemical Society*, *102*(2), 59–66. Available from https://doi.org/10.1149/1.2429995.

Wang, D., Tang, X., Qiu, Y., Gan, F., & Chen, G. Z. (2005). A study of the film formation kinetics on zinc in different acidic corrosion inhibitor solutions by quartz crystal microbalance. *Corrosion Science*, *47*(9), 2157–2172. Available from https://doi.org/10.1016/j.corsci.2004.10.003.

Zehra, S., Mobin, M., Aslam, J., & Parveen, M. (2017). Assessment of glycine derivative N-benzylidine-2((2-oxo-2-(10H-phenothiazine-10yl)ethyl)amino) acetohydrazide as inhibitor for mild steel corrosion in 1 M HCl solution: electrochemical and theoretical approach. *Journal of Adhesion Science and Technology*, *32* (3), 317–342. Available from https://doi.org/10.1080/01694243.2017.1354669.

Zehra, S., Mobin, M., & Aslam, J. (2022). Chromates as corrosion inhibitors. Elsevier BV. Available from https://doi.org/10.1016/b978-0-323-90410-0.00014-3.

21 Corrosion protective coatings

21.1 Introduction

The most extensively utilized practical strategy for controlling corrosion is undoubtedly the application of protective coatings. They offer long-term defense against various corrosive conditions, from air exposure to the most challenging chemical processing conditions (Stoye & Freitag, 1998).

Although protective coatings do not offer much in the way of structural support, they do shield other materials so that they can maintain their strength and integrity. The primary purpose of a protective coating is to shield corrosive structural elements from the environment. Any flaw in a coating might serve as a focal point for the deterioration of the substrate and corrosion, so it must provide a continuous barrier to the substrate.

In addition to the composition of the materials, other elements affect the quality of the coatings. For a coating to be effective against corrosion, it must have several characteristics, including a high level of adhesion to the substrate, a minimal amount of coating discontinuity (porosity), high resistance to the flow of electrons, a sufficient thickness (the more significant the thickness, the more excellent the resistance), and a slow rate of diffusion for ions like Cl^- and H_2O.

Although coatings have been used for esthetic and identification purposes for thousands of years, it was not until World War II that their industrial significance was realized. There are billions of gallons of paint sold worldwide each year. Metal surfaces receive about one-third of the paint produced for protection and decoration. All modes of transportation, including buses, cars, ships, and airplanes, as well as buried underground structures like tanks, oil and gas pipelines, offshore installations, iron structures, and all metallic equipment, call for coatings. As a result, the coating sector has grown to be one of the biggest in terms of production. The fact that coating is almost never disregarded in any corrosion protection strategy speaks volumes about the significance of coating. At a compound annual growth rate of 5% throughout the forecast period, it is expected that the worldwide paints and coatings market will increase from $167.04 billion in 2022 to $235.06 billion in 2029.

When a metal is otherwise mechanically and physically appropriate for the service circumstances, it is coated primarily to protect it from a corrosive environment. When it comes to cost and necessary life, coating metal with strong mechanical properties, typically

Handbook of Corrosion Engineering. DOI: https://doi.org/10.1016/B978-0-323-95185-2.00021-6

steel, is frequently more practicable than opting for a more expensive corrosion-resistant material.

There are four ways to achieve protection, and many coatings work in more than one of these ways:

a A *barrier coating* that prevents the corrosive environment from contacting the metal.
b A *cathodic layer* that protects the underlying metal while providing a corrosive sacrificial coating.
c An *inhibitor coating* that slows electrode reactions.
d An *electrically resistive coating* that stifles electrochemical corrosion cells.

Paints fall into *electrically resistive coating* category.

Appropriate surface preparation is the secret to adequate protection, which is even more crucial than the characteristics of the coatings. If the metal surface is rusted or contaminated with oil, salt, or other substances, no coating will remain functional for an extended period. Most of the coatings include many layers. Each layer adds one or more qualities to shield the base metal.

Fig. 21–1 depicts the complexity of the coating system as well as some of the factors that have an impact on the effectiveness and longevity of an anticorrosive coating system.

This chapter aims to develop a basic understanding and cover major principles involved in the field of anticorrosion coating.

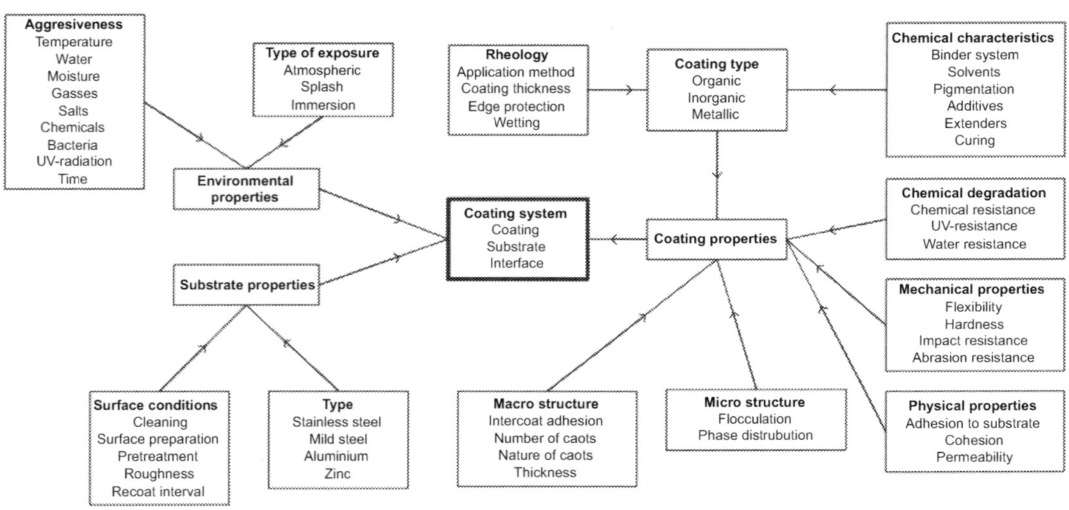

FIGURE 21–1 Factors affecting the durability of an anticorrosive coating system. *From Sørensen, P. A., Kiil, S., Dam-Johansen, K. & Weinell, C. E. (2009). Anticorrosive coatings: A review.* Journal of Coatings Technology and Research, *6(2), 135–176. https://doi.org/10.1007/s11998-008-9144-2.*

21.2 Major goals to apply anticorrosion coatings

The following are the main goals of applying anticorrosion coatings:

- Equipment and structures are protected from the elements by serving as a barrier between the substrate and the aggressive environment, such as the maritime and industrial environments.
- Control of solvent losses.
- Marine fouling is controlled by coating components that prevent the mildew growth and marine fouling in seawater.
- Reduction in friction; coating reduces friction between two contacting surfaces.
- Pleasant appearance; certain coatings provide a pleasant appearance and produce attractive surroundings.
- The intensity of light can be changed by choosing the right coatings for rooms and structures to meet specific needs.
- Visibility: to warn aircraft, TV, and radio towers use a variety of color combinations that are visible from great distances.
- By altering the material's chemical, mechanical, thermal, electrical, and optical properties.
- Thin coatings are applied on inexpensive substrates to boost efficiency and save costs.

21.3 Classification of the coatings

Based on how they retard metal from corrosion, anticorrosive coatings are often categorized into three categories: sacrificial protection, passivation of the substrate surface, and barrier protection (galvanic effect). The general classification is discussed in Fig. 21–2.

Using a coating system with low permeability for liquids, gases, and ions, one may create a barrier that will stop harmful species from penetrating the surface of the substrate. The incorporation of inhibitive pigments or a chemical conversion layer into the coatings can passivate the surface of the substrates. Sacrificial protection, or when a more electrochemically active metal in electrical contact with the substrate corrodes to give protection, has been widely utilized to shield metals against corrosion via metallic, organic, and inorganic coatings.

21.3.1 Barrier coatings

By fabricating an insulating and physical barrier, a barrier coating prevents corrosive substances like the electrolyte from encountering the substrate. Barrier coatings are used to create an efficient physical barrier, as well as a smooth foundation for finishing. Barrier coatings feature short drying periods, excellent dimensional stability, exceptional toughness, and abrasion resistance. They also offer outstanding resistance to corrosive chemicals, heat, and ultraviolet radiation. Additionally, they offer powerful adhesion to various substrates, including glass, metals, fibers, and several other materials. When metals and ceramics cannot resist highly abrasive working conditions, barrier coatings are often used.

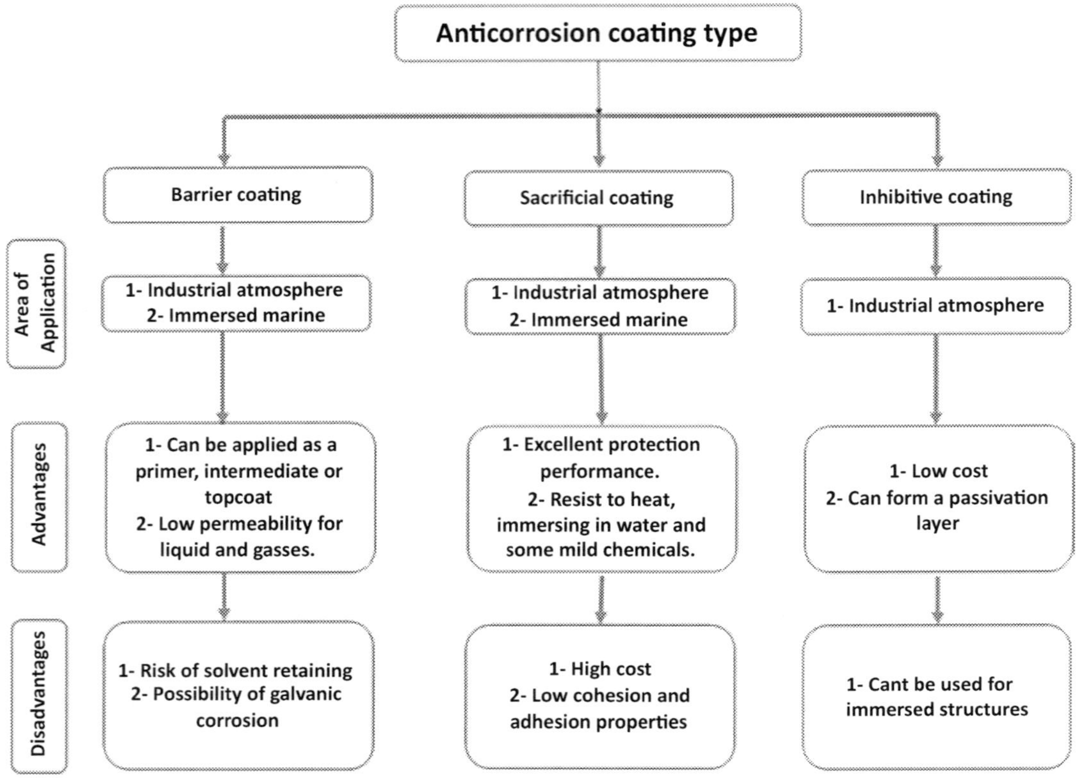

FIGURE 21–2 Outline of the classification of anticorrosion coatings.

Many industrial operations occur at extremely high temperatures while surrounded by corrosive, harsh gases. Metals are frequently prone to corrosion, resulting in catastrophic failure during operation. Metal and ceramic components are coated with a different substance that can survive these harsh conditions to prevent deterioration and increase life and operational efficiency. The propeties of the material of barrier coating must match those of the underlying substrate. Finding a method to bind the top coat chemically or physically to the substrate is a difficult task. Anodic oxides, inorganic coatings, inhibitive coatings, and organic coatings are the four different forms of barrier coatings.

Anodic oxides: By electrolysis, a layer of Al_2O_3 is formed on the surface of the aluminum. The oxides are sealed with a potassium dichromate solution since they are permeable. Reducing porosity is the goal of sealing. Chromate use is restricted in various countries due to health risks.

Inorganic coatings: Coatings made of ceramic and glass are among them. Water is essentially impervious to glass coatings. As long as they are not mechanically harmed, cement coatings are impenetrable to damage.

Inhibitive coatings: Inhibitors are frequently applied to create surface layers that act as environmental barriers. Inhibitors, such as cinnamic acid, are added to paint coats to retard steel from corrosion in alkaline or neutral conditions.
Organic coatings: In the industrial context, coatings made of epoxy, polyurethane, chlorinated rubber, and polyvinyl chloride are frequently utilized. They operate as a barrier against oxygen and water and stop the cathodic reaction from happening beneath the coating. Impregnating the corrosion inhibitor in the primer further improves the barrier qualities.

21.3.2 Conversion coatings

The surface of a metal can experience a surface reaction that changes it into an inorganic compound that is particularly adherent since it was created from the metal itself. Given that the metal has already reacted, this surface is incredibly corrosion-resistant. Examples of conversion coatings are phosphate and chromate coatings. Conversion coatings are so-called because the surface metal is converted into a compound having the desired porosity to act as a good base for paint.

21.3.3 Anodic coatings

An anodic coating is a coating material that uses anodizing to provide enhanced thickness, color, and protection to provide aluminum or any other type of substrate. This coating is made up of the oxide layer that electrolysis produces on metal, with the metal acting as the anode. The natural protective layer on ductile and soft metals like aluminum can be pretty thin, rendering it vulnerable to harm. As a result, increasing the anodic coating provides the surface of the aluminum several advantageous qualities.

Anodizing, also known as reversed electroplating, is the method used to create anodic coatings. During anodizing, the surface acts as the anode, or positive electrode, within the electrolytic cell. The aluminum is immersed in an acid electrolyte or water solution. The regulated temperature will affect the uniformity of the coating. The coating will be transparent and pliable at 20°C. The eventual effect of exposure to 50°C is a thick, gray, and dull covering known as hard anodizing. A direct current then connects the anode or aluminum and the cathode. After application, the water molecules of the solution breaks, releasing oxygen into the anode. The oxide coating that is present on the surface is produced when oxygen reacts with aluminum to make oxide.

When the required anodic layer thickness is reached, the aluminum or anode is removed from the electrolyte and thoroughly rinsed to remove the acids. The anodic coating is ready for coating at this point if necessary. The anodic coating has the potential to be very porous and will either trap or absorb nearly all substances that pass through its pores. The current properties may benefit or suffer as a result of this. This is avoided by hydrolyzing or adding water to the oxide, which seals the coatings or pores. The final coating will be homogenous, clear, durable, and smooth.

21.3.4 Cathodic coatings

The metals that are deposited in this kind of coating are electropositive to the substrate. For instance, copper (£° = +0.337 V) is positive for steel (F = -0.440 V) with a copper coating. The coatings must be thick and devoid of pores. In general, discontinuities and pores are not visible in electroplated coatings. On the other hand, the coating functions as the anode concerning the substrate if it has a fault (crater). Electrons, therefore, go from the crater to the noble covering.

21.3.5 Other coatings

Some other coatings include glass coatings, porcelain coatings, and high-temperature coatings.

21.3.5.1 Glass coatings

Alkyl resins are used to distribute chemically resistant glass flakes that are 3–4 μm thick. Utilized as a pigment is TiC > 2. Five microns of glass flakes are obtained. The impermeability and erosion resistance are increased by the application of glass flakes.

21.3.5.2 Porcelain enamels

In the wet method, clay and a mixture of 15% solids are suspended in the water together with powdered frit (ground glass). Melting combinations of B_2O_3, SiO_2, Al_2O_3, ZrO_2, $Na_2CC > 3$, and PbO create the mixture. The amount of SiO_2 increased for acid resistance. Onto the metal surface that needs coating, the mixture is sprayed. The workpiece is heated to dry the coating before being fired for 5 minutes at 850°C. Up until the desired coating thickness is achieved, this procedure is repeated. In the dry method, the workpiece is covered with the powdered frit after being heated to the frit's fusion temperature. The coating for iron and steel is fired at a temperature of 700°C. Up to 1000°C, porcelain enamel-coated steels are suitable for usage.

21.4 Metallic coatings

Metallic coatings offer a layer that converts the workpiece's surface characteristics to those of the metal being applied. The workpiece transforms into a composite material with characteristics that, when employed separately, are often not possible for either material. The coatings offer durable, corrosion-resistant covering, while the load-bearing ability comes from the core material. Because metallic coatings are significantly more expensive than organic coatings, their usage is only justified by their extended lifespan. Metallic coatings shield the substrate by serving as a barrier between it and the environment or by corroding more quickly in relation to the substrate. The main techniques for applying metallic coatings are as follows:

a electroplating and electrodeposition;
b diffusion;

c thermal spraying;
d flame spraying;
e hot-dipped galvanized;
f vapor deposition (chemical and physical);
g welding, cladding, bonding;
h miscellaneous methods.

Cadmium, chromium, nickel, aluminum, and zinc are a few crucial coatings. Additionally, electrical equipment and occasionally applications for specialized fasteners employ copper, gold, and silver. Silver is utilized to prevent fretting; gold is occasionally used to promote electrical conductivity in waveguides and at contacts; and copper serves as the base layer in multiple-plate electroplating. In order to attain the appropriate surface condition, metal objects are dipped into and then retrieved from baths containing various chemicals during plating and surface treatment procedures, which are often batch operations. The procedures entail passing the item being coated through a number of baths that are intended to yield the desired result. Depending on the complexity and amount of automation used, these procedures can be highly automated or manual.

21.4.1 Electroplating

Electroplating is accomplished by running an electrical current through a solution containing dissolved metal ions and the metal item to be plated. In an electrochemical cell, the metal item acts as the cathode, drawing metal ions from the solution. Various metals, such as aluminum, brass, bronze, cadmium, copper, chromium, iron, lead, nickel, tin, zinc, and precious metals like gold, platinum, and silver, are plated on ferrous and nonferrous metal items. Controlling a number of factors, such as voltage and amperage, regulates the process, temperature, lengths of residency, and bath solution purity. Only those metals that can be reduced from aqueous solutions of their salts may be electrodeposited since plating baths are nearly invariably aqueous solutions. Aluminum is the one significant exception since it may be plated using organic electrolytes. A typical electroplating operation's unit activities include many cleaning processes, peeling off old plating or paint, electroplating procedures, and washing before and after each of these phases. Similar procedures are used in electroless plating, which involves depositing metal on a substrate without the use of external electrical energy. There are three approaches used:

21.4.1.1 Vat-plating

Thousands of liter-capacity tanks or vats are used to execute the electrodeposition process. The workpiece is hung in the electrolyte while being fastened in a jig. A few millimeters away from the workpiece, inert anodes are hanging since they do not dissolve in the electrolyte. A transformer-rectifier supplies the DC (4–8 V). The major application for the acquired thin coatings is decoration. Vat plating results in dense coatings. The relationship between the coating's thickness and current density is linear.

21.4.1.2 Selective plating

With this method, components may be electrodeposited without masking or immersion onto the targeted, localized locations. An absorbent pad coated with an electrolyte-soaked electrolyte anode is fitted in an insulated handle. The circuit is completed by the contact of the absorbent pad with the workpiece, which is coupled to the negative side of a DC power source. The method eliminates the need for masking. Compared to vat plating, the deposition rate is greater, and this method has the advantage of mobility.

21.4.1.3 Electroless plating

The catalytic reduction of nickel ions in an aqueous solution (including a chemical reducing agent) is the basis for the chemical reduction process known as electroless nickel plating, which results in the deposition of nickel metal without the use of electrical energy. For instance, the reduction chemistry for nickel deposition is as follows:

$$Ni^{2+} + H_2PO_2^- + H_2O \overset{\text{Catalyst}}{\rightarrow} Ni\,(\text{metal}) + 2H^+ + H_2PO_3 \quad (21-1)$$

The procedure is carried out in a stainless steel tank with polytetrafluoroethylene lining at 90°C. Deposits are made consistently. Thus a chemical reducing agent in solution provides the driving force for the reduction of nickel metal ions and their deposition during the electroless nickel plating process. As long as there is enough agitation to guarantee a homogeneous concentration of metal ions and reducing agents, this driving potential is practically constant across the component's surface.

Because of this, the thickness of the electroless deposits is fairly consistent across the size and shape of the item. Thus when plating irregularly shaped items, holes, recesses, interior surfaces, valves, threaded pieces, etc., the technique offers specific benefits. In many industrial settings, electroless (autocatalytic) nickel coating offers a hard, homogeneous, corrosion-, abrasion-, and wear-resistant surface to safeguard machine components. Because electroless nickel is chemically deposited, the coating's thickness is remarkably consistent. Thorough process control can accurately recreate the surface finish, obviating the requirement for expensive postplating machining. Only the surface of a catalytic substrate in contact with the plating solution experiences a reduction of metal ions during a genuine electroless plating process. The plating process continues because the deposited metal is also catalytic after it has completely covered the catalytic substrate. Because of this, electroless plating techniques are frequently utilized in industry to satisfy practical end-use requirements and are rarely used for ornamental purposes. Unique benefits of electroless nickel plating include the uniformity of the deposits, even on intricate geometries. The electroplated coatings are thicker on parts that protrude and are thinner in locations that are recessed.

Deposits frequently offer substantially better barrier corrosion protection for steel substrates than electroplated nickel and hard chrome because they are frequently less porous. Deposits generate approximately a tenth as much hydrogen absorption as hard chromium and about a fifth as much as electrolytic nickel. Deposits may be plated under compressive or zero stress. Thus the substrate material's debit for fatigue strength is either zero or positive.

Unlike electrolytic nickel, deposits have intrinsic lubricity and nongalling properties. Oils can wettably penetrate deposits. Deposits are easily brazed, are weldable, and may be soldered.

Deposits are substantially harder, with as-plated microhardness ranging from 450 to 600 HV; this value may be raised to 1000–1100 Vickers Hardness (HV) with the right heat treatment. This increase is the result of Ni_3P precipitating, which hardens the alloy generally. Deposits have distinctive magnetic characteristics. In the as-plated form, electroless nickel deposits with a P content of more than 8% are often regarded as being largely nonmagnetic. The nonferromagnetic behavior of as-plated Ni-P deposits is caused by the presence of phosphorus in Ni-P coatings as a supersaturated solution in a fine microcrystalline solid solution that borders on an amorphous or liquidlike (glasslike) metastable structure (with P–8%). By co-plating silicon carbide particles that are smaller than a millimeter with nickel, the second generation of electroless nickel plating has been created that is exceptionally resistant to wear and corrosion.

Nickel: For air corrosion resistance and usage in bright, esthetic finishes, nickel is frequently plated on copper. Nickel plating is most typically used as an undercoat for chrome plating. Changes in current density, temperature, plating duration, and bath chemistry can affect the coat's thickness, ductility, and hardness.

Nickel electroless: Metal components can be plated by submerging them in a specific solution that contains an unstable nickel salt, which is broken down by the metal surfaces. Due to the nickel deposit's 4%–12% phosphorus content, which hardens it, the plating has high corrosion resistance and wear resistance. The uniformity of the coatings makes plating inside holes and protected regions significantly superior to electroplating. However, the coatings are brittle, and the chemicals are expensive.

Chromium plating: Unlike other plating methods, electroplating employs insoluble lead alloy anodes, necessitating the addition of chromic acid to the bath. Anodes made of chromium would passivate and halt the current flow. Although chromium is a reactive metal, it passivates so efficiently that it is plated cathodically on steel with a layer of nickel in between. Microcracked chromium electrodeposited at circumstances that leave 400–800 cracks/ern (1000–2000 cracks/in) or more per inch provides the highest resistance to air corrosion. Microcracks in the plate lower stresses without encroaching very deeply; as the number of fractures increases, so does the depth of penetration.

Zinc: Zinc serves as a barrier and a sacrificial layer for steel since it is anodic to zinc. Because it may be applied more smoothly than hot dipping, a thin (5–10 μm or 0.2–0.4 mil) electroplated zinc coating is frequently used as a paint foundation, particularly on the lower portions of automotive bodywork. Zinc is given a chromate dip to keep its bright, shining appearance if it is not painted (such as on screws, nuts, and bolts). Because of health risks, zinc-plated steel should not be used for prolonged immersion or in contact with food.

Cadmium: Due to its high ductility and electrical conductivity, cadmium, like zinc, is anodic to steel and is thinly coated (5–15 μm or 0.2–0.6 mil) onto steel springs, fasteners, television chassis, and electronic equipment. It is toxic, has equal atmospheric corrosion resistance to zinc, and is not chemically resistant. Cadmium plating is not often painted, unlike zinc. High-strength steels can become brittle when exposed to cadmium at temperatures exceeding 230°C (450°F). High-strength steels must be baked at roughly 200°C (400°F) right away after plating because hydrogen that is charged into the metal during electroplating, particularly cadmium plating, might cause embrittlement. For steels with tensile strengths around 1500 MPa (220 ksi), baking might take as little as 3 hours, although portions thicker than 25 mm (1 in.) may need up to 24 hours.

Tin: Tin coatings on steel are nearly typically electroplated, with coat thicknesses as low as 0.4 μm (0.02 mil), and then they undergo a flash meeting process to decrease porosity. Tin is superior to steel in water; however, it is sacrificed to steel when food is sealed inside airless tin cans. It is commonly used in food handling equipment since it is not harmful and seems to have no flavor. Because it solders well and does not tarnish, tinplate is also often employed in the electronics sector.

21.4.2 Diffusion

In this method, the surface of the metal to be coated is modified by diffusing it at a high temperature into a metal or an element, which would provide the required resistance when combined with the parent metal. Such coatings are called "diffusion coatings." Diffusion coatings can be applied to a range of metals and alloys, such as nickel, titanium, and molybdenum, but the widest use is on ferrous metals. Examples are zinc diffusion coatings and aluminum diffusion coatings. Diffusion coatings are formed by depositing a layer of metal on the metal surface and then heating the component in a furnace for a period of time. During this heat treatment, the aluminum and metal atoms migrate, or diffuse, into each other, which is the reason these coatings are called diffusion coatings. This processing is usually performed by a pack cementation process in which the aluminum deposition and the heat treatment occur simultaneously.

21.4.3 Pack cementation

Pack cementation is widely used to confer oxidation resistance on ferrous alloys. Usually, relatively expensive aluminum or binary alloy grade reagent is used during the packing process with aluminum as a source. Pack cementation processes include aluminizing, chromizing, and siliconizing. Components are packed in metal powders in sealed heat-resistant retorts and heated inside a furnace to precisely control temperature–time profiles. In the aluminizing process, a source of aluminum reacts with a chemical activator on heating to form a gaseous compound (e.g., pure Al with NaF to form AlF). This gas is the transfer medium that carries aluminum to the component surface. The gas decomposes at the substrate surface, depositing aluminum and releasing the halogen activator. The halogen activator returns to the pack and reacts with the aluminum again. Thus the transfer process continues until all of the aluminum in the pack is used or until the process is stopped by cooling. The coating forms at temperatures ranging from 700°C to 1100°C over a period of several hours. Pack cementation is the most widely used process for making diffusion aluminide coatings.

Diffusion coatings are primarily aluminide coatings composed of aluminum and base metal. A nickel-based superalloy forms a nickel-aluminide, which is a chemical compound with the formula NiAl. A cobalt-based superalloy forms a cobalt-aluminide, which is a chemical compound with the formula CoAl. It is common to incorporate platinum into the coating to improve corrosion and oxidation resistance. This is called a platinum-aluminide coating. Diffusion chrome coatings are also available. Diffusion aluminide coatings protect the base metal by forming a continuous, aluminum oxide layer, Al_2O_3, which prevents further oxidation of the coating (Actually, oxidation continues but at much slower rates than without a continuous aluminum oxide scale.). When part of the Al_2O_3 scale spalls off, the underlying aluminide layer is exposed to form a new Al_2O_3 scale. Thus the coating is self-healing. Pack cementation can also be used to produce chromium-modified aluminide coatings. The addition of chromium is known to improve the hot corrosion resistance of nickel-based alloys. Although chromium can be co-deposited with aluminum in a single-step process, a duplex process is frequently used to form the chromium-modified aluminide. The component is first chromized using either pack cementation or a gas phase process, and this is then followed by standard aluminizing treatment. The final distribution of the chromium in the coating will depend on whether a low- or high-activity aluminizing process is employed. For a platinum-aluminide coating, a thin (typically 8-m) layer of platinum is first deposited onto the substrate, usually by a plating process. The second step involves aluminizing for several hours using the conventional packed cementation process to form the platinum-aluminide coating. Conventional pack cementation processes are unable to effectively coat internal surfaces such as cooling holes. The coating thickness on these internal surfaces is usually less than on the surface due to limited access by the carrier gas. Access can be improved by pulsing the carrier gas, or by the use of vapor phase coating process.

Another method of coating both the internal and external surfaces involves generating the coating gases in a reactor that is separate from the vessel the parts are in. The coating gases are pumped around the outside and through the inside of the parts by two different distribution networks. Internal passages can be coated by filling them with the powder used in the pack. Slurry processes can also be used to deposit the aluminum or the aluminum and other alloying elements. The slurry is usually sprayed on the component. The component is then given a heat treatment, which burns off the binder in the slurry and melts the remaining slurry, which reacts with the base metal to form the diffusion coating. After coating, it is usually necessary to heat treat the coated component to restore the mechanical properties of the base metal.

21.4.4 Spraying

Coatings can be sprayed from rod or wire stock or powdered materials. The material (e.g., wire) is fed into a flame, where it is melted. The molten stock is then stripped from the end of the wire and atomized by a high-velocity stream of compressed air or other gases, which propels the material onto a prepared substrate or workpiece. Depending on the substrate, bonding occurs either due to mechanical interlock with a roughened surface, due to localized diffusion and alloying, and/or by means of Van der Waals forces (i.e., mutual attraction and cohesion between two surfaces).

The basic steps involved in any thermal coating process are substrate preparation, masking and fixturing, coating, finishing, inspection, and stripping (when necessary). Substrate preparation usually involves scale and oil and grease removal, as well as surface roughening. Roughening is necessary for most of the thermal spray processes to ensure adequate bonding of the coating to the substrate. The most common method is grit blasting, usually with alumina. Masking and fixturing limit the amount of coating applied to the workpiece to remove overspray through time-consuming grinding and stripping after deposition. The basic parameters in thermal spray deposition are particle temperature, velocity, angle of impact, and extent of reaction with gases during the deposition process. The geometry of the part being coated affects the surface coating because the specific properties vary from point to point on each piece.

In many applications, workpieces must be finished after the deposition process, the most common technique being grinding followed by lapping. The final inspection of thermal spray coatings involves verification of dimensions, and a visual examination for pits, cracks, and so forth. Nondestructive testing has largely proven unsuccessful.

There are three basic categories of thermal spray technologies: combustion torch (flame spray, high-velocity oxy-fuel, and detonation gun), electric (wire) arc, and plasma arc. Thermal spray processes are maturing, and the technology is readily available. Environmental concerns with thermal spraying techniques include the generation of dust, fumes, overspray, noise, and intense light. The metal spray process is usually performed in front of a "water curtain" or dry filter exhaust hood, which captures the overspray and fumes. Water curtain systems periodically discharge contaminated wastewater. Noise generated can vary from approximately 80 dB to more than 140 dB. With the higher noise-level processes, robotics are usually required for spray application.

The use of metal spray processes may eliminate some of the pollution associated with conventional tank plating. In most cases, however, wet processes, such as cleaning, are necessary in addition to the metal coating process. Therefore the complete elimination of tanks may not be possible. Waste streams resulting from flame spray techniques may include overspray, wastewaters, spent exhaust filters, rejected parts, spent gas cylinders, air emissions (dust, fumes), and wastes associated with the grinding and finishing phases.

21.4.4.1 Flame spraying

In general, the coating material is melted and blown onto the surface to be coated. Very few molten particles or droplets make up the coating substance. Based on the shape of the coating material, four ways are often used: an electrical arc and an additional gas stream that blows the molten metal out of the arc in droplets.

a Oxyacetylene flame melts the coating material, which is in the shape of a rod, and blows it onto the surface that needs coating.

b An oxyacetylene flame heats the powder coating material before atomizing it and blowing it onto the surface.

c The coating is shaped like a wire. The wire is fed into a nozzle's central aperture. It is sprayed over the metal surface after going through an oxyacetylene flame.

d By putting the coating material, which is in the shape of a wire, through the plasma of an electric arc, it is heated. An additional gas stream blows the molten metal out of the arc in droplets.

The method is adaptable and requires little upfront money. The particles discharged from the spray gun during the electrostatic spraying process are electrostatically charged and pushed at low speed by air or a rotating spray head. Avoid using too much air pressure. With this process, placing the workpiece is not necessary to generate a proper wrap-around. The method works well with tubular objects.

21.4.5 Hot-dip galvanizing

Iron and steel can be treated using this procedure to retard corrosion. An item is dipped into a molten zinc bath where the zinc interacts with the iron to generate a coating. First, hot alkali and cold alkali solutions are used to remove thick greases, paints, and lacquers. Grit blasting is used to eliminate surface contaminants. H_2SO_4 or HC1 pickling solution is used to remove oxides and scales. The item to be galvanized is submerged in a 450°C molten zinc bath. Both iron and steel are instantly wetted by the zinc when submerged in a galvanized bath, where they react to produce an iron zinc coating. The initial rate of reaction is quite quick at the typical galvanizing temperature, which is 445°C–465°C. During this time, the primary thickness is produced. The average immersion time is 1–2 minutes.

An inner layer of zinc and iron in contact with the base metal and an exterior layer of unalloyed zinc make up the majority of a hot-dipped galvanized coating. To decrease alloy layer formation and increase ductility, alloying additives are done. Gray coating refers to a coating comprised completely of an alloy layer. First, thick greases, paints, and lacquers are removed using hot and cold alkali solutions. Surface contamination is removed by grit blasting and pickling removes the oxides.

To alter the coating, other metals including aluminum and zinc are added to the zinc bath. Tin and aluminum additions increase adherence and generate a uniform coating, respectively. Fluxing is required to remove surface contaminants and guarantee that only spotless steel interacts with molten zinc. The workpiece is initially degreased in a hot alkaline solution before being washed with hot and cold water in the dry process. To get rid of the mill scale, oxides, and rust, it is rinsed and then pickled in HC1 or H_2SO_4. To get rid of iron salts, it is washed once more in water. The workpiece is dried after being submerged in a flux tank after degreasing. Drying may be done at 120°C in ovens or on hot plates. The dry workpiece is subsequently placed, as previously said, in the molten zinc bath. The workpiece is not fluxed or dried before being submerged in the molten zinc bath during the wet galvanizing process. It is submerged immediately in a bath of molten zinc that has a layer of flux covering the zinc's surface. Zinc ammonium chloride is the main component of the flux cover utilized in the wet process.

The workpiece is removed using a flux sheet. As the workpiece moves through the flux and into the molten bath, its surface is cleaned and moistened. With the dry process, dross quality is reduced and the industrial environment may be kept cleaner. As the workpiece moves through the flux blanket, less space is needed for the wet process and a better surface for galvanizing is created. Iron and aluminum in trace amounts (0.005% each) slow down oxidation and increase brightness.

21.4.6 Electrogalvanizing

In electrogalvanizing, steel strip is passed through a series of electrolytic cells instead of the hot-dipped techniques' furnaces, galvanizing pots, and cooling towers. An electrical current travels through a zinc solution from the anode to the cathode in each plating cell (workpiece). Typically, zinc slab anodes that are soluble in acid are employed. Zinc is produced by dissolving these anodes, and it is then deposited on the cathode (workpiece). In contrast to the hot-dipping procedure, electrochemistry is used to link zinc to steel. In the electrogalvanizing process, no alloy is formed.

21.4.7 Vapor deposition

The workpiece is removed using a flux sheet. As the workpiece moves through the flux and into the molten bath, its surface is cleaned and moistened. With the dry process, dross quality is reduced and the industrial environment may be kept cleaner. As the workpiece moves through the flux blanket, less space is needed for the wet process and a better surface for galvanizing is created. Iron and aluminum in trace amounts (0.005% each) slow down oxidation and increase brightness.

$$2AlCl_3 + 3Fe \rightarrow 3FeCl_2 + 2Al \qquad (21-2)$$

21.4.8 Cladding

Corrosion resistance can be improved by metallurgically bonding to the susceptible core alloy a surface layer of a metal or an alloy with good corrosion resistance. The cladding is selected not only to have good corrosion resistance but also to be anodic to the core alloy by about 80–100 mV. Thus if the cladding becomes damaged by scratches, or if the core alloy is exposed at drilled fastener holes, the cladding will provide cathodic protection by corroding sacrificially. Cladding is usually applied at the mill stage by the manufacturers of the sheet, plate, or tubing. Cladding by pressing, rolling, or extrusion can produce a coating in which the thickness and distribution can be controlled over wide ranges, and the coatings produced are free of porosity. Although there is almost no practical limit to the thickness of coatings that can be produced by cladding, the application of the process is limited to simple-shaped articles that do not require much subsequent mechanical deformation. Among the principal uses are lead and cadmium sheathing for cables, lead-sheathed sheets for architectural applications, and composite extruded tubes for heat exchangers. Because of the cathodic protection provided by the cladding, corrosion progresses only to the core/cladding interface and then spreads laterally, thus helping to prevent perforations in thin sheets. The cut edges of the clad product should be protected by the normal finish or by jointing-compound squeezed out during wet assembly.

For aluminum-copper alloys (2000 series) dilute aluminum alloys such as 1230, 6003, or 6053, containing small amounts of manganese, chromium, or magnesium, may be used as cladding material. These have low-copper contents, less than 0.02%, and low-iron content, less than 0.2%. However, these alloys are not sufficiently anodic with respect to the Al-Zn-Mg-Cu alloys of the 7000 series, and they do not provide cathodic protection in these cases. The 7000 series alloys are substitute cleaning and coating removal. The advanced coatings provide excellent adhesion between the substrate and the coating; as a result, these coatings are much more difficult to strip than conventional coatings. Many coating companies have had to develop proprietary stripping techniques, most of which have adverse environmental or health risks.

21.4.9 Paint coatings

The paint coating system comprises (Painting, 1999):

- The primer
- The intermediate coat
- The top coat

The illustration of the layers of paint is represented in Fig. 21–3.

21.4.9.1 The primer

The primer is the most crucial part of the coating system. Primers serve the following purposes:

1. The substrate must be firmly attached.
2. It must be able to inhibit corrosion in corrosive situations.
3. It must firmly adhere to either the top coat or the intermediate coat.

Before any coating is applied, primers are given enough time to stand before being covered.

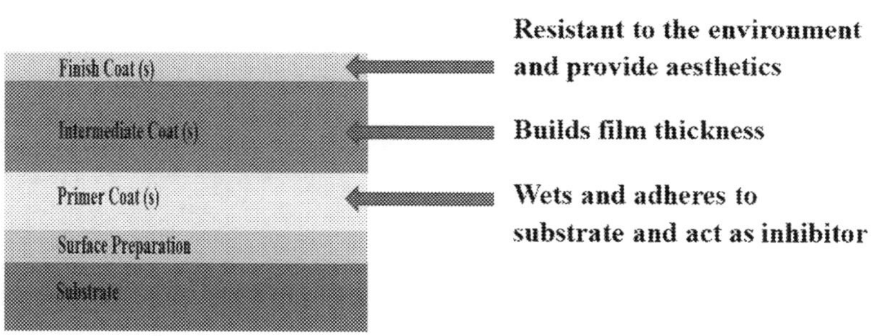

FIGURE 21–3 General representations of the layers of the paint coatings.

21.4.9.1.1 Types of primers

Inhibitive primers Inhibitory, impermeable, or cathodic primers, as detailed below, can be used depending on the environment in that a metallic construction will be exposed. A passive surface is created on the substrate, such as steel, by the reaction between the pigment in the primer and moisture absorbed in the coating. Red lead and chromate salts are two examples of pigments that are inhibitive yet have health risks.
Impervious primer This primer increases the coating's resistance to the passage of CO_2, oxygen, air, ions, and electrons. It is utilized in impermeable coating systems. Conventional thin film coatings cannot stop oxygen and water permeability. The thick coal tar enamel and primer in an impervious system combine to provide a highly impenetrable finish. The pigments are often metal salts, strontium, lead, and zinc chromates; however, the presence of Cr and Pb poses health risks.
Cathodically protective primer A zinc-based primer is an excellent illustration of this. Zinc corrodes preferentially to steel because it is anodic to steel, preventing steel corrosion. Experience has demonstrated that a zinc-rich primer may increase the lifespan of epoxy or chlorinated rubber top coat twofold. Three crucial coating systems are built on the three types of primers mentioned above: (a) impermeable coating systems for equipment that must be submerged, (b) inhibitive systems for use in marine conditions, and (c) cathodically protective systems for extremely corrosive situations.

21.4.9.2 Components of paints

The major components are discussed below (Fig. 21–4):
Major components of the paint.

- *Vehicle*: This is the liquid portion of the paint in which the pigment is dispersed. It is composed of a binder and the solvent or thinner or both.
- *Binder*. This binds the pigments in the coating in a homogeneous film. The binder also binds the total coating to the substrate. The binders provide the basis for the generic terminology of paint. The binders determine the physical and chemical properties of paints.
- *Pigment*. A pigment not only provides a pleasing color but also protects the binder from the adverse effect of ultraviolet radiation on the coating.
- *Solvent*. The purpose of the solvent is to provide the surface with a coating material in a form in which it can be physically applied to the surface.
- *Additives*. These are used to modify the properties of the coating, such as reducing drying time and enhancing the desired properties. One example is a plasticizer which makes the film flexible. Similarly, a drier may be added. It is a substance, such as a compound of lead, manganese, which, when introduced in drying oils, reduces their time. A drying oil, such as tung oil, forms a tough solid film in the air.
- *Extenders*. They are added to improve the application properties of the paint.

Paints and coatings may be categorized into two types: convertible and nonconvertible. Convertible paints and coatings need chemical reactions like oxidation or polymerization to

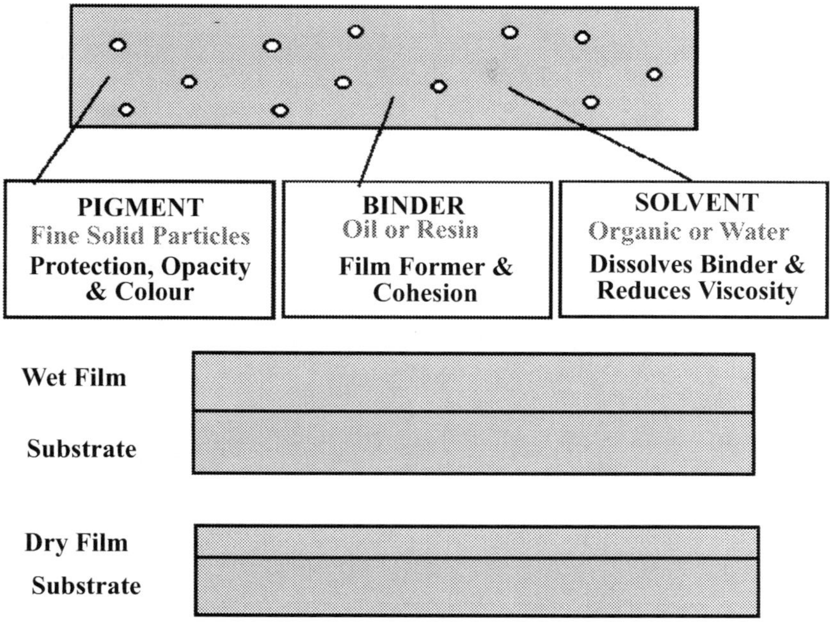

FIGURE 21–4 Major components of the paint.

form, whereas nonconvertible paints and coatings are created by solvent evaporation. Alkyds, epoxy, esters, polyesters, urethanes, silicon, and other resins fall within the first group.

21.4.9.2.1 Binders or resins

Binders can be classified in various ways. One of the classifications is given:

- **a** Drying oil, alkyd resins
- **b** Epoxy, polyurethane, and coal tar (two-pack chemical resistance)
- **c** Vinyl and chlorinated rubber (one-pack chemical resistance)
- **d** Bituminous coatings
- **e** Lacquers

a) Drying oil and alkyd resins

Drying oils These are natural binders that may be used either directly or after some processing. Natural drying oils included in this category include linseed oil, soybean oil, and coconut oil. To provide them with beneficial characteristics, synthetic resins are added to them. Based on fatty acid content, oils are categorized as drying, semidrying, and nondrying. Saturated fatty acids are those that have no double bonds, unsaturated fatty acids have one double bond, and polyunsaturated fatty acids have more than two double bonds.

A minimum of 50% of the fatty acids in a drying oil must be polysaturated. For instance, linolenic acid [$CH_3 \cdot CH_2CH = CH\ CH_2CH = CH_2CH2CH = CH(CH_2)_7\ COOH$] is a polyunsaturated fatty acid. Linseed oil is a good drying oil since it contains more than 50% linolenic acid. All oils undergo the same chemical processing and heating process to create alkyd resins.

Heating oils create oleoresinous binders with either natural or synthetic resins that have already been pre-formed. Alkyds and other synthetic resins have essentially taken the role of oleoresinous vehicles.

Alkyd resins The reaction of the oil with an acid and an alcohol produces the alkyd resin. The most widely used synthetic polymers in the coatings business are alkyd resins. Oils, fatty acids, polyhydric alcohols, and dibasic acids make up the bulk of the raw ingredients.

Compared to the original drying oil, the alkyd offers certain significant advantages. Alkyds are frequently used with other types of generic binders to improve characteristics. Vinyls, silicones, urethanes, and chlorinated rubber are the most often used binders. They have commonly used paints that may be applied successfully in a variety of settings. They are also considered to be convertible coatings.

b) Epoxy, polyurethane, and coal tar (two-pack chemical resistance)

Epoxy esters They also fall within the category of coatings that are changeable. Epoxy-based coatings come in a variety of forms, including heat-cured (epoxy-phenolic), air-drying (epoxy esters), catalyzed (amine or polyamide), epoxy copolymers (coat tar, phenolic, amine or polyamide), and high-build epoxy-based coatings. Some of the most potent adhesives now in use are made with the category of synthetic resins. Epoxy ester paints, one of its components, are cured by oxygen.

Epoxy mastics, which are materials without a solvent, are a unique family of epoxies that have just been developed. They may be cured down to 10°C and need minimal surface preparation. They produce highly uniform coatings. Coal-tar epoxy is a different class; some epoxy is changed for tar. To line tanks, utilize these materials. To synthesize epoxy esters, epoxy resins, and drying oils are chemically mixed. Diphenylolpropane (bisphenol A) and epichlorohydrin (DPP + ECH) combine to create epoxy resins. Both ends of the chain contain the epoxy group $H_2C{-}O{-}CH_2$. Epoxy resins employ amines or polyamides as catalysts to produce molecular cross-linking. These epoxy coatings that have been catalyzed exhibit good resistance to acids, solvents, and alkalies.

Epoxy ester can be created by reacting fatty acids with the terminal epoxide group and secondary hydroxy groups of solid epoxy resins. The polymer's C–C bond and ether linkage, as well as the molecule's terminal epoxide group, are what cause the resistance. The polar nature of the polymer gives it adhesive qualities. They outperform alkyds in terms of all their qualities. Their resistance to water and hydrocarbons has improved. They do, however, have a propensity to "chalk."

Polyurethane About half of all alkyds products are urethane oils and alkyds made of urethane. The sole difference between them and the ordinary air-drying alkyds outlined earlier is that instead of phthalic acid, which is utilized in dibasic acid in air-drying alkyds, isocyanates ($R{-}N=C=O$) are employed. Toluene diisocyanate, often known as TDI, is the most widely used isocyanate. TDI combines with an active hydrogen atom from hydroxy-containing substances such as polyethers and vegetable oils. The properties of the isocyanate and hydroxy compound utilized determine performance. With dampness, they do not respond. The fatty acid groups are still present in the alkyd molecule. Since the oil goes

through oxidation, this alkyd is cured by oxidation of oils. These alkyds offer a high level of abrasion resistance. Water, solvents, organic acids, and alkalis are all things that polyurethane resins can withstand with ease. They are highly resistant to abrasion.
c) Bituminous materials Asphalt is a part of crude oil, while coal tar is the byproduct of distilling coal tar. To create a lacquer, the aforementioned two are often mixed with a solvent, such as aliphatic or aromatic hydrocarbons. They are dried using solvent evaporation. The process of making water emulsions involves floating tiny bits of bitumen in the liquid and emulsifying them with additives like coal dust, limestone dust, manganese, and aluminum silicates. For use with hot metals, the bitumen is heated to a temperature of 175°C–245°C. They offer the best resistance to chemicals and dampness. They are cost-effective and are utilized in foundries, vehicle chassis, radiators, and waterproofing. They are damaged by nonpolar solvents, notwithstanding their strong resistance to nonoxidizing acids. Coal breaks down into gas and coke when heated to 1095°C without air. Gas condenses, and coal tar is produced. All coal tar coatings are susceptible to cracking and UV radiation-induced deterioration.

Asphalt is a brittle, naturally hard substance. Complex polymeric aliphatic hydrocarbons make up asphaltic compounds. They are relatively resistant to water and UV rays. Four main categories may be found in bituminous resin-based coatings:

- *Solid asphalt*: Solid asphalt bricks are melted and spread over the base.
- *Solvent cutbacks*: Bitumen is dissolved in a solvent, and the solvent's evaporation creates a nonconvertible coating. Cutback asphalt is what it is known as.
- *Emulsion*: Water is used to emulsify the bitumens before they are applied as a top coat.
- *Varnishes*: In oleoresinous paints, the bitumen substitutes some or all of the resin (oils combined with resin to obtain paints, such as alkyd and phenolics). In blast tanks, bitumen and tar have both been utilized extensively.

d) Lacquers: These are mixtures of substances that create organic films in organic solvents, and the solvent evaporates when the mixture is applied to a substrate. After evaporation, a solid film is created. The type of resin used affects the characteristics of lacquers. The lacquers dry up quite fast. The main categories of lacquers are as follows:

- cellulose derivatives,
- polyvinyl chloride,
- chlorinated rubber, and
- acrylics

The outline of these categories is discussed in Table 21–1.

21.4.10 Pigments

A component called pigment is mixed with the binder to give it certain physical and chemical qualities. The following are the roles that pigments play:

- *Esthetic appeal:* color or special effects.

Table 21–1 Detailed outline of the different types of lacquers.

S. no.	Name	Structure	Source
1.	Cellulose derivatives		The sources of cellulose include wood pulp or cotton linters, which are used as raw materials. The cellulose is nitrated, then heated in water to restore the required properties. It serves as a lacquer film former.
2.	Polyvinyl chloride		A vinyl chloride gas is created when hydrochloric acid, ethylene, or acetylene react. In the process, ethylene loses one hydrogen atom and gains a chlorine atom, rendering it noncombustible. Polyvinyl chloride is produced during polymerization. A linear chain polymer with substantial chlorine side groups is polyvinyl chloride.
3.	Chlorinated rubber		It originates from natural rubber, an isoprene polymer. Tetrachloroisoprene is created when natural rubber is chlorinated.
4.	Acrylics		For usage in both land-based and maritime industries, vinyl and chlorinated rubber paints have largely taken the role of acrylic paints. R and R' are either hydrogen atoms or alkyl groups in their overall chemical structure. Because of their superior strength, chemical resistance, and weather resistance, acrylic polymers are mostly used. They are created by the polymerization of acrylic and methacrylic acid esters. The primary resins are copolymers with ethyl, butyl, or methacrylate, as well as polymers of methyl and esters of acrylic and methacrylic acid.

- *Surface protection*: It offers corrosion and weathering resistance, such as iron oxide and chrome oxide.
- *Corrosion inhibition*: Numerous pigments, including chromates, operate as inhibitors by assisting in the formation of a barrier layer.
- *Film reinforcement*: The paint coating is robust thanks to various pigment particles, including oxides and silicates.
- *Increased coverage*: Many pigments expand the number of solids present without reducing the pigment's efficacy.
- *Adhesion*: Numerous pigments, including aluminum flakes, improve adhesive properties.

Criteria for the selection of the pigments:

As the pigment used must not be impacted by the solvent, diluent, or any other paint component, it must be compatible with the type of resin employed.

Paint performance: A pigment must work with how the paint is applied. For instance, carbon black is not suitable for industrial facilities, although titanium dioxide is.

Transparency versus opacity: The choice of color is determined by the desired level of opacity. Black, brown, and green pigments can be utilized if opacity is needed. Iron oxide paints in the colors yellow and red can be used to create transparency.

Ability to combine pigments: Various organic and inorganic colored pigments may be blended to create various color combinations. For instance, they are mixing yellow and blue results in a green pigment and mixing red and white results in a pink pigment.

Durability: This refers to how well the pigments can endure environmental changes without losing their capacity to retain color over time.

Gloss: The pigments need to give the surface an excellent glossy finish. Gloss is the degree to which a surface may reflect incident light in a way that resembles a perfect mirror image.

21.4.10.1 Types of pigments

Different types of pigments are discussed in Table 21–2.

21.4.10.1.1 Solvents

It is a material used to convert a solid into a fluid form. Low molecular weight organic substances are known as solvents. A solvent breaks down the paint's binder. A solvent should be removed from the paint film after application because it is no longer necessary. However, the solvent must dissolve the binder to give the paint the right application qualities.

Hydrocarbons, aliphatic and aromatic chemicals, and chemical solvents, including alcohols, ethers, ketones, esters, etc., are all examples of solvents. C, H, and O molecules are present in chemical solvents.

Types of solvents

Hydrocarbon solvents For example, normal paraffin, hexane (C_6H_{14}), boiling point 69°C; iso-paraffin, 2-methyl-pentane, boiling point 60.3°C; cyclo-paraffin (naphthene), cyclohexane, boiling point 81°C.

Chemical solvents The following are the major groups: methyl iso-butyl carbinol, boiling point 130°C–133°C; diacetone alcohol (keto-alcohol): DAA, boiling point 166°C with decomposition to acetone; ethanol, boiling point 78.3°C; hexylene glycol, boiling point 192°C–200°C.

Properties of solvents The following are the main properties of solvents:

Solvency The stronger the solvent needed to dissolve a polymer, the more complicated it is. Toluene is an excellent solvent for chlorinated rubber because of its increased solvent power, whereas white spirit is appropriate for oil and alkyd resins.

Viscosity and solid content To maintain a high solid content, less solvent is required for a given viscosity, the greater the solvent power.

Ease of application Solvents with a rapid evaporation rate frequently produce a dry mist when sprayed. The evaporation must be gradual when using a brush.

Drying A slower rate of evaporation slows down the physical drying of the paint. Rapid evaporation may lead to the formation of several defects.

Boiling points and evaporative rates The rate of evaporation and the boiling point is connected. Generally speaking, the rate of evaporation increases with decreasing boiling point.

Flash point A flash point signals a potential fire. Paints are thought to be very flammable if their flash points are lower than 23°C.

Maximum acceptable concentration value It represents the level of risk the body can withstand without endangering itself. The concentration is stated in terms of air cubic meters. A lower number than a greater value denotes higher toxicity. Maximum acceptable concentration (MAC) value of certain significant solvents is summarized as follows:

- *Toluene*: Flash point - 6C; MAC- 100 ppm,
- *Ethanol:* Flash point - 12C; M.A.C.- 1000 ppm

Table 21–2 Outline of various types of pigments used.

S. no.	Type	Description
White pigment		
	Titanium dioxide	Rutile and anatase are the two mutations that are produced. Inert, noninhibiting white pigment. Provides high gloss, tinting strength, and opacity. At high temperatures, unaffected by organic and mineral acids.
	White lead [$2PbCO_3$-$Pb(OH)_2$]	It offers a robust, flexible, and long-lasting paint film. Reacts with acidic binders. Health risk.
	Zinc oxide	An excellent UV radiation absorber that prevents the resin from deteriorating. It is mildew resistant and has high fungicidal properties.
	Lithophone [ZnS 30%/$BaSO_4$ 70%]	Excellent resin dispensing characteristics. An intense pigment. It has strong tinting strength and good opacity but poor resistance to chalking.
Black pigment		
	Carbon black	Flue gases are produced by burning carbonaceous materials and contacting an acidic surface. Depending on the degree of darkness, blacks are categorized as high, medium, and low color blacks. Fine particle size (0.010–0.25μm) is utilized for painting automobiles. Paints for ornamental purposes employ the coarser particle size. They are readily dispersed in paints.
	Graphite	Possess low tinting strength and imparts low-intensity blacks.
	Black iron oxide [ferroso-ferric oxide ($FeO.Fe_2O_3$)]	Pigment with coarse particle size is produced by black iron oxide. It is an innocuous, inexpensive pigment. UV radiation-resistant.
Brown pigment		
	Iron oxides [Fe_2O3]	These have a weak tinting strength and are translucent.
	Calcium plumbate	Lime and lead oxide are heated in a stream of air to create it. Not prone to corrosion. Provides outstanding toughness and adherence.
	Monoazo-brown and diazo-brown	Both pigments have good corrosion resistance. Inexpensive, light-resistant.
Blue pigment		
	Ultramarine blue [$Na_6Al_4Si_6S_4O$]	Heat-resistant. Low opacity and poor tinting strength. Have excellent alkali resistance but weak acid resistance.
	Phthalocyanine [copper-phthalocyanine]	These need to combine with other colors to flocculate. High tinting strength, outstanding resistance to chemicals, UV light, and weathering.
Green pigments - Brunswick greens		
	Chrome greens	Cheap and highly opaque. Toxic in nature. May become blue on exposure to the environment. Offer good heat resistance.

(Continued)

Table 21–2 (Continued)

S. no.	Type	Description
Yellow pigments		
	Lead chromate [pigment yellow 34 – $PbCrO_4$]	The following are the important chromate pigments: Primrose chrome (greenish-yellow lead sulfochromate) Lemon chrome (yellow lead sulfochromate). Middle chromes (reddish-yellow chromates). They feature strong tinting and high opacity. Pale primrose to crimson color beams is produced using lead chromates and lead sulfate. inexpensive and has deterrent qualities utilized in coatings on automobiles.
	Yellow iron oxide [hydrated ferric oxide – Fe_2O_3-H_2O]	These exhibit outstanding resistance to deterioration and chemical assault.
	Monoazo yellow (2-nitro-*p*-toluidine coupled with aacetoacetanilide) or (hansa) yellow	They are chemically resistant and have a high tinting strength.
Orange pigments		
	Basic lead chromate [x$PbCrO_4$, yPbO with somelead sulfate $PbSO_4$]	Toxic like other chrome pigments. Get darkened on exposure to sunlight. Resistant to alkalies.
	Red pigments	Except for red lead, these are made of mineral-based pigments and contain Fe_2O_3, which is responsible for the red hue.
	Red iron oxides (natural and synthetic ferric oxide) (Fe_2O_3)	The earth's crust contains the red iron oxide known as hematite. Excellent chemical resistance is included in the synthetic variety. Good lightfastness and opacity.
	Diazo-redpigments (pyrazolone)	It is exceptionally solvent resistant and lightfast.
Special effect pigments		
	Aluminum flakes (leafy and nonleafy type)	These metallic pigments well protect the metal substrate exposed to the sea environment.
	Fluorescent pigments (rhodamine).	It is produced by adding fluorescent dyes to resins. They fluoresce by retransmitting energy at lower frequencies after absorbing UV light at a particular frequency. They lack transparency and are pretty pricey. They are not very weather-resistant.
	Red Lead [Pb_3O_4]	Frequently used as a steel-inhibiting primer. Inexpensive and climate-resistant. Health risk.
	Corrosion inhibiting pigmentsFor example, Zinc phosphate [$Zn(PO_4)_2 \cdot 4H_2O$]	Creating a barrier to the environment will stop rusting.

- *Butyl acetate*: Flash point - 25C; MAC- 200 ppm
- *Xylene*: Flash point - 27C; MAC- 100 ppm

21.4.10.1.2 Solvent retention

Any solvent in the coating will weaken adhesion and reduce its ability to withstand chemicals.

21.4.10.1.3 Compatibility
Industrial coatings used to be solvent-borne before the 1960s. The primary solvents used to dissolve the resins were weak organic acids. Then, these solvents were evaporated.

However, their evaporation increased environmental contamination. Environmentalists have thus strongly emphasized reducing the solvents' volatile organic content (VOC) to lessen their potentially harmful environmental impact. Commercially available coating compositions with fewer or no solvents are a recent development. The environmental protection agency's VOC standards have severely limited the formulation options for coatings.

To produce a high-quality, extremely resistant film, the solvent that is utilized must be compatible with the composition of the coatings. A suitable solvent combination must be chosen since the formulation may contain many resins. This will increase the compatibility of the resins employed in the formulation. The solvents may impact the coating by causing porosity, discoloration, poor gloss, and poor adherence even after dissipating. Therefore choosing suitable solvents is crucial.

21.4.10.2 Diluents or thinner
They are added before the application to achieve the right viscosity for the paints. These thinners must have the same solvent power as the most volatile element of the particular paint. The rate of evaporation must be compatible with the application method.

21.4.10.3 Additives
These are anticorrosive pigment enhancers, antifoams, antisettling agents, antiskidding agents, driers, and preservatives, which are added in a small amount to improve the properties of the paints. Examples of additives are calcite ($CaCO_3$), quartz (SiO_2), and kaolin clay ($Al_2Si_2O_5(OH)_4$).

21.4.11 Conversion coatings

Metal can go through a surface reaction that changes its surface into an inorganic substance that was created from the metal itself and is very adherent. Considering that the metal has already reacted, this surface is exceptionally corrosion-resistant.

21.4.11.1 Anodizing
Anodizing is the process of electrolytically oxidizing a surface to create an oxide scale that is thicker than the naturally existing coating and securely adhering. Aluminum serves as the anode during the electrochemical process known as anodizing. The metal surface is changed into strong aluminum oxide by the electric current flowing through an electrolyte. The oxide coating is a part of the metal substrate, as opposed to being a metallic coating deposition, which is the distinction between plating and anodizing. The oxidized surface offers some corrosion protection and is tough and abrasion-resistant. An additional layer of protection, such as painting, is typically needed since anodizing cannot be depended upon to offer corrosion resistance to metals that are prone to corrosion. Fortunately, the anodic coating offers

a superb surface for adhesive bonding as well as painting. Anodic coatings degrade chemically in both very alkaline (pH > 8.5) and severely acidic (pH < 4.0) conditions. Additionally, due to their greater brittleness and potential for stress cracking, stress corrosion-prone metals require additional protection, such as painting.

Chromic, sulfuric, phosphoric, or oxalic acid solutions can create anodic coatings. The 7000 series metals are frequently anodized with chromic acid to increase corrosion resistance and paint adherence, and unsealed coatings serve as a solid foundation for structural adhesives. These coatings frequently get discolored, hence sulfuric acid anodizing could be recommended in situations where esthetic appeal is crucial. The Al_2O_3 coating created by anodizing is generally 2–25 m thick and is made up of a porous outer layer that can be sealed by hydrothermal treatment in steam or hot water for several minutes, as well as a thin nonporous barrier layer adjacent to the metal. This produces a hydrated oxide layer with improved protective properties. If the sealing is carried out in a hot metal salt solution, such as a chromate or dichromate solution, increased corrosion resistance is produced. It is also possible to dye the oxide coatings to provide ornamental surface colors, and this can be done either during or after the anodizing process. The International Standards Organization has developed global guidelines for the anodic treatment of aluminum alloys, which address both colored and colorless coatings. Anodizing a component is done for a variety of reasons. Several factors and the industries that use them are listed below:

- *Appearance*: Products have a better, polished, cleaner appearance that lasts longer. Color softens the harsh metal aspect of metal and promotes a robust, well-built appearance. Any piece of metal may have its surface colored.
- *Corrosion resistance*: Weathering is slowed down while a smooth surface is kept. Useful for managing maritime items and food. Any anodized product will remain cleaner for a longer period of time and is simpler to clean when it does become filthy.
- *Abrasion resistance*: The treated metal is strong—harder than many abrasives—and is perfect for air cylinder applications, tooling, and caul plates.
- *Nongalling*: Wear in these places is reduced and screws and other moving components will not jam, drag, or seize. Common uses include screw threads, instrumentation, and gun sights.
- *Heat absorption*: This can give aluminum for use in the food processing sector homogenous or selective heat-absorption capabilities.
- *Heat radiation*: This technique is used to complete radiators and heats sinks for electronic devices.

Additionally, anodizing is hygienic and flavorless, is good as a paint foundation, will not rub off, and eliminates small blemishes.

There are several variations of the anodization process. The following are examples of some of the practices used in the industry:

a) *Hard-coat anodizing*: A hard-coat finish is utilized when crucial features like surface hardness, abrasion and corrosion resistance, and durability are required. In order to create the hardened surface, the electrolyte concentration, temperature, and electric current parameters are changed throughout the hard-coating process, which is essentially an anodizing in sulfuric acid. Case-hardened steel or hard chrome plate have been shown to have worse wearing properties.

b) *Bulk anodizing*: Small, atypically shaped items are electrochemically anodized by the bulk anodizing process and processed in perforated aluminum, plastic, or titanium baskets. This method is very cost-effective due to the enormous number of pieces that can be finished in a very short period of time. The uniformity in color and quality that results from processing such big quantities at once is another benefit. Utilizing the bulk method, finishing components like rivets, ferrules, medical hubs, and so forth enable production to be economically viable.

c) *Sulfuric acid anodizing*: The most popular technique for anodizing is this one. The component is given a specific electric current to run through an electrolyte of sulfuric acid, which changes the surface into the aluminum oxide coating that can absorb dyes in a variety of hues. The surface's resistance to abrasion and/or corrosion is improved, and it can also serve as a foundation for other coatings like paint, teflon, and adhesives. Any specification may be filled with custom coloring, and surfaces with a matte, satin, or highly reflective finish can be provided using prefinish processes. Magnesium and titanium alloys can also get anodizing treatments. The treatments frequently used with magnesium alloys involve a variety of processing options to produce either thin coatings, up to a thickness of 30 m, for maximum corrosion and abrasion resistance, or thick coatings, up to a thickness of 5 m for flexibility and surfaces suitable for paint adhesion. When anodizing is used to treat titanium and titanium alloys, it can offer some protection against galvanic corrosion to the less noble metals, and when combined with solid film lubricants, it aids in preventing galling. The procedure results in a uniformly smooth covering that is blue-to-violet in color and texture.

21.4.12 Phosphate coatings

An insoluble coating of metal phosphate forms on the surface of iron, steel, galvanized steel, or aluminum when it is sprayed on or submerged in a diluted solution of phosphoric acid with zinc, iron, or manganese phosphates. The phosphate layer serves as a substrate for coatings on items like car bodywork and refrigerator cabinets, and it also offers some corrosion protection, helps with cold forming, and boosts wear resistance. The phosphate enhances paint adherence and prevents corrosion under the paint coat from spreading. It shields the nearby anode and cathode locations on the metal surface due to its dielectric properties. In order to increase corrosion resistance at the coating's weakest points, the pores in phosphate coatings are frequently sealed using chromate rinses. With its multilayer protection system, galvanized sheet steel phosphate for paint adherence offers outstanding corrosion resistance. Zinc phosphate with fluoride can be sprayed over aluminum to coat it quickly and readily. Calcium phosphate is increasingly being used to cover medical implants. The least costly and thinnest protective coating is produced by iron phosphating, which also offers an excellent paint foundation. In order to effectively remove rust, it is applied at the greatest acidity, skipping the pickling process.

Zinc phosphates often have an intermediate thickness and serve as a paint foundation or a framework for holding oils or wax. Although paint provides considerable corrosion resistance when put over zinc phosphate, the layer is typically too thick for coil coating, where the painted sheet must be produced. Although heavy manganese phosphate coatings are not painted, they are sufficiently porous to allow the absorption of corrosion-inhibiting waxes or oils, which are the typical finish used on steel nuts, bolts, and screws.

Phosphate coatings are of three types: (a) zinc phosphate, (b) iron phosphates, and (c) manganese chromium phosphates.

21.4.13 Zinc phosphate conversion coatings

The metal reacts with phosphoric acid producing iron phosphate.

$$3Fe + 2H_3PO_4 \rightarrow Fe_3(PO_4)_2 + 3H_2 \qquad (21-3)$$

Formation of an insoluble tertiary zinc phosphate from soluble primary zinc phosphate.

$$3Zn(H_2PO_4)_2 \rightarrow Zn_3(PO_4)_2 + 4H_3PO_4 \qquad (21-4)$$

Formation of sludge (ferrous phosphate) by addition of NaNO2.

$$H_3PO_4 + Fe\ _3(PO_4)_2 + 3HNO_2 \rightarrow 3FePO_4 + 3NO + 3H_2O \qquad (21-5)$$

Sludge is the precipitation suspended in the bath. It settles down at the bottom.

21.4.14 Iron phosphate conversion coatings

The basic composition is modified by the addition of chemicals, like fluoride. They do not offer as good resistance as zinc phosphate coatings. Following are the reactions:

$$2Fe + 3NaH_2PO_4^- \rightarrow 2FeHPO_4 + Na_3PO_4 + 2H_2 \qquad (21-6)$$

$$4FeHPO_4 + O_2 \rightarrow 4FePO_4 + 2H_2O \qquad (21-7)$$

21.4.15 Chromium phosphate conversion coatings

The coating bath consists of an acidic mixture of phosphate, chromate, and fluoride. The fluoride acts as an accelerator. It is applied to aluminum. The mechanism is given below:

$$Al + 3HF \rightarrow AlF_3 + 3H^+ \qquad (21-8)$$

$$CrO_3 + 3H^+ + H_3PO_4 \rightarrow CrPO_4 + 3H_2O \qquad (21-9)$$

The coating is characterized by the formation of $Al_2O_30.2CrO_40.8H_2O$.

21.4.16 Chromate coatings

Aluminum, zinc, cadmium, steel, and magnesium are routinely given a chromate treatment, usually by immersion or spray, to provide a protective chromate surface on the metal, seal further coats of phosphate, oxide, and the like, or provide an excellent, nonporous bond for paint. The solutions include chromic acid or Cr^{6+} ions. The end products, which range in thickness from 10 to 1000 nm (5×10^{-4} to 0.05 mil), are mostly made of hydrated chromium chromate and also contain considerable amounts of other metals and anions. Thicker coatings often have yellow undertones, and color varies from transparent to iridescent blues and greens as thickness increases. They may also be colored. Chromate coatings provide exceptional corrosion resistance when immersed because they are less soluble in water than other conversion coatings. The coatings' ability to provide a barrier layer, passivate metal, and hinder the cathode process are all directly inversely connected with corrosion resistance. Cadmium plating has to be stretched out or baked prior to chromating in order to avoid the protective chromate from being damaged by heat. Galvanizing cannot be successfully painted or phosphated if it has been chromated beforehand. To produce "tin-free" steel sheet that is later coated for tin cans, the majority of the steel is subjected to chromating.

21.4.17 Oxide coatings

Aluminum is routinely oxidized using solutions containing chromate or dichromate to increase corrosion resistance or as a foundation for lacquers and varnishes. The produced films are gray in color and can be sealed with hot dichlornate solution. In the past, Fe_3O_4, which has a good appearance and was thought to give corrosion resistance, was mostly used to make the surface layer on steel tools and weapons that were blued. The steel was heated to roughly 300°C (575°F) on a bed of hot charcoal or to around 330°C (625°F) in a salt bath until the desired color was obtained. By immersing the metal in a hot alkali solution that contains an oxidizing agent, the metal can be chemically blued.

21.4.17.1 Nitriding and carburizing

The surface hardening techniques of carburization and nitriding improve the steel's resistance to wear. They distribute carbon or nitrogen atoms into the surface to harden it in addition to applying large compressive pressures to the surface. Untreated steel is not more or less resistant to corrosion than steel that has been carburized or nitrided, but treated steel is more resistant to SCC, which requires tensile stress at the surface to induce a crack to develop.

21.4.17.2 Rust converters

A variety of chemical formulations, many of which contain tannin, can be applied on corroded steel to penetrate, and react with the rust, converting it into stable, nonporous compounds that act as a barrier layer to protect the steel. They do penetrate and do transform the rust into a mostly protective covering that, in some situations, covers the steel.

However, they do not remove moisture from the rust and do not counteract the effects of the rust's aggressive salts, especially chlorides, thus corrosion can continue unabatedly beneath the converted rust. Rust converters only function with pure, uncontaminated rust that is less than 2.5 μm (0.1 mil) thick and under low service conditions.

21.5 Linings

The difference between a polymer coating and a lining is the difference in thickness and, therefore, in service conditions and a difference in the mechanism of protection (Coatings & Linings, 2003). A coating is thin (by definition <0.5 mm or 20 mils), typically 50–100 μm (2–4 mils), while a lining will usually be around 3 mm (0.1 in.). A coating is not a full barrier between the metal and its surroundings, but a lining is. While coatings are applied to the exterior of tanks and vessels, where they may be exposed to weathering, condensation, fumes, splashes, and other similar conditions but not immersion, which is a particularly harsh one, linings are applied to the interior of tanks and vessels where they are exposed to immersion conditions.

Both at the mill and on the job site, coatings and linings can be used. If they can be put into a mill, they will function significantly better since the mill offers much greater surface preparation. Additionally, field treatments will often cost at least twice. To minimize damage during transportation, storage, and handling on the building site, mill-applied coatings must be substantially tougher and more resistant to weathering and sunshine.

21.6 Surface preparation and pretreatments before coatings

According to the coatings and paint system specifications, the surface must be cleansed to the required level of surface preparation. Before wire brushing or blasting the surface, contaminants that impede the removal of rust, such as salts, oils, and greases, must be removed from the surface. This is the most crucial phase of the surface treatment process. The main goal is to obtain a surface with excellent cleanliness and suitable roughness. Degreasing is the initial step in the pre-treatment, followed by removing mill scale and rust and roughening the surface.

Degreasing: degreasing can be done using any of the following ways: (1) organic solvents like white spirit and paraffin; (2) alkaline cleaning agents like soap or other wetting agents in solutions containing sodium phosphate, silicate, carbonate, or hydroxide; (3) an emulsion of organic and alkaline cleaning agents; or (4) high-pressure steam containing small amounts of a cleaning agent. The cleanest surface is produced by alkaline cleaning, which is frequently employed as the final phase of pretreatment, for example, right before electrolytic metal plating. Before applying new coatings, steam can be used to degrease surfaces after using emulsions to remove existing coatings and after doing an alkaline clean (Pludek, 1977). Before coating, the surface must be clear of grease according to all coatings and application techniques.

Removal of rust and mill scale: Either chemical pickling or mechanical blast cleaning is preferred.

Pickling: Steel is pickled by submerging it in a pickling acid bath that contains a pickling inhibitor. Pickling is typically done using a solution of 3%–10% by weight H_2SO_4 in water, and it takes place at 65°C–90°C for 5–20 minutes. Other pickling acids, such as cold hydrochloric acid, phosphoric acid, nitric acid, or a combination of different acids, are also employed for a variety of reasons (Shreir, Jarman, & Burnstein, 1994). Pickling removes scale in two ways: by dissolving the mill and rust scale and by causing the oxide scale to be displaced by hydrogen gas created by corrosion beneath the scale. Inhibitors are introduced to stop the metal from corroding severely. Pickling is done prior to applying organic coatings, inorganic coatings like enamel, and metallic coatings (by electroplating or hot dipping).

Blast cleaning: It is the ideal preparatory procedure for painting (in addition to necessary degreasing). Centrifugal blasting (also known as sling cleaning), compressed air blasting with a dry blasting agent, or possibly a sand and water mixture (ideally with an inhibitor) that is blown against the object under high pressure can all be used to complete the operation. The following are employed as blasting agents: steel shot or wire cuts (in centrifugal blasting facilities), olivine sand, iron slag or copper slag, iron silicate, or aluminum silicate, the grit of aluminum oxide, steel, or white iron (mainly when the blasting agent can be recirculated). Due to the risk of silicosis, silica sand is banned in various nations. Water blasting at extremely high pressure (without solid blasting medium) has become customary for huge structures like offshore platforms in recent years. Demands must be made with regard to the caliber of a surface that has been blast cleaned, as was previously described. Many people use the Swedish Standard SS 055900 (ISO 8501-1) to describe the quality of blasting. Color photographs of the various grades are included in a book (ISO-Standard 8501-1, 1994) to be used as a tool for evaluating the blasting quality in real life (together with the rust grade before blasting and the quality of wire-brushed and scraped surfaces).

It is challenging to completely clean badly rusted steel surfaces with blast cleaning, especially if the rust has absorbed salts from the sea or marine or industrial environments. For removing salt traces, pickling is preferable. After that, the surface needs to be gently cleansed with water. It should be remembered that a reasonably fine-grained blasting agent is most effective when blast cleaning is advised for rust removal. The ideal pretreatment for thermal spraying of metals, ceramics, or ceramic-metallic coating materials is typically blast-cleaned. Rough grinding is employed for specific purposes. The adhesion of arc-sprayed and flame-sprayed aluminum and zinc on steel depends on Sa-grade and blasting medium, the latter of which is characterized by material and grain size (Bardal, Molde, & Eggen, 1973), as an illustration of the significance of blasting quality in general, and for thermally sprayed coatings. Hammering, scraping, wire brushing, grinding, and tumbling (mechanical procedures), as well as flame cleaning and induction heating with subsequent quenching, are further techniques used to remove rust and mill scale from steel surfaces in addition to pickling and blast cleaning (thermal methods).

21.7 Coating failures

A mechanistic understanding of failures of paints and coatings is important for engineers. The failures can be divided into four categories:

a formulation-related failures,
b adhesion-related failures,
c application-related failures, and
d design-related failures.

An account of major coating failures is given below. Only selected failure categories are described here:

21.7.1 Formulation-related failures

These failures arise due to improper coating formulations. Some of the major formulation-related failures of coatings are:

21.7.1.1 Chalking

It is the failure of a coating by the formation of a powdery layer on the coating surface. Chalking is a loose, white powdery pigment residue on the surface of the coating accompanied by a loss of sheen and color. This can be expected to occur over time and is a normal way for a paint film to wear. In some cases, this is a self-cleaning feature designed into the coating. However, excessive film erosion can result from heavy chalking and result in the need for premature repaint.

The binder of the coating disintegrates due to severe sunlight and leaves the pigments which were held by it. The binder continues to disintegrate until corrosion starts to develop on the substrate and the coating surface is worn out. A pigment may interact with ultraviolet radiation and breakdown. A change in the color of a pigment indicates chalking. Previous coatings may not have satisfied the porosity of the substrate.

Chalking can be prevented by following opting strategies:

a by the selection of an appropriate combination of binder and pigment on the basis of experience,
b by the selection of pigments with the least tendency for chalking, and
c by the removal of loose chalk deposits and application of a new coating system.

21.7.2 Checking, alligatoring, and cracking

Although different in appearance, all three failures mentioned above are related to the deterioration of the coating with time Fig. 21–5 (Table 21–3).

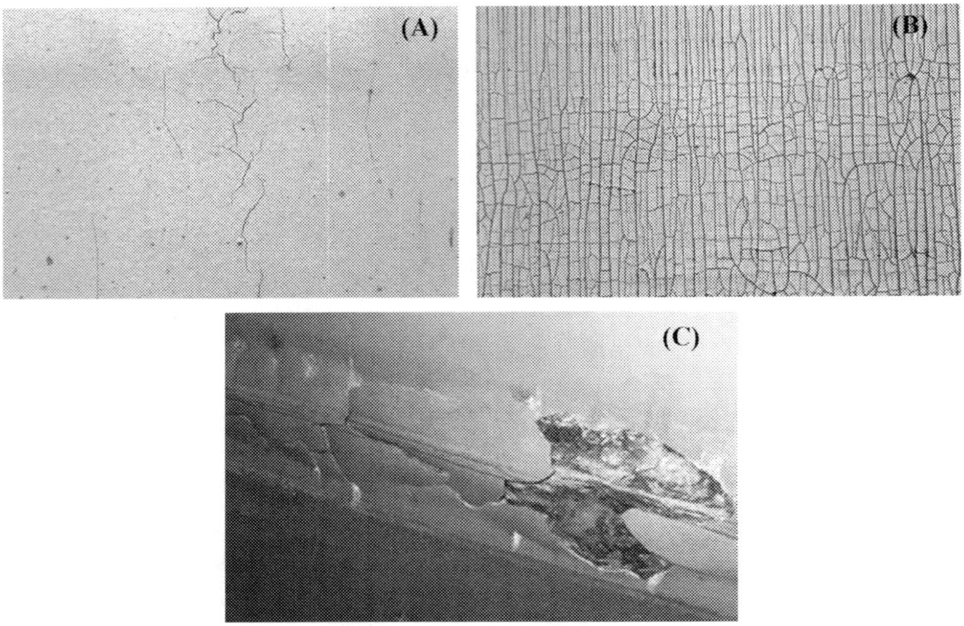

FIGURE 21–5 Failures related to the deterioration of the coating with time (A) checking, (B) alligartoring, and (C) cracking.

Table 21–3 Failures related to the deterioration of the coating with time.

	Checking	Alligatoring	Cracking
Definition	Small check-board patterns are formed on the surface of a coating as it ages. The coating becomes harder, and the brittleness increases as it ages.	It is a film rupture caused by the application of a hard, dry, brittle film over a more softer and extensible film. The failed surface of the coating resembles the hide of an alligator, hence the name alligatoring. It is also called crazing.	The surface of a coating cracks due to weathering and aging are similar to alligatoring, however, in this case, the cracks reach the substrate and cause a more serious defect than either checking or alligatoring. Cracking may also be caused by a fast curing rate.
Mechanism	The primary reason for the above failures is the introduction of stresses in the coating. In the case of checking, stresses are caused by shrinkage of the coating due to weathering, whereas in the case of alligatoring, the stress is caused by the shrinkage of the surface of the coating at a higher rate than the rate at which the body of the coating shrinks. In the case of cracking, stresses are set by the continuous weather of the coating. The end result is the rupture of the coating by stress.		
Prevention	Alligatoring can be prevented by selecting the proper formulation and allowing a specified drying time between coats. The ingredients selected must show a minimum difference in the rates of expansion and contraction so that stresses created by the above processes are minimized. Cracking can be minimized by selecting a coating that is resistant to weathering and oxidation. Remove all dust and contaminations and prime, fill, and then reapply the full paint system. The coating selected must also be flexible. Similar prevention is required for checking. The pigments selected for the latter must be inert and reinforcing.		

21.7.2.1 Erosion

It is the failure of a coating due to erosion by high winds and sand particles, mostly encountered in deserts. In erosion, it is the high winds or sand particles which hit the coating and disintegrate it. Once the coating is disintegrated, and the substrate is exposed, corrosion proceeds rapidly on the substrate, particularly in a humid environment. The prevention of erosion coating failure is similar to that of chalking.

Wrinkling: It appears as the wrinkles on the coatings' surface. This type of failure arises when the surface of the coating is cured more rapidly compared to the body of the coating. Due to this, a serious stress is created in the coating due to the differences in the expansion rates. This results in wrinkles formation. This type of coating failure mostly arises in oil-based coatings. This failure can be avoided by utilizing slow-drying solvents and a mixture of silicates (in order to check the drying rate of the coatings). Furthermore, a thin film of coating material should be employed in order to avoid this type of defect.

21.7.2.2 Failures related to improper application

21.7.2.2.1 Holidays or misses

This is a particular area in which the coating has been left uncoated while the rest of the surface has been coated. "Holidays" includes pinholes. Holidays can also be caused by dust particles being incorporated during painting. These types of failures can be neglected by checking none of the areas must remain uncoated.

Pinholes: These are minute holes that are formed during the application of paint, which exposes the underlying substrate. The flow of applied paint is not continuous due either to poor mixing of two-pack materials or insufficient agitation leading to insufficient wetting of the substrate. These types of failures can be avoided by applying coatings evenly and paying proper attention to the surface, or by applying coating by the help of brush uniformly and then allowed to dry. After the drying processes another coat of paint should be applied.

21.7.2.3 Spattering

This is the defect caused when only droplets of the paint reach the substrate and a gap of bare metal exists between the droplets. The bare metal serves as the "holiday area." Spattering can be avoided by applying paint with a spray gun adjusted properly.

21.7.2.4 Cratering

It is the formation of small bowl-shaped depressions in the paint or varnish film. It is caused by the falling of impurities in the wet paint. It can be removed by the addition of antifoams and bubble-releasing agents (for solvent-based paints).

21.7.3 Failures related to welding and design

Weld areas can cause the failure of coatings by several causes: interference of the welding flux with the coating creating a lack of proper adhesion; splatters (droplets) of welds creating cavities that expose the bare metal surface; underfilm rusting may proceed with the creation of cavities. Welds are the potential holidays in underground pipelines. Weld areas are generally the first regions to show coating failures in a nonaqueous environment. This failure can be prevented by adopting a proper cleaning process by removing the first cause (flux interference).

21.7.3.1 Edge failure

The failure of coating on the edge is a design-related failure because the coating is thinner at the edge compared to the rest of the surface. An edge provides a break in the coating. The coat fails by the development of rust in the coating which penetrates the substrate. It can be avoided by the application of an extra coat on the edges.

21.7.4 Failures related to the lack of adhesion

The following may be the reasons for the lack of adhesion.

Poor surface preparation: Good surface preparation is a primary prerequisite for coating. Any dirt, grease, rust, or scale, if present, must be removed. Before the application of the new coating, the old coatings must be removed completely. The mill scale must be completely removed from the surface. If they are not removed, the oxygen and moisture may penetrate under the scale and cause the surface to corrode. Once the surface corrodes, the substrate and the coating are destroyed. Loss of adhesion may not occur immediately. However, it is dependent on how severe the environment is.

Poor application: Porosity can result from improper application.

Film thickness: If the film thickness is very high, stresses are introduced by shrinkage and the coating peels off.

Some common failures arising due to the adhesion are discussed below (Fig. 21–6; Table 21–4):

21.7.5 Failure of immersed organic coatings

Blistering and delamination (Fig. 21–7) are two of the most severe and frequent types of visual failure in submerged organic coating systems (discussed above). The actions that take place after the hydroxyl ions engage with the metallic substrate are what distinguish cathodic blistering from cathodic delamination. Due to the high-water solubility of the reaction products from the cathodic reaction, blistering is the result of an osmotic pressure that occurs. The alkalinity of the cathodic reaction products causes bonds to break at the coating-metal contact, which leads to delamination from damage. The fundamental corrosion processes of steel in saltwater are connected to the alkaline environment found under a disbonded covering and in cathodic blisters. Steel will be exposed to the environment if the coating contains flaws or faults. As a result, a galvanic cell is created with anodic and cathodic areas. The cathodic reaction, which is mediated by oxidized metals and does not significantly take place in the absence of solid metal, counterbalances the anodic reaction. The reduction of oxygen will be a part of the cathodic reaction in the majority of naturally occurring circumstances. The mechanisms involved in the deterioration are comparable even if the modes of

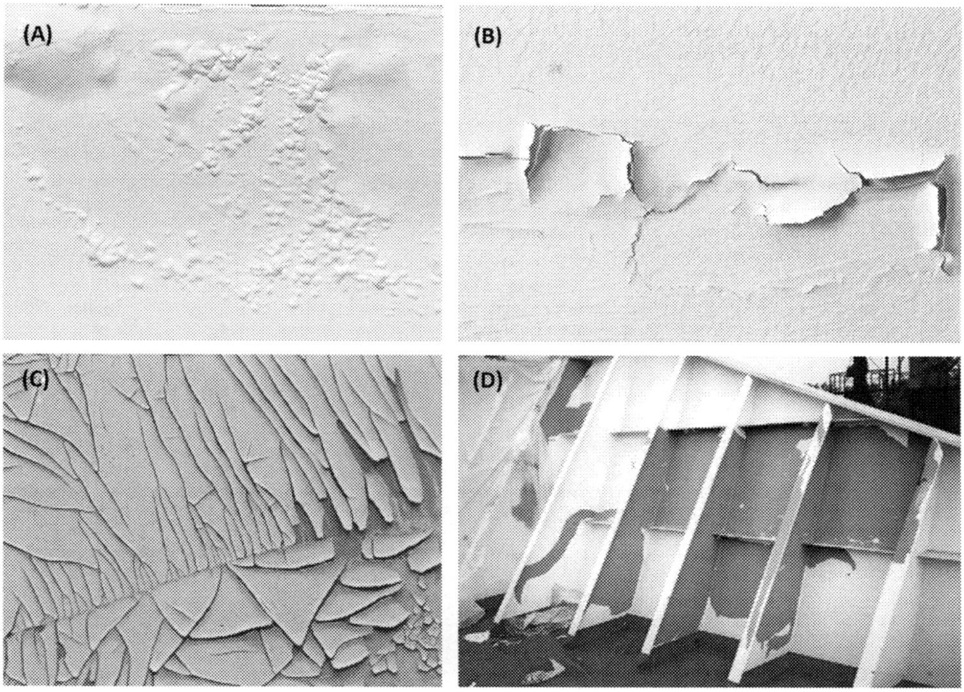

FIGURE 21–6 Failures related to the adhesion (A) blistering, (B) peeling, (C) flaking, and (D) intercoat delamination.

degradation for intact and defective organic coatings exposed to a neutral electrolyte solution differ in theory.

21.8 New technologies in the field

Using environmental changes to trigger a reaction in the coating or substrate that alters surface composition to increase corrosion prevention, nanostructured materials are thought to expand the potential for building "smart coatings." Other facets of the coatings business already use coatings' capacity to alter their physical features. Numerous researchers have focused on electroactive conducting polymers for corrosion control during the past few decades. They could be beneficial in a variety of applications due to their distinctive electrical conductivity. The majority of investigations have demonstrated that these coatings can offer significant protective effectiveness, particularly for mild steel. However, they are only partially effective in preventing the corrosion of metals in very hostile situations. The strongest options have been polypyrrole (Tan & Blackwood, 2003) and polyaniline (Ahmad & MacDiarmid, 1996; Kinlen, Silverman, & Jeffreys, 1997; Wesseling, 1994), both of which offer some benefits for this use. Steel may reportedly be protected against acidic and neutral conditions by having coatings with pinholes and scratches that re-passivate exposed portions. Since electroactive polymer coatings have been claimed to offer both anodic protection and

Table 21–4 Failures arising due to the adhesion.

Name of failure	Definition	Mechanism	Prevention
Blistering	This is the failure of a coating by the development of large or small round hemispherical pimples from the surface, either dry or filled with a liquid.	The blister may develop by one of the following mechanisms: (1) Absorption of moisture by the soluble pigments: When the soluble pigments absorb moisture, the solution becomes concentrated and water is pulled into all the areas where the pigment is dissolved, by the process of osmosis. Osmosis is the transfer of liquid through the coating from a lesser concentration to a higher concentration. Blisters filled with liquid are, therefore, formed. If some soluble salts are present between the intercoats or in the substrate, absorption by osmosis would occur and blisters would be formed. (2) Inadequate solvent release: If the coating is a rapid drying type, some of the solvents may not be released and is trapped in the coating. When the temperature of the coated surface is raised, the solvent may create a significant vapor pressure and cause the formation of blisters. A similar condition may arise when a top coat is applied over a porous undercoat. (3) Poor coating adhesion: In the areas of poor adhesion, adhesion of the coating to substrate, or adhesion of inter-coats, the liquids and gases may be trapped and this can lead to the development of blisters. Salt contamination on a surface may also cause blistering. They may occur in extremely wet environments. (4) Improper film thickness: The thickness of the coating must be strictly in accordance with the specifications for a particular coating. (5) Proper choice of primer: As primer is the foundation of a coating, it is important to select a proper primer according to the intended application of the coating. For instance, if the coating is to be applied in a marine environment, an inhibitive primer must be selected. For immersion service, an impermeable primer must be selected. (6) Weather: High humidity, fluctuating temperatures, and wind speeds can lead to the failure of coatings by either creating conditions inappropriate for curing or by introducing contamination during coating application. Inhibitive primers can be usefully applied to overcome the effect of moisture adsorption. (7) Cathodic protection: If a structure is cathodically protected, hydrogen gas may be formed in sufficient volume. It may also be formed by stray currents. Hydrogen vapor pressure pushes the coatings up	Blistering is a serious type of failure. The only practical remedy is to remove the coating and repair it, if blistering is localized. If blistering occurs all over the surface, completely change the coating which is more expensive but provides better adhesion. Water-soluble pigments in humid environments must be avoided. Ensure that prior to application of coating the substrate is completely dry and free from any contaminations. Sometimes blistering may be harmless, for example, blistering of subsea coal tar epoxy paints on oil rigs can be caused by cathodic protection but can be a harmless effect.

		and blisters are formed. The size of the blisters depends on the degree of adhesion of the coating, the salt content in the blisters and the internal pressure of either the liquids or gas inside the blisters. Generally, blistering leads to rust formation. The blister may expand to a sufficient degree to cause the lifting of the coating.	
Peeling	Peeling is the loss of adhesion resulting in detachment and curling out of the paint film from the substrate or between intercoats. It may occur by a number of different causes given below: (1) If the tensile strength of a coating exceeds the bond strength between coating and substrate. (2) If the solvent used for one coating softens the other coating. (3) If corrosion proceeds under the coat. (4) Painting of a surface when it is wet with rain or dew. (5) Inadequate removal of the old coat before the application of a new coat. (6) Ingress of moisture to the substrate.	One or more than one of the factors given above may cause a loss in adhesion of the coating. The paint film is curled from the surface.	Remove the old coat completely before applying the new coat. Never coat the surface under wet conditions. Do not allow moisture to penetrate.
Flaking	Flaking is the pulling away of the coating from the substrate because of the hard and brittle nature of the coating	It is similar to the mechanism of peeling. The coating that flakes is generally hard and brittle.	Avoid contamination between coatings and do not apply very thick coatings. Never paint a wet surface and use coatings that are compatible.
Intercoat delamination	The loss of adhesion between two coats. May be caused due to: (1) Not using a compatible coating with the old coating during repairs. (2) Lack of removal of contamination from the substrate before applying a new coat. (3) Application of very thick coatings. (4) An increase in the rate of the curing by sunlight, such as in case of coal tar epoxy coatings.	One or more of the above factors may come into play and cause delamination.	If the delamination is localized, the coating may be repaired by sanding off and recoating with a compatible paint. However, when the delamination spreads over a very large area, the best remedy is to reapply the coating all over again, after removing the old coating.

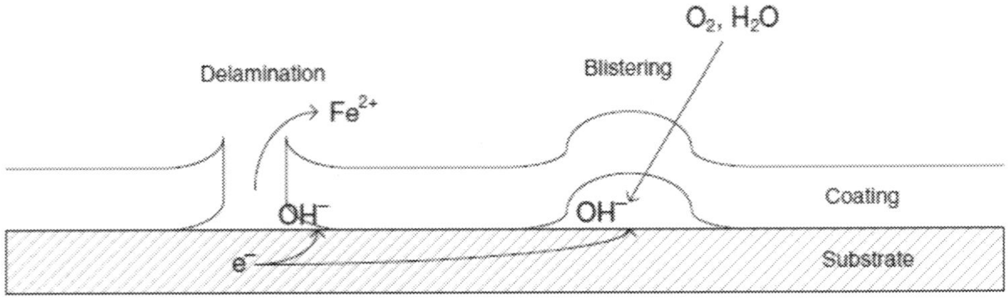

FIGURE 21–7 Idealizedsketch of delamination and blistering.

a barrier effect to prevent assault by a corrosive environment on steel, the precise protective mechanism of conducting polymers in anticorrosive coatings is up for debate (Tansug, Tuken, Ozyilmaz, Erbil, & Yazici, 2007).

When developing smart corrosion-inhibiting coatings, electroactive polymers may be used. These coatings will only produce or release an inhibitor when necessary to halt the onset of corrosion. As a result, coatings could be able to spontaneously release corrosion inhibitors in response to, for instance, mechanical or chemical stress or an electrical or mechanical control signal. In biological systems, damage results in an autonomic healing response, which serves as inspiration for the idea of self-healing polymers. Prior to manufacture, functional components (a healing agent and a catalyst) are incorporated into microcapsules or nanocontainers that are then placed into a polymer matrix to create self-healing materials (Keller & Sottos, 2006). They will be made to mechanically burst when the coating is broken by impact or abrasion, releasing their substance. When the healing agent comes into touch with a catalyst built into the polymer matrix, polymerization begins, and the damage is cured. Subsequently, the contents of these micro- or nanocontainers are carried to the injured regions. The next generation of anticorrosive coatings may rely on corrosion-generated electrical signals to start self-healing. But because self-healing technology is still relatively new, there are still a number of obstacles to overcome, even though some preliminary findings suggest that it may be used to prevent corrosion. However, it is evident that several years of study are required to perfect the technology for commercial usage.

21.9 Future prospect and discussion

Innovations in anticorrosive coatings are primarily driven by the need to comply with environmental standards and save production costs. Finding a solution with a higher environmental profile while maintaining the tried-and-true performance characteristics of conventional formulas is a problem for coating providers. This difficulty is more remarkable for the protective coatings industry than for less demanding parts of the coatings market. This explains why emerging technologies like water-borne coatings have not yet been widely adopted in the protective segment. The lack of clarity on the physical and chemical mechanisms causing the breakdown of anticorrosive coatings while in service is a significant problem for the development of the next generation of high-performance anticorrosive coatings. A thorough comprehension and quantification of the degradation mechanisms using mathematical models may be employed to ensure a correlation between accelerated exposure and natural exposure testing and serve as a beneficial tool in creating novel, intriguing goods. Additionally, greater research into the mechanisms underlying coating system deterioration as well as advancements in the creation of binders and pigments, may lead to the creation of anticorrosive coating systems that can offer superior corrosion protection. Though they take time to develop, novel concepts like self-healing coatings might contribute to the answer. However, high anticorrosive performance must be attained without sacrificing other system features that already exist. The upcoming decade will provide light on whether the objective will be accomplished.

References

Ahmad, N., & MacDiarmid, A. G. (1996). Inhibition of corrosion of steels with the exploitation of conducting polymers. *Synthetic Metals, 78,* 103.

Bardal, E., Molde, P., & Eggen, T. G. (1973). Arc and flame sprayed aluminium and zinc coatings on mild steel. Bond strength, surface roughness, structure, and hardness. *British Corrosion Journal, 8,* 15–19, January.

Coatings and Linings. (2003). *ASM handbook* (Vol. 13A). Materials Park, OH: ASM International.

ISO-Standard 8501-1 (1994). *Preparation of steel substrates before application of paint and related products.* Part 1, 1988/suppl.

Keller, M. W., & Sottos, N. R. (2006). Mechanical properties of microcapsules used in a self-healing polymer. *Experimental Mechanics, 46,* 725.

Kinlen, P. J., Silverman, D. C., & Jeffreys, C. R. (1997). Corrosion protection using polyaniline coating formulations. *Synthetic Metals, 85,* 1327.

Painting. (1999). (9th ed.). *ASM handbook,* (Vol. 5). Materials Park, OH: ASM International.

Pludek, V. R. (1977). *Design and corrosion control.* The MacMillian Press.

Shreir, L. L., Jarman, R. A., & Burnstein, G. T. (Eds.) (1994). Corrosion: Vol. 2 (3rd ed.). Oxford: Butterworth Heinemann.

Stoye, D., & Freitag, W. (1998). In L. D. Vincent (Ed.), *Corrosion prevention by protective coatings* (2nd revised ed.). Texas: Houston, USA: NACE.

Tan, C. K., & Blackwood, D. J. (2003). Corrosion protection by multilayered conducting polymer coatings. *Corrosion Science, 45,* 545.

Tansug, G., Tuken, T., Ozyilmaz, A. T., Erbil, M., & Yazici, C. (2007). Mild steel protection with epoxy top coated polypyrrole and polyaniline in 3.5% NaCl. *Current Applied Physics, 7,* 440.

Wesseling, B. (1994). Passivation of metals by coating with polyaniline—Corrosion potential shift and morphological changes. *Advanced Materials, 3,* 226.

PART 7

Corrosion failures, corrosion maintenance & monitoring

22

Corrosion failures

22.1 Introduction

Corrosion is one of the deteriorating processes of metallic equipment (among overload fracture, fatigue, creep, wear, and so on), leading to either "slow deterioration" in functionality (thickness reduction, insulating scale build-up, choking, and so on) or unexpected "sudden" failures (stress corrosion cracking [SCC], hydrogen cracking, and so on). All failures, on the other hand, are not caused by corrosion, and all corrosion events are not failures. One cannot, and should not, conclude about the method of failure unless they do a systematic exercise that probes into all facets of the experienced failure; this analysis is known as failure analysis. Every unexpected failure must be thoroughly investigated in order to determine the root cause and take preventative measures to avoid future failures. Replacement would not ensure that subsequent failures would not occur. Immediate replacement, on the other hand, is an essential and financial need.

System failures and subsequent failure investigations have become increasingly essential in modern civilizations. Aside from liability concerns, one of the main reasons for conducting a failure inquiry is to determine the mechanisms and causes of a problem in order to prevent it from happening again. The failure analysis process is incomplete without the advice of corrective action. Failure to identify and address the underlying causes of corrosion failures can lead to litigation, lawsuit, and a loss of consumer and public confidence in a company's product(s). In today's worldwide competitive corporate environment, such risks are unacceptable. The task of conducting a failure analysis is not easy. Because much valuable corrosion information might be lost if a failure scene is altered or changed before proper observations and testing are done, early discovery of corrosion as a factor in a failure is critical.

Specific systematic approaches have been offered to lead an investigator through the failure analysis process in order to avoid these traps. However, failure analysis is best learned by experience, and a failure analyst must gain proper qualifications by participating in genuine investigations and solving a range of problems. In failure analyses, it is critical to employ precise and consistent terminology. Inattention to this aspect reduces the value of information in reports, especially if the information is to be kept, accessed, and processed by computers later.

A thorough failure analysis begins with the gathering of relevant background information and service history, followed by visual inspection, photographic documentation, material evaluation, data analysis, and conclusion procurement. When it comes to studying corrosion failures, it is important to understand the various types of corrosion mechanisms that can occur. The characteristics of any corrosion failure provide important hints as to the corrosion's most likely origin. The characteristics of any corrosion failure provide significant

Handbook of Corrosion Engineering. DOI: https://doi.org/10.1016/B978-0-323-95185-2.00022-8

hints as to the corrosion's most likely origin. Understanding the standard corrosion mechanisms will help you interpret the traits and findings that every corrosion research will disclose. General corrosion, pitting, galvanic corrosion, dealloying, crevice corrosion, microbiologically influenced corrosion (MIC), corrosion fatigue, SCC, intergranular corrosion, fretting, erosion-corrosion, and hydrogen damage are all common corrosion mechanisms. This chapter outlines a tried-and-true method for finding the root cause and mechanism of failure.

Unfortunately, in practice, key terminology is frequently misused. The first section of this chapter covers the fundamentals of recognizing corrosion damage. A review of guides for performing corrosion-failure investigations follows. The chapter finishes with some basic corrective measures.

22.2 Significance of corrosion failure analysis

The goal of a corrosion failure investigation is to figure out what caused an unexpected failure in the field. Failure of equipment is defined as the unanticipated loss of functioning while in use. The major cause could be due to a design flaw, a material flaw, mechanical forces or environmental interactions, or insufficient process parameters and maintenance control. Corrosion is a common environmental interaction that frequently occursin chemical process equipment. The goal of the investigation is to determine the precise reason for the failure and give recommendations for corrective action.

22.3 Mechanisms and forms of corrosion failures

The characteristics of any corrosion failure provide essential hints as to the corrosion's most likely origin. Understanding the standard corrosion mechanisms will help one interpret the traits and findings that every corrosion research will disclose. General corrosion, pitting, galvanic corrosion, dealloying, crevice corrosion, MIC, corrosion fatigue, SCC, intergranular corrosion, fretting, erosion-corrosion, and hydrogen damage are all common corrosion mechanisms. Part I went over many types of corrosion in great depth.

22.3.1 General corrosion

Millions of microscopic and submicroscopic anodic and cathodic sites form on the surface when a metal is immersed in an electrolyte. The polarity of a particular site is determined by a number of elements, including chemical composition, residual stress, and microstructure. The anodic site corrodes in general but then produces a protective coating that effectively protects it from future corrosion. As a result, the polarity of the site reverses, and it becomes more electrochemically noble than the surrounding sites. As a result, certain areas corrode, but a protective coating forms almost quickly. Because of the continual polarity reversal, the entire exposed surface deteriorates uniformly. Because of the homogeneity of the attack, this type of corrosion is easily predicted by laboratory testing, and corrosion rates can be directly determined using weight loss measurements.

22.3.2 Pitting corrosion

In certain circumstances, many metals and alloys have a tendency to build a tenacious passive layer on their surfaces, which shields them against general corrosion. Pitting happens when anodic sites form on the surfaces of such metals and stay anodic due to the robustness of the surrounding protective coating. When exposed to chloride-containing chemicals, stainless steel, for example, has been observed to pit, especially under acidic circumstances. Pitting is a more severe issue than ordinary corrosion for a number of reasons. To begin with, laboratory testing makes pitting difficult to forecast. Although several electrochemical test methods have been established to detect if various metals are prone to pitting in specific settings, these methods do not always yield findings that are representative of actual components in real-world situations. Second, due to localized galvanic effects and acidic conditions, corrosion rates within pits are often relatively high; as a result, pits can swiftly pierce vessel walls and create leaks without notice. Third, even if pits are discovered before they create leaks, they can result in costly and time-consuming repairs. Pitted sections may require substantial grounding to properly remove localized damage of the pits, resulting in lengthy shutdowns and costly restorations.

22.3.3 Galvanic corrosion

When two dissimilar metals are electrically linked and immersed in an electrolyte, the more electrochemically active metal will corrode at a faster pace, while the more noble metal will be protected to some extent. It is crucial to remember that moisture from medium- to high-humidity surroundings can act as an electrolyte on its own, so even if components are not immersed, they might suffer galvanic corrosion. The relative potentials of dissimilar metals, relatively exposed regions of the anode and cathode, electrolyte conductivity, and the polarization tendencies of the metals in the cell are all elements that influence the pace of galvanic corrosion in an electrochemical cell. In most applications, however, careful design may easily reduce the risk of galvanic corrosion. If appropriate materials with similar electrochemical potential must be utilized, nonconductive gaskets can be used to prevent electrical contact between the metals, preventing galvanic corrosion.

22.3.4 Dealloying

Dealloying is a type of corrosion that affects tiny phases in metal alloys due to chemical composition differences that cause localized variances in electrochemical potential. Dezincification is one of the most prevalent types of dealloying. Brasses containing at least 15% zinc are prone to dezincification, which occurs when the dissolved zinc is electrochemically more anodic than the copper surrounding it. When exposed to sufficiently corrosive settings, the zinc in the solid solution is leached away and dissolved, leaving just the pure copper matrix material remaining. Visual inspection and metallography may easily detect dezincification in brass components as the afflicted surfaces change color from orange to red.

22.3.5 Crevice corrosion

When fissures exist between a metal surface and another component, the environment inside the crevice can differ dramatically from the environment outside the crevice, owing to changes in the amount of accessible oxygen. The localized electrochemical potential is effectively altered by environmental fluctuations, causing the material within the fissure to have a higher anodic potential than the material exposed to the general environment. In addition, the lack of oxygen prevents the formation of a protective oxide film within the crevice. As a result of these variables, the metal surface within the crevice suffers from localized corrosive attack.

22.3.6 Microbiologically influenced corrosion

Untreated or inadequately treated water can contain a wide variety of bacteria, resulting in MIC. This type of corrosion is most frequent in stainless steel; however, it has been observed in carbon steel under certain circumstances. Such microscopic species have a tendency to settle onto metal surfaces and produce slime deposits when the flow is somewhat slow and stationary. They tend to colonize and exude corrosive chemicals once connected to the metal surface, which can pit stainless and carbon steel. Anaerobic bacterias have a proclivity for migrating to low-oxygen conditions. These organisms prefer the lower oxygen conditions within corrosion pits to the environment in the surrounding sections of the surface when pits grow. These organisms generate subsurface caverns as they congregate within the pits, which can be easily discovered using metallographic techniques.

22.3.7 Stress corrosion cracking

While exposed to slightly corrosive conditions, passive film-forming metals that are subjected to continuous tensile stress may experience rapid catastrophic SCC. Because this cracking mechanism is relatively quick, it can result in unexpected failures. Stainless steels, carbon steels, copper alloys, aluminum alloys, and titanium alloys are among the materials susceptible to SCC. Austenitic stainless steels that are subjected to persistent tensile stress and exposed to aqueous chloride-containing chemicals, notably steam, typically develop transgranular SCC. SCC is distinguished by multiple branched cracks that extend deep into the cross-section of the material.

22.3.8 Corrosion fatigue

Corrosion fatigue, like SCC, is caused by stress and a corrosive environment. Corrosion fatigue, on the other hand, occurs when there is a cyclic stress rather than a constant load. Corrosion fatigue cracks normally begin in corrosion pits on the surface, and as the cracks spread, corrosion occurs to the fractures. Corrosion-fatigued components often have traits comparable to fatigue fractures, as well as significant corrosion of the fracture surface. Stress intensity, load frequency, ambient variables, and metallurgy all influence the pace at which cracks propagate.

22.4 The failure analysis process

A rational and systematic approach is critical in finding the root cause of any failure, whether due to corrosion, fracture, wear, or any other mechanism, . Important details and features will be less likely to be neglected during the failure investigation if this approach is used, as such details may not be recovered once the failed component has been dissected. In a failure inquiry, several guidelines have been proven to be helpful in maximizing the chances of an accurate prediction. The detailed procedure for corrosion failure can be referred to in “Failure Analysis and Prevention” (ASM Handbook, 2002).

22.4.1 Site visit

After being briefed over correspondence about the broad generic nature of the breakdown, the site visit is the first significant step in the failure analysis process. The significance of a site visit cannot be overstated. The main goal of the site visit is to get firsthand information on the design, the material of construction, specified and operating parameters, failure incident, and operating history, to physically see the failed component/equipment/location, and to formulate and get approval for a detailed failure analysis procedure before beginning the latter. According to the authors’ experience, any shortcut taken to avoid a site visit owing to time, money, or any other cause is likely to result in incorrect interpretation, analysis, and remedial efforts, resulting in major multiple failures. A poor experience was had when a sample for failure analysis was sent through courier to avoid a “time-consuming expensive site visit.” It was discovered that the sample was taken from a high-pressure conduit without paying close heed to the analyst’s instructions. (1) Specifics of the required orientation of the samples, (2) cutting procedure, (3) location of the sample, etc. The entire courier operation resulted in not only the destruction of the critical fingerprint of the breakdown, but also the additional time lost and pipeline replacement without identifying the true reason of failure.

22.4.2 Background data collection

The first logical step in any failure study is to collect as much background information about the failed component as possible. Part information, identification, relevant drawings, and specifications, should be gathered and stored for future reference. The component’s known service history is also crucial and should contain information about how it was built, as well as the stresses and service conditions it endured during its lifetime. Details on the maintenance schedule and repair history, as well as any unusual service events, should be noted in cases involving machines and equipment. The frequency of the observed failure is particularly relevant since it can help determine whether the failure is related to a production lot or exceptional service conditions. The frequency of the observed failure is particularly relevant since it can help determine whether the failure is related to a production lot or exceptional service conditions.

22.4.3 Initial inspection and scope definition

The initial inspection can be carried out once a clear understanding of the background information and component history is made. In most circumstances, the failure analyst should undertake this inspection directly; however, in cases requiring major components or machinery, it may be more cost-effective to have it completed by fairly skilled employees who are familiar with the equipment. The initial inspection includes documentation of wear patterns, atypical surface damage, corrosion features, and other factors, as well as an evaluation of the failed component and neighboring components. Photographic documentation should also be done to clearly explain the nature of the problems and provide visual aids for the analyst to analyze and debate. The scope of the study is defined during this step. The extent of the material evaluation will be determined by factors such as the amount of time available to conduct the study, the quantity and condition of representative failed and nonfailed components, and the amount of money available to do the material evaluation.

22.4.4 Material sampling and shipment

Failed components that are relatively tiny can be easily delivered to a laboratory for in-depth analysis. Large components, on the other hand, will frequently need to be sectioned to make shipping and handling easier. It is critical that the evidence is not harmed throughout the sectioning process and that the failed portions are kept as much as possible in their original condition. To avoid harming the fracture surface in cases involving broken components, cautious sectioning procedures should be used. It should be noted that some cracks may be caused by corrosion; as a result, cutting fluids and coolants should be used sparingly, as they may contaminate the corrosion deposits. Clean water should be used instead of cutting fluids if cooling is required to avoid affecting the part microstructure. The size of the samples size should be taken into account, and enough material should be removed to allow for mechanical testing, if necessary. It is also a good idea to photograph the failed component before and after the sectioning procedure, so you can see exactly where it was in service. Before sending the samples to the laboratory, they should be identified and wrapped in protective packaging materials.

22.4.5 Visual inspection

The analyst can begin the extensive visual inspection process after the defective component has arrived at the laboratory. During this stage, careful observations of the condition of components and the appearance of the damage that appears to have caused or contributed to the failure are made. Surface damage outside of the failed areas should be evaluated as well since it could indicate misalignment or unusual service circumstances. At this point, low and high-magnification photographic recordings should be done to show the morphology and color details. If laboratory cleaning is required, the failed regions should be documented both before and after the cleaning operation, as the cleaning process might disguise the original color of any rust deposits. Before and after any dismantling or sectioning, the component should be photographed. Physical samples of corrosion products should be taken in circumstances when adequate amounts of corrosion products are present, either by light scraping or by using acetate replicas. It is important to keep the corrosion deposits as close to their original state as possible because they can sometimes reveal microscopic details that reveal how they developed. It is critical that the corrosion products are not contaminated, regardless of the sampling method utilized. In most cases, it is a good idea to check the elemental makeup of the underlying material of the failing component. This procedure can be done using a variety of ways, the best of which relies on the material grade and component shape. Techniques like glow discharge spectroscopy and optical emission spectroscopy have been established as accurate and reasonably cost chemical analysis procedures for very large solid metal samples. Inductively coupled plasma-atomic emission spectroscopy can be employed for small components. This method of the analysis entails dissolving clean machining chips or drillings from a metallic component in acid, then analyzing the resultant solution (Raj et al., 1999). When a totally nondestructive examination is necessary, portable X-ray fluorescence, which involves the use of hand-held analyzers, can be used to provide estimated chemical compositions. Although this approach is widely used for positive material identification, it has limits in terms of accuracy and the ability to analyze specific elements. Energy dispersive X-ray spectroscopy (EDS) or wavelength dispersive X-ray spectroscopy (WDS) can be used in conjunction with scanning electron microscopy (SEM) to perform semiquantitative elemental characterization of corrosion deposits, where the emitted X-rays from the electron beam provide information on the approximate elemental composition of the target material. By examining crystal plane spacing, X-ray diffraction (XRD) can be utilized to detect chemicals in corrosion deposits. Characterization of organic compounds can be done using Fourier transform infrared spectroscopy, which involves bombarding the material with light in the infrared (IR) spectrum and analyzing the reflectance or absorbance pattern to determine the material's type. To more clearly identify the material, further characterization employing thermogravimetric analysis (TGA) and differential scanning calorimetry (DSC) is frequently required. TGA is a technique that measures the weight loss profile as a function of temperature and can be used to quantify the material's distinct components. The sample is steadily heated in the latter approach, DSC, and the temperature and heat flow associated with material changes are measured. This gives quantitative and qualitative data on physical and chemical changes including endothermic and exothermic processes, as well as heat capacity variations.

22.4.6 Mechanical testing

Measuring the mechanical parameters of a failed component can be very useful, especially when fracture or wear is involved. The durability of a material is determined using very common test procedures under a variety of stress and temperature situations. In most cases, hardness can be used to approximate the ultimate tensile strength at ambient temperature; however, tensile testing can be done at a variety of temperatures to directly evaluate tensile strength, yield strength, and ductility at low strain rates. Impact testing, on the other hand, can be done at a specific temperature or a series of temperatures to measure fracture toughness at a high strain rate. If the part is too small for tensile or normal hardness testing, it can be mounted in a metallographic mount and tested for microhardness, which can also be used to estimate ultimate tensile strength. In cases involving formed parts, Olsen Cup testing can be used to determine the raw material's formability, and specific tensile testing techniques can be used to determine the material's strain hardening exponent (n-value) and plastic strain ratio (r-value).

22.4.7 Scanning electron microscopy

SEM is used to characterize the component surface at high magnification . In cases when there are fractures, this device provides for positive identification of the fracture mode, as well as a detailed analysis of corrosion features that are difficult to see with light microscopy. In most cases, this microscope allows the analyst to observe the sample at magnifications of up to 5000 × . Magnifications of 20,000 × can be achieved with acceptable instrumental conditions and no residual vibration while keeping high imaging quality; however, these magnifications are rarely required for proper analysis of fracture or corrosion characteristics. EDS or WDS capabilities are available on some SEMs, allowing elemental analysis of inclusions and corrosion deposits. This is useful for figuring out what is causing the corrosion and what the most likely corrosion mechanisms are.

22.4.8 Cross-section examination

Evaluating a failing component in a cross-section can provide a wealth of information. Cross-section investigation is routinely performed at the fracture origin in cases involving fractures in order to identify any material anomalies that may have functioned as crack initiation sites. In the case of corrosion failures, examining the surface profile at the corroded areas can indicate what caused the corrosion as well as any localized microstructure features that may have contributed to the problem. Metallographic techniques are used to identify and characterize features such as pits, cracks, inclusions, and intermetallic precipitates, and etching is frequently required to show the heat treat state as well as potentially harmful metallic phases. Higher magnifications of 1000 × to 2000 × can be used to describe carbide dispersion and appearance, while lower magnifications of up to 100 × can be utilized to characterize overall microstructure homogeneity. Furthermore, plated parts can be evaluated for plating thickness by looking at the surface profile; however, in specimens that will be studied along the surface, significant care must be taken to ensure adequate edge retention.

22.4.9 Data review, conclusions, and recommendations

The analyst must draw logical conclusions when all of the data has been collected and analyzed. The conclusions should be based on sound engineering principles and take into account all of the background information, service history, and evaluation data that has been provided. To minimize misinterpretation of the results, it is critical to take into account all of the test data when drawing conclusions. After reaching the correct conclusions, the analyst can suggest corrective steps that, if followed correctly, should considerably decrease or eliminate the likelihood of future failures. This step requires the analyst to use common sense and appropriate experience in order to minimize excessive costs and, more significantly, the emergence of new problems. The simplest option is frequently the best solution.

22.5 Failure analysis report

The failure analysis report should be prepared in a clear, deliberate, and logical language. For corrosion failures, it should be clearly comprehended by a practicing nonspecialist who is neither a corrosion chemist, engineer, or metallurgist. An overabundance of scientific evidence is unnecessary; it would obscure the precise diagnosis in a sea of data. The following should be included in the report:

- The necessity of conducting the aforementioned failure analysis in order to prove that the failure is premature.
- A brief description of the chemical reaction at hand.
- Description of the failed component and failure incident, including previous chronological events and service history.
- Actual service conditions at the moment of breakdown, with any departures from design parameters highlighted.
- The material of the equipment/construction component, its specification, its production sequence, and the metallurgical treatment it received during manufacture.
- A factual account of the observations and conclusions reached after discussions with plant staff on the site. Actual results of all tests performed on the cut and acquired samples, including both the failed and nonfailed baseline samples if tests were performed on the latter.
- Specific mention of the component's metallurgical quality as evidenced by the test findings.
- All of the input data is interpreted, discussed, and analyzed.
- Specific failure diagnosis, if necessary, stage-by-stage mechanisms/ events that resulted in the ultimate failure.
- The given diagnosis is used to explain all of the observed symptoms.
- Conclusions on the causes of failure are clear.
- Practical solutions that are simple to put into action in order to avoid such failures in the future. The recommendations should be succinct and detailed, and they should be doable in a reasonable amount of time.
- If a better building material (one of the corrosion-resistant alloys) is used, and quick coupon testing in the running plant is desired, the recommendations should include the procedural specifics of such coupon tests in order to avoid corrosion occurrences.
- Finally, the report must include an executive summary that highlights the diagnosis and recommendations.

Short questions

1. What causes corrosion failure?
2. What are the type of corrosion failures?
3. How can corrosion failures be prevented?

References

ASM Handbook. (2002). (10th ed.). *Failure analysis and prevention*, (Vol. 11). Materials Park, OH: ASM International.

Raj, B., Jayakumar, T., & Thavasimuthu, M. (1999). *Practical nondestructive testing*. New Delhi: Narosa Publications.

23

Corrosion maintenance and monitoring

23.1 Introduction

Corrosion causes pipes, tanks, steel, and other structures to corrode, as it is already a well-known challenge. It has severe economic and other consequences (Bullard et al., 2003). As a result, corrosion management must be prioritized both during the design process and during the operating period. Whenever a structure, a plant, or any type of equipment is placed into service, it requires regular maintenance in order to operate effectively and reliably. This is especially true for aging systems and buildings, since many of them may continue to function much beyond their initial design life. The level of damage caused by corrosion can be widely predicted and analyzed with the aid of various corrosion inspection and monitoring techniques (Balakrishnan, 1994). It is important to carry out corrosion inspections and monitoring to assess the state of the system, as well as the effectiveness of corrosion management and maintenance actions. Corrosion monitoring encompasses a variety of methodologies, ranging from basic coupon exposure to automated sensor systems for smart structures. The knowledge offered has benefited the development and execution of effective corrosion-related loss-reduction solutions.

The procedure of the assessment of the damage caused by corrosion over a period of time and the evaluation and finding of the reason for the variation in corrosion rate is defined as the process of corrosion monitoring. It is not always easy to differentiate between the process of corrosion inspection and corrosion monitoring. The inspection usually refers to one-time, short-term measures carried out in compliance with the appropriate plans for both maintenance and corrosion inspection. When corrosion inspection and monitoring are used together, they are the most helpful and cost-effective procedures. Both of these procedures are interdependent and should not be deemed replaceable. One seemingly simple method is the employment of visible or infrared (IR) cameras, which are used in order to conduct periodic visual inspections of assets. These methods were used to monitor the hidden sides of the metallic structures, which include the inner portion of the transformers, pipelines, storage tanks, and bulky building structures. In order to monitor and predict the corrosion, other probes like ultrasonic and X-ray can also be employed; however, these methods involve a hectic process shutdown. Furthermore, because these sensors produce local measurements, the time required in order to monitor the large structures may be unnecessarily prolonged.

The corrosion coupon technique or the weight loss measurement is still considered one of the most used offline methods for checking the integrity of metallic structures.

Handbook of Corrosion Engineering. DOI: https://doi.org/10.1016/B978-0-323-95185-2.00023-X

This technique comprises subjecting or exposing a material specimen, called the coupon, to the environment (similar conditions) for a predetermined amount of time. This is followed by the removal of the subjected specimen or coupons, and estimation of the results of weight loss for the laboratory evaluation (Bruyn, 1996; Lazzari, Ormellese, & Pedeferri, 2005; Herts, 2006). In order to calculate corrosion rate, one can divide weight loss by material density, coupon area, and exposure duration. A strategy such as this is frequently employed if the rate of deterioration is slow. Furthermore, it is best suited for the estimation of the corrosion potential of the metallic structure rather than erosion. Apart from these, several other techniques like manual or offline monitoring, comprehensive online, noninvasive, and completely autonomous monitoring techniques of metallic structures remain a considerable problem that continues to entice researchers due to their importance in numerous sectors. Online monitoring is anticipated to help in the assessment of corrosion damage, the definition of corrosion rates, the understanding of corrosion behavior, the evaluation of the efficacy of corrosion control techniques, and the provision of prompt corrective actions to avoid failures. Transducer type, fit in the position, reaction time, resolution, and measurement precision are all important factors to consider when choosing a fit sensor. Online monitoring becomes vital not only for the protection and avoidance of any accidents that could endanger persons or materials but also for the coverage of enormous networks of metallic structures. Fig. 23–1 presents how to integrate data from diverse sources to provide managerial information for the sake of strategic decision-making.

Individual plant inspection and monitoring systems have significant limitations when it comes to the various complicated forms and causes of corrosion degradation. Because of the

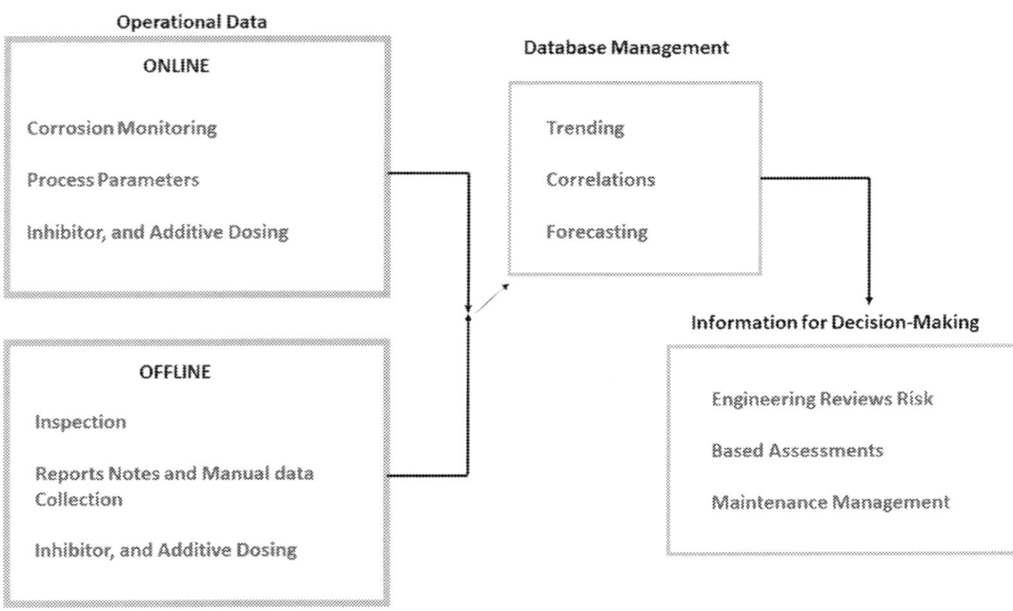

FIGURE 23–1 Review of how corrosion inspection and monitoring programs are integrated to bring strategic management information to the forefront.

huge number of variables involved, no one solution can be anticipated to be suitable in all settings and environments. As a result, the purpose of this chapter is to provide a brief overview of the various approaches used for corrosion inspection and monitoring.

23.2 The cost of poor maintenance

Most industrial sectors spend a substantial part of their operating budgets on maintenance, particularly when older structures or equipment are involved. Modern maintenance management (also known as profit-centered maintenance) aims to cut costs while enhancing plant and equipment reliability and availability. Maintenance is viewed as an investment rather than a cost in this scenario. The maintenance role, on the other hand, is typically expected to be completed with limited technical and financial resources, necessitating a focus on the most vital components as part of the overall simplification process. Maintenance is carried out to either prevent or fix problems in manufacturing equipment. It is important to remember that both preventive and corrective maintenance can be linked to either conformance or nonconformance expenses. A simple CoPM (corrective and preventive maintenance) model is established by putting up a matrix with CoPM set in proportion to costs of conformance and nonconformance (as illustrated in Fig. 23–2; Salonen & Deleryd, 2011). Four cost categories may be determined from the matrix:

- indispensable corrective maintenance cost;
- valid preventive maintenance cost;

	Corrective maintenance	Preventive maintenance
Cost of Conformance	Indispensable corrective maintenance: Corrective Maintenance due to: - Failures with random distribution and no measurable deterioration - Failures which are not financially justified to prevent	Valid preventive maintenance: Preventive Maintenance, nacessary to uphold necessary dependability Improvements intended to increase the reliability of equipment
Cost of Non-conformance	Non-accepted corrective maintenance: Corrective Maintenance due to: - Lack of preventive maintenance - Poorly performed preventive maintenance - Poor equipment reliability	Poor preventive maintenance: Unnecessary Preventive Maintenance Poorly performed Preventive Maintenance

FIGURE 23–2 Corrective and preventive maintenance (CoPM) model in which corrective and preventive maintenance is separated into the cost of compliance and cost of nonconformance. *From Salonen, A. & Deleryd, M. (2011). Cost of poor maintenance: A concept for maintenance performance improvement.* Journal of Quality in Maintenance Engineering, *17(1), 63–73. https://doi.org/10.1108/13552511111116259.*

- nonaccepted corrective maintenance cost; and
- poor preventive maintenance cost.

When it comes to meeting new expectations, "old" corrective repair, and time-based preventative maintenance techniques are usually insufficient. The following are the implications of poor maintenance practices and/or a lack of investment in the maintenance function:

1. The capacity of production gets lowered. There will be more downtime, and additionally, assets will also operate poorly during uptime.
2. Costs of production got increased. Asset performance is affected by real and opportunity costs.
3. Products and services of poor quality. Customer dissatisfaction and, most likely, lost revenue will be the end result.
4. There are risks to safety. Failures can result in fatalities, injuries, and significant financial losses.

Table 23–1 shows examples of expenses associated with the major categories in the CoPM model. Some of the costs in the idea are aggregated for convenience of usage, such as the cost of lost production time, which may be separated further into various fixed production expenses. This aggregation was done on purpose to make the notion as simple as feasible to implement inside enterprises with minimal resources.

Table 23–1 Categories of costs related to CoPM.

S. no.	Cost of conformance		Cost of nonconformance	
	Cost for indispensable corrective maintenance	**Cost for valid preventive maintenance**	**Cost for nonaccepted corrective maintenance**	**Cost for poor preventive maintenance**
Manhours for maintenance (including administration)	×	×	×	×
Exchange part	×	×	×	×
Lubricants	×	×	×	×
Other materials for maintenance (rags, cleaning solvents, etc.)	×	×	×	×
Lost production time	×	×	×	×
Logistics (for spare parts and entrepreneurs)			×	
Breakdown related to scrap and/or re-work due to poor maintenance	×		×	
Breakdown-related overtime for recovery				×
Production	×		×	
Speed losses due to poor maintenance				×

23.3 Corrosion maintenance strategies

The four types of maintenance procedures that may be distinguished are (1) corrective, (2) preventive, (3) predictive, and (4) reliability-cantered maintenance. The most recent advancement is predictive maintenance. In practice, all of these categories are employed in engineering system maintenance. To maximize profits, the objective is to strike the best feasible balance between them. Whenever the maintenance requirements are very high, the corrective maintenance strategy is typically the least cost-effective option.

23.3.1 Corrective maintenance

Corrective maintenance is when intervention is made only after a system or component has failed. As a result, it is a plan that is being implemented after the fact. In this situation, the maintenance team's job is normally to make repairs as quickly as feasible. Repair expenses (replacement components, labor, and consumables), missed output and lost revenue are all costs connected with corrective maintenance. The expansion of the size of maintenance crews, deploying backup systems, and implementing emergency procedures might be explored to reduce the consequences of lost output and speed up repairs. Regrettably, in the short term, such interventions are only cost-efficient and/or successful. Considering an example, if heat-exchanger tubes have spilled due to pitting corrosion and production must continue immediately, it may be able to temporarily seal the leaking tubes. Obviously, such measures do not guarantee the long-term functioning of the heat exchanger tubes.

23.3.2 Preventive maintenance

Preventive maintenance involves repairing and servicing equipment before it fails. Schedules govern the frequency of maintenance actions. The bigger the potential repercussions of failure, the more preventative maintenance is justified. In the end, this means a trade-off between the expense of undertaking preventative maintenance and the cost of using the equipment until it fails. Of course, safety, environmental, insurance, and other regulatory issues can also influence preventative maintenance duties. Inspection is an important part of preventative maintenance plans. At certain intervals, components are examined for corrosion and other deterioration in order to take remedial action before problems occur. Preventive maintenance performed at regular intervals is frequently associated with lower failure rates. As preventative maintenance entails large expenditures, particularly in terms of scheduled downtime, careful planning is essential. The primary goal of maximizing asset value and performance is to do preventative maintenance just before major damage occurs. Furthermore, the relevance of the equipment to the process and the required degree of dependability must drive the number of preventative maintenance activities. Computerized preventative maintenance systems are employed in facilities of all sizes to achieve these goals in current complex systems. If the preventive tasking needs to be reviewed regularly to ensure it is still relevant and to determine whether it can be replaced by predictive tasks, then a dynamic preventive maintenance system should have that capability.

23.3.3 Predictive maintenance

Maintenance that is based on the actual state of a component is known as predictive maintenance. Maintenance is carried out when a specific change in features is noticed, rather than on a predetermined preventative schedule. Corrosion sensors, which provide diagnostic information on the status of a system or component, are crucial to this maintenance method. Preventive maintenance attempts to reduce the number of superfluous inspection and maintenance jobs, perform extra maintenance tasks when needed, and concentrate efforts on the most vital objects. It is possible to make a helpful parallel to change the oil in a car. A preventative maintenance method is to replace the oil every 5000 km to extend engine life, regardless of whether the oil change is actually required. Change the oil depending on changes in its qualities, such as the building of wear debris, as part of predictive maintenance. A longer service interval may be determined by a vehicle's oil analysis when it is driven responsibly and only used for long-distance highway trips. Through proactive planning and scheduling, breakdown maintenance will be reduced and utilization will improve, thus providing some of the resources that preventative maintenance needs. In order to identify recurring issues and problem areas that have the largest impact, proper record-keeping is critical.

23.3.4 Reliability-centered maintenance

The development or modification of a maintenance program in the most cost-effective and technically viable manner is referred to as reliability-centered maintenance (RCM). It adopts a deliberate, disciplined approach that focuses on failure repercussions. As a result, it emphasizes the functional value of system components, as well as their failure and maintenance histories, rather than time-based maintenance. As opposed to preventive maintenance, RCM helps determine if maintenance is needed for a specific system component. RCM has existed since the early 1960s. RCM procedures for commercial aircraft were established when widebody jets were initially placed into commercial airline service in the late 1960s (Stern & Method, 1958). Airlines were concerned that existing time-based preventative maintenance procedures would put larger, more complicated planes out of business. Components are frequently refurbished after a specified amount of flight hours under the time-based maintenance strategy. RCM, on the other hand, establishes maintenance intervals based on the criticality of a component and its performance history, as previously mentioned. Airlines using the RCM strategy found that maintenance costs stayed relatively constant but plane availability and dependability increased (Douglas, 1995). The majority of the world's airlines currently use RCM as a regular procedure. The first development work was done by "maintenance steering groups," (MSGs), in the North American civil aviation sector. The MSGs were formed to review all that was being done to maintain planes in the air. Representatives from aircraft manufacturers, airlines, and the federal aviation administration (FAA) made up these organizations. Washington, D.C.-based Air Transport Association published the first attempt at creating maintenance policies using a logical, zero-based approach in 1968. Based on the first letters of the MSG, it was called MSG 1. It was approved in 1970 to perform a refinement called MSG 2.

23.4 Corrosion monitoring

With the rapid development of industrial corrosion monitoring systems in Europe and the United States over the last decade, several definitions of corrosion monitoring have evolved. According to one definition, "Corrosion monitoring is the systematic measurement of the corrosion or degradation of an item of equipment, to assist the understanding of the corrosion process and/or obtain information for use in controlling corrosion and its consequences" (Hines, 1968). The second definition of corrosion monitoring defines as "The assessment of the degradation of a material caused through environmental, chemical, electrochemical, or biological action or reaction" (Rothwell & Tullmin, 2000). And the third definition of corrosion monitoring defines as "Any technique used to observe or measure the progress of corrosion." Among these three definitions, the first one is preferred by all. It clearly describes the primary goal of corrosion monitoring and gives crucial information about the advantages gained from a corrosion monitoring program's functioning. The second definition provides some details on potential corrosion mechanisms as well as environmental factors. Although true, the third definition does not indicate the probable aims or benefits of online monitoring in connecting corrosion data with plant conditions and corrosion management strategies.

Oil and gas (production and refining), petrochemical and chemical manufacturing, and electricity production are all capital-intensive sectors where corrosion monitoring is especially helpful, as these have high production volumes and long intervals between scheduled shutdowns, resulting in significant financial penalties. Monitoring corrosion regimes during start-ups and shutdowns can help manage corrosion during transient operating conditions. If the steady-state conditions are altered beyond their original specifications, such as temperature or composition, it might indicate corrosion is imminent. Corrosion monitoring can demonstrate where corrosion damage has occurred and/or track the effectiveness of suggested treatments in the event that corrosion damage is unexpectedly discovered during inspections.

Currently, corrosion risk reduction methods are primarily based on corrosion inhibitor addition at various stages of oil and gas production and refining, and at utilities (boilers, cooling water systems, fire water circuits) and power generation. Continuous monitoring can help determine the best inhibitor doses, resulting in lower chemical expenditures. In most cases, a plant trial of candidate corrosion inhibitors would be conducted after laboratory screening, in which corrosion rates would be observed as a way of determining its efficacy and optimal dose rate. It may be possible to control the injection of inhibitors automatically and cost-effectively by using output data from monitoring systems connected with dosing pumps. The corrosion monitoring has evolved from offline to online, and online to real-time measurement (Fig. 23–3).

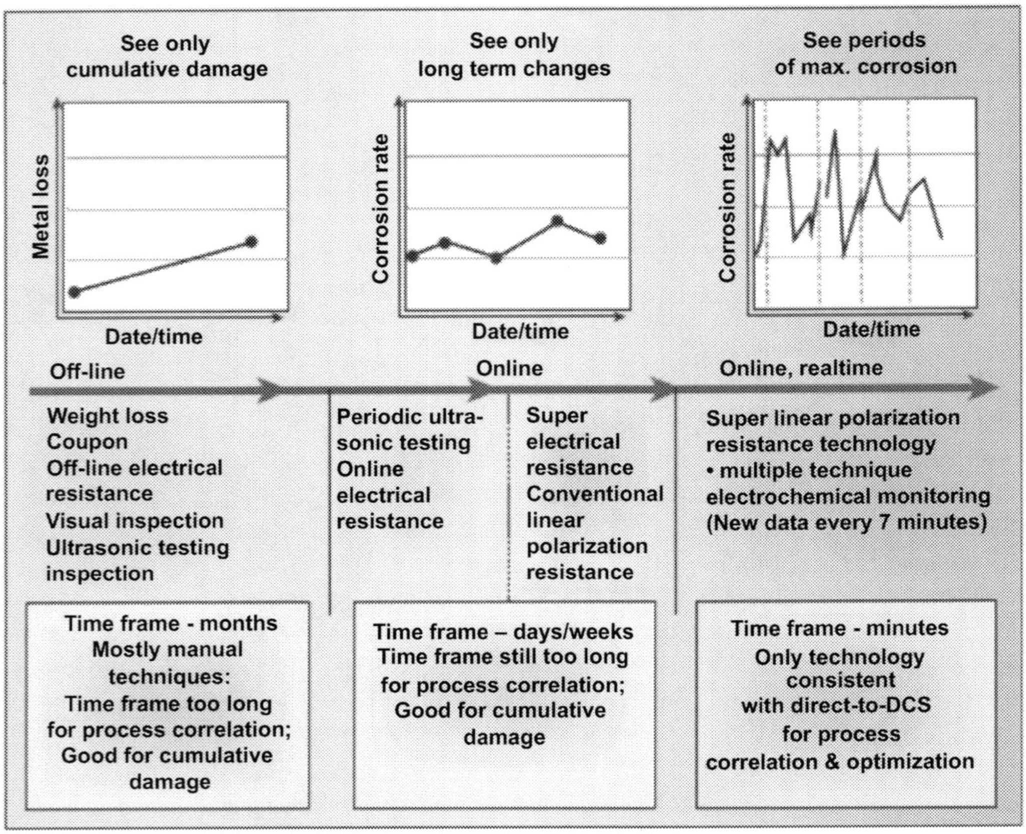

FIGURE 23–3 Corrosion monitoring has evolved from offline to online, and online, real-time measurements. *From Kane, R. D. (2007). A new approach to corrosion monitoring,* Chemical Engineering *(http://www.che.com).*

23.4.1 Objectives of the corrosion monitoring

The objectives of considering the corrosion monitoring undertaken are as follows:

1. To aid in the diagnosis of problems related to the corrosion in the operational equipment.
2. Adding inhibitors to operating equipment is one way to monitor and manage the efficacy of corrosion risk reduction techniques.
3. As part of a risk-based inspection approach to make shutdown scheduling easier.

23.4.2 Legislation and standards for the corrosion monitoring

Many countries have enacted corrosion engineering regulations, and organizations such as the National Association of Corrosion Engineers (NACE) and the European Federation of Corrosion (EFC) have imposed corrosion monitoring standards and guidelines. Corrosion monitoring guidance has been published by numerous professional organizations, including the Health and Safety Executive (HSE) in the United Kingdom (Appendix A).

23.4.3 Elements of corrosion monitoring

There are many types of corrosion monitoring systems, ranging from simple coupon exposures to hand-held data loggers to fully integrated surveillance devices that can access and manage data remotely. Corrosion monitoring programs are shown to provide cost savings in lockstep with their sophistication (and cost). Simpler monitoring systems, however, can still provide considerable financial benefits.

Corrosion sensors (probes) are a necessary part of any corrosion monitoring system. The nature of the sensors varies depending on the monitoring techniques utilized, however, a corrosion sensor is frequently referred to as an instrumented coupon. The most common types of retrievable corrosion probes can be inserted into a single high-pressure access fitting. Sensor insertion and extraction under pressured working conditions may be accomplished with specialized instruments (and courageous expert operating staff).

In most cases, the signal from a corrosion sensor must be processed in some way. Filtering, averaging, and unit conversions are examples of signal processing. Furthermore, in certain corrosion sensing systems, an input signal must disrupt the sensor surface in order to create a corrosion signal output. Electronic sensor leads were commonly used in older systems for these functions, as well as to relay sensor signals to a signal-processing unit. Microchips, which may effectively be regarded as intrinsic to the sensor units, are enabling sensor signal conditioning and processing thanks to advances in microelectronics. The microelectronic revolution has also resulted in wireless data transmission with such sensor devices. For online and real-time corrosion monitoring, regardless of the sensor specifics, a data-gathering system is necessary. The data collecting system for numerous facilities is contained in transportable labs that may be made inherently safe.

As the corrosion rate fluctuates, real-time corrosion measurements are extraordinarily sensitive, with a signal response appearing virtually quickly. The amount of corrosion damage is rarely (if ever) stable throughout time, according to several real-time corrosion monitoring programs in diverse sectors. When operating parameters have diverged “abnormally,” serious corrosion damage is more likely to occur. These undesirable operation windows can only be identified via a real-time monitoring method.

A computer system is typically used to gather data, process it, and manage it all at the same time. A data-processing technique is used to convert corrosion monitoring data (which has a lower inherent value) into information (which has a greater intrinsic value). For input into the management information system, corrosion sensors can be supplemented with additional data from other relevant sources, such as process parameter records and inspection reports. Using more advanced database management and data presentation technologies, this system converts primary corrosion data into management information.

23.4.4 Corrosion-monitoring techniques

The techniques to be used for corrosion monitoring and inspection should be defined, including whether they will be invasive (penetrating equipment/process) or noninvasive (measurements made outside of equipment), and whether they will be continuous (data available continuously from probes/sensors exposed to the corrodent) or periodic (data available on a regular basis from probes/sensors exposed to the corrodent) (sensors exposed to the process but which need to be extracted for analysis such as corrosion coupons or bio-studs). According to a variety of factors, corrosion monitoring techniques can be classified into various groups, such as direct or indirect, intrusive or nonintrusive, online or offline. If the monitoring technique measures a direct result of corrosion, it is direct, otherwise, it is indirect.

Various corrosion monitoring techniques can be classified into different categories based on various parameters, such as direct versus indirect, intrusive versus nonintrusive, and online versus offline. Corrosion is measured directly if it results directly from the monitoring technique, otherwise indirectly. A number of factors to take into account include any proposed monitoring objective, the appropriateness of new techniques in relation to the corrosion expected, the compatibility of these techniques with existing inspection and monitoring methods, and the types of sensors (portable or automatic) that can be used. A variety of monitoring techniques can be used to monitor uniform or general corrosion. The monitoring of localized corrosion, such as pitting corrosion, has traditionally been difficult, but recent advances in electrochemical methods, such as electrochemical noise (EN), have made it more viable. Few engineers, on the other hand, have any electrochemistry training, and as a result, they are less amenable to electrochemically based monitoring systems. Electrochemical impedance spectroscopy (EIS) and EN techniques are particularly problematic in this regard, and while software can produce simple corrosion rates from complex responses, a lack of understanding of the technique's fundamentals prevents its adoption, particularly by new users and in remote areas.

It is forecast that by 2025, the global corrosion monitoring equipment market (excluding inspection) will reach about $25 million, inclusive of ancillary fittings and tools. In the United States, the first corrosion monitoring equipment producers were specialty chemicals companies (particularly corrosion inhibitors), which viewed corrosion monitoring as complementary to their products. Instruments were touted as offering parameters for determining inhibitor efficacy and optimal dosing rates. Instrument production and supply activities were not considered profit centers at the time, and they were underfunded and undersupported in general. The chemical industry was one of the first industries to use corrosion monitoring. Several large chemical companies that had in-house corrosion expertise used electrochemical potential measurements to monitor passive materials. While the method did not provide a corrosion rate, it did allow operators to alter process conditions to maintain passivity. Additionally, electrical resistance (ER) technology enabled the development of sensors/probes that were eventually licensed for the use on a commercial level. An industry notorious for conservatism, the oil and gas industry, was initially skeptical of equipment purporting to deliver internal corrosion data over the internet. The corrosion monitoring systems, both for existing facilities and for new facilities, have gradually been used, even though some teething issues did arise (wrong approach, wrong sensors, probes that were not appropriate for the environment, sensors that were not robust enough). Due to the American origin of most commercial monitoring equipment, European delivery took longer, causing delays, limited repair facilities, no spares, and little or no general application guidance and support. Several sectors in the United Kingdom have used corrosion monitoring since 1981 and 1984. By the time the 1984 study was conducted, several of the issues raised in the 1981 study had been resolved. It is noteworthy that the 1981 study was conducted 10 years after UK oil production began. Overall, users of integrated systems that included various strategies had the most success. A survey found that the typical package of techniques for monitoring industrial plant corrosion at the time included corrosion coupons, and online probe system (ER or linear polarization resistance [LPR]), and a nondestructive test, most commonly ultrasonography.

23.4.4.1 Coupon technique

Specimens also known as the coupons are subjected to exposure to an environment for a period of time before being removed and weighed, as well as undergoing a more complete examination. Despite the concept's simplicity, there are a number of possible risks that may be avoided by following the recommendations of an ASTM guide in its entirety (ASTM G4 standard).

Coupons are metal samples that have been subjected to a corrosive environment in order to determine the rate of corrosion. The corrosion rate is measured using gravimetric methods each time they are removed from the plant.

Coupons are frequently used as the baseline criterion in a corrosion monitoring program, which may then be compared to instrumented techniques such as ER or different electrochemical techniques. The coupon numbers can be lowered once confidence has been gained. Coupons are normally fixed in appropriate racks or holders (Fig. 23–4) and are inserted in a shutdown and removed at the end of it. Coupons that are put in a bypass line or installed in a running plant still require full-bore valves and high-pressure access systems.

It is important to electrically isolate coupons from support systems after they have been exposed to the corrosive environment for a period of time. They are available in a variety of materials and shapes from commercial providers and are usually surface-treated, weighed, and stamped with an identifier. To minimize residual stresses from the cutting process (typically guillotining), it is recommended to anneal coupons created at factories. Coupons are susceptible to corrosion when exposed to liquid corrosives, high-temperature gases, soil, and the environment. To assess corrosion, it is necessary to remove corrosion products after exposure. By incorporating corrosion inhibitors in the cleaning solutions, errors can be reduced when metal is removed during the cleaning process. A blank coupon can be cleaned to compensate for any metal loss during the cleaning process. Corrosion inhibitor vendors frequently offer coupon assessments as part of their inhibitor supply contracts; independent laboratories can also carry out this service for smaller businesses. Coupon assessments are typically done in-house by large businesses, but can also be accomplished by independent laboratories. The rate of corrosion is measured in millimeters per year (Europe/Middle East/Asia) or mils per year (USA), with 1 mm equal to 39.37 mils. To determine the penetration

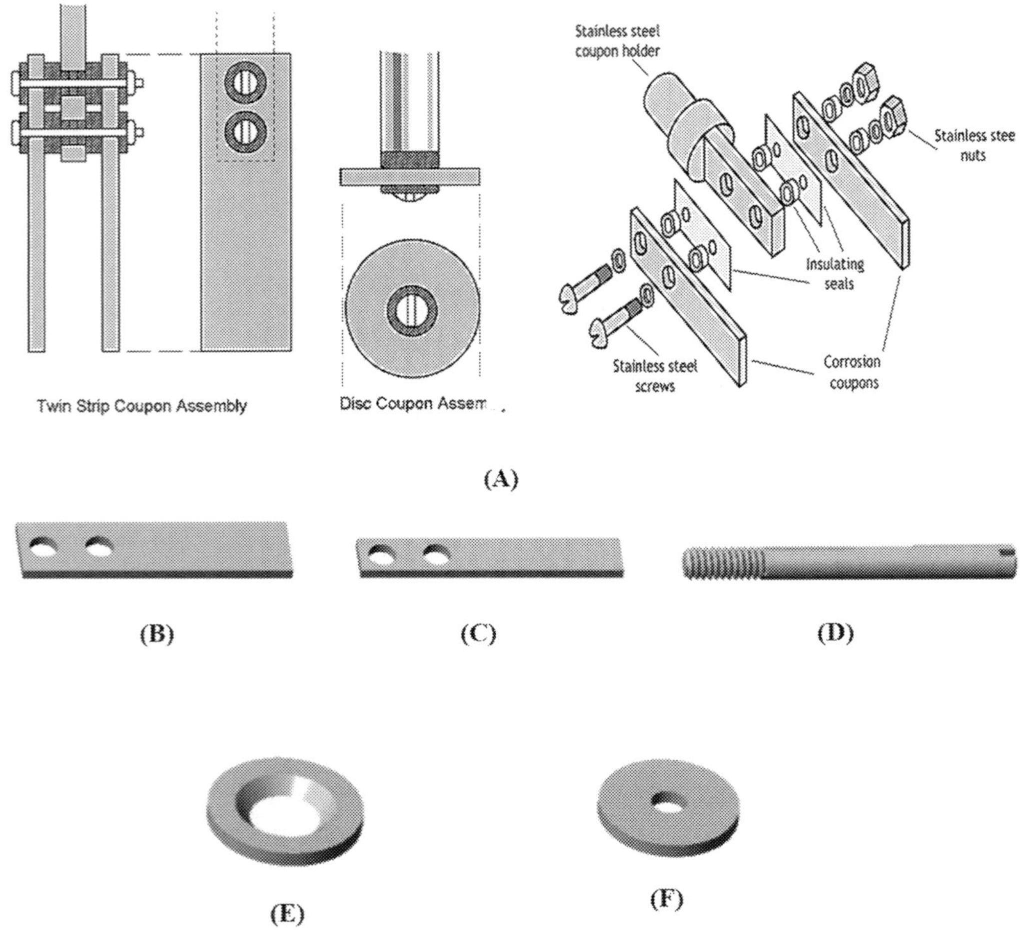

FIGURE 23–4 (A) Corrosion coupons assembly, and corrosion coupons: (B) strip coupon 3/4″ × 1/8″ × 3″ lg, (C) strip coupon 1/2″ × 1/16″ × 3″ lg, (D) rod coupon, (E) flush disc coupon, and (F) disc coupon.

rate, the weight lost after exposure to the plant or field after the corrosion products are removed is used. Additionally, the coupon surface area, the metal density, and exposure duration must be considered, and they are related as follows:

$$\text{Corrosion rate} = \frac{10.W}{\rho A t} \tag{23.1}$$

where Corrosion rate: mm/year; W, weight loss; ρ metal density(g/cm^3); A, area(cm^2); t, time (years).

There are coupons that can be used for welding corrosion testing, crevice corrosion testing, and stress corrosion testing (U-bends, C-rings) that can be purchased from vendors or created by users with design guidance.

Coupon assessment, like any other monitoring approach, is susceptible to errors. Only by being conversant with all of the standards, codes of practice, and counsel from co-workers can they be avoided. Localized attacks, like pitting and crevice attacks, can occur on coupons. There is no purpose in reporting weight loss data if an excessive localized attack is present. Localized corrosion will very likely nullify the weight loss assessment; nonetheless, representative reporting of coupon appearance is required. Pitting assaults on coupons are reported according to specified guidelines.

Without a standard mechanism for tracking coupon appearance, evaluating coupon outcomes created by diverse employees is extremely difficult. It is advised that you get familiar with the assessment literature, such as that produced by NACE, the American Society for Testing Materials (ASTM), and other organizations (Field Corrosion Evaluation Using Metallic Test Specimens, 2008; Standard Guide for Conducting Corrosion Tests in Field Applications, 2005; Standard Guide for Examination & Evaluation of Pitting Corrosion, 2005).

Monitoring coupons has several advantages, such as a low cost and the ability to provide information on localized corrosion, such as pitting, crevice corrosion, stress corrosion, and galvanic corrosion (bimetallic corrosion). In a coupon ladder rack, coupon strips can be displayed as duplicates or more than one array, making it possible to display several types of coupons at once.

In addition to being labor-intensive (especially for offshore applications), this technique averages measurements across the time period, which may obscure significant differences in corrosion rates over shorter periods. One way to minimize the possibility of this is to expose a series of coupons at one site and recover individual coupons at predetermined intervals.

23.4.4.2 Electrochemical Methods

a) Electrical resistance probes

In relation to Fig. 23–5, online corrosion repair involves the use of a metal element made from the metal or alloy of interest, mounted inside a suitable housing, and placed in corrosive surroundings. The electrical resistance (ER) probe is made up of a metal coupon made of the same pipe material that allows the corrosion rate to be calculated by measuring its electrical resistance. According to the second Ohm's law, as corrosion progresses, the coupon's electrical resistance increases because the corrosion attack lowers its cross-section area. A straight line with a slope proportionated to corrosion rate is created by charting ER versus time. The approach has several advantages: there is no need for a retrieve operation because the measurement can be done externally; the corrosion rate (i.e., the slope) can be computed after only a few days of exposure; and it works in any H_2S-free environment. The creation of iron sulfide, which has an electronic conductivity, influences the ER in the presence of hydrogen sulfide, resulting in unpredictable readings. Because resistance varies with temperature, the gadget has a calibration mechanism that takes temperature into account. The fundamental disadvantage of this technology is that when the projected corrosion rate is low, the coupon thinning and, as a result, the ER fluctuation is small, necessitating high electrical instrumentation precision.

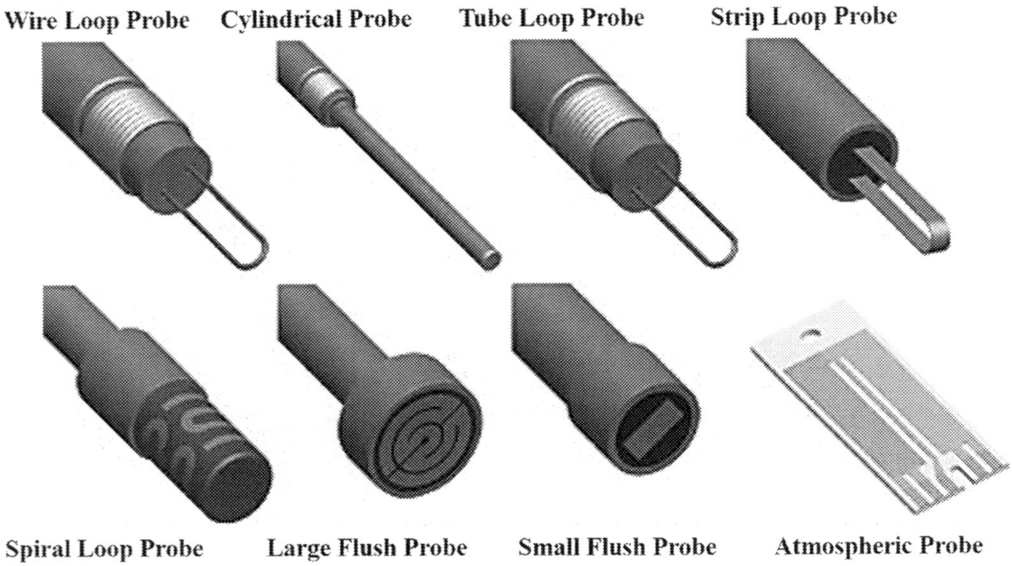

FIGURE 23–5 Different electrical resistance probes. *From http://www.alspi.com/erintro.htm.*

b) Potential Measurement

Potential measurement: The potential of passive films and their fluctuation over time are indications of their state, however, corrosion penetration rates are not obtained. Typically, this method is employed in the chemical industry to check the state of passive materials. A high-impedance voltmeter must be used, as well as a suitable reference electrode, as well as a working electrode, which might be the plant component of interest. In pioneering work on this technology, Edeleanu (1955) revealed that the potential of UNS S30400 stainless steel varied with pumping speed, temperature, and impurities. A laboratory assessment of polarization properties can determine the ranges of passive and active potentials for a specific material/environment combination. Choosing the right reference electrode can be challenging for those unfamiliar with corrosion. It is preferred to use a reversible reference electrode, such as calomel or silver–silver chloride, though nonreversible electrodes can also be employed in practice provided the potential remains constant. Commercially available polarization resistance probes can be used for this.

c) Potentiodynamic polarization/cyclic polarization

With these approaches, the polarization properties of the metal–environment system can be measured, and laboratory procedures have been outlined and described (Publication EFC4. Guidelines for Publication EFC4, 1990). By using external current through a counter electrode, polarization curves are created in traditional electrochemical cells by polarizing the potential of a working electrode with respect to a reference electrode. When it comes to potential measurement, selecting a reference electrode can be difficult. Tafel constants, b_a and b_c, can be calculated by using the slopes of the linear sections of the anodic and cathodic polarization curves affected by corrosion. In order to determine corrosion rates, the linear sections of the curves can be extrapolated to cross at the natural corrosion potential. A cyclic polarization is similar to potentiodynamic, except that the potential is reversed in a negative direction following an anodic scanning. Anodic breakdown potentials and hysteresis signatures in the reverse scan's anodic section may indicate a tendency toward pitting and crevice corrosion, respectively. An ASTM standard contains instructions for performing these tests (Standard Test Method for Conducting Cyclic Potentiodynamic Polarization Measurements for Localized Corrosion Susceptibility of Iron-, Nickel-, or Cobalt-Based Alloys, 2003). There are portable commercial potentiostats that can be used to assess polarization in real-time in running facilities. However, polarization features are typically complicated, and these approaches need careful, informed interpretation. As a result, they are not commonly utilized for monitoring of corrosion monitoring.

d) Linear polarization resistance

In principle, since water corrosion is an electrochemical process, using Faraday's law and measuring the current involved in corrosion reactions should yield a quantitative measure of corrosion rate. The internal corrosion current of a corroding metal cannot be measured effectively with an ammeter when the cathodic and anodic sites are inseparable. Stern and Geary's relationship (Stern & Geary, 1957; Stern & Method, 1958) describes how an applied potential changes a material's polarization characteristics in response to an environment with a corroding electrode close to the corrosion potential. This relationship can be used to calculate the current with the environment's polarization characteristics:

$$R_p = \frac{\Delta E}{\Delta I} = \frac{b_a b_c}{2.3 i_{corr}(b_a + b_c)} = \frac{B}{i_{corr}} \tag{23.2}$$

where R_p is the Polarization Resistance; ΔE (equal to $E - E_{corr}$) is the potential shift from the corrosion potential, E_{corr}; ΔI is the measured external current flowing due to this potential shift; i_{corr} is the corrosion current at the free corrosion potential, E_{corr}; b_a and b_c are Tafel constants.

Stern–Geary's relationship only holds when specific experimental and theoretical circumstances are met, one of them being a relatively low DC potential (generally around 20 mV). Readings are rapid and in effect immediately, so they may be used to control procedures for corrosion mitigation in operational equipment, such as adding corrosion inhibitors to recirculating cooling water systems in refineries or cutting additives to crude distillation columns. Employing a probe access fitting is the preferred method of inserting probes into plants. It is possible to use two- or three-electrode probes with the LPR technique. Current is measured by applying a voltage (up to 30 mV) between identical electrodes in a two-electrode probe. In most cases, a cycle period of roughly a minute is used, and steady-state conditions should then dominate as a result of a second application of the voltage with reversed polarity. In three-electrode probes, the potential difference between the working electrode and the reference electrode is measured and then polarized by 10 mV. With the three-electrode probe, you can test more resistive solutions, but it poses the problem of inaccuracies in measurements if the current selected is too high and the electrodes are polarized beyond 10 mV. Metal rod electrodes come in a variety of metallic compositions, and they are connected to threaded rods within the probe. The devices may be fixed or retractable, or they may be fitted with a pressure gauge through a full-bore opening valve, and they may be paired with an automated retrieval system. Instrumentation is available in various shapes and sizes, from portable devices to fully automated retrieval systems. For a wide variety of material/environment combinations, B values are used to measure penetration rates (mils per year/m year^{-1}/mpy, or millimeters per year/mm year^{-1}/mmpy, respectively). The method's main benefit is that it can estimate corrosion rates almost instantly, allowing it to be linked to process parameters almost immediately. Besides the limitation to conducting solutions, which can often be compensated for with resistances, and the required values of B for the relevant reactions, the approach has two other drawbacks. When deposits build up on the probes, readings become distorted, and oxidizing or reducing processes that aren't part of corrosion can cause reading errors. As part of the validation and calibration of the methodology, other corrosion monitoring techniques should be included.

e) Electrochemical Noise (EN)

This method examines spontaneously occurring potential or current transients on or between corrosion-prone metal electrodes. Iverson (1968) was the first to recognize the EN method, and there is now a large body of literature on its uses in corrosion monitoring. The theoretical foundations and evolution of EN are covered in depth elsewhere in this book. EN uses in power generation plants, cooling water systems, and concrete reinforcement was documented in the latest edition of this book, published in 1994. It was also said that probes comprising combination elements have been created and used for measurements in condensing acid conditions using a variety of electrochemical methods (AC and DC). Electrical equipment advances have enabled EN to advance rapidly thanks to its ability to distinguish between true voltage/current transients and electronic background noise. There has been interest in the technique due to its unique sensitivity to localized corrosion processes such as pitting crevice corrosion and stress corrosion cracking (SCC) which is unavailable by traditional electrochemical techniques. This makes EN a very good candidate for corrosion/process control because it is also very sensitive to changes and disturbances in a system. In this method, the electrode surfaces are not perturbed by the natural voltages and currents created in the environment, as they are with techniques using polarization, which are electrochemical. When compared to LPR, measurements can be made in a wider range of ambient resistance. Potential noise and/or current noise are measured between two identical electrodes in the standard setup. A zero-resistance ammeter is used to measure current. The corrosion rate and mechanism of corrosion are determined by the data analysis. Corrosion rates can theoretically be inferred using EN data in a similar way as LPR measurements. The noise resistance (R_n) is equal to the LPR polarization resistance (R_p), and is computed using the equation (Bullard et al., 2003):

$$R_n = \frac{\sigma V_n}{\sigma i_n} \tag{23.3}$$

where V_n and i_n are the standard deviations of the EN potential and current signals, respectively.

The temporal basis of the data acquired, statistical analysis, which can offer mechanistic information, frequency domain transformation to define localized corrosion, and fractal and neural analysis are all approaches that can be utilized to study the data (Hladky & John, 1984). The technology has the benefit of being able to be used in running plants using ordinary, commercially available LPR probes and traditional access systems. The intricate instrumentation, like EIS, can be costly, and the verification of instrument/computer outputs, as well as the interpretation and analysis of data, require specialist knowledge. Unless a bigger organization has in-house competence, the approach is usually given by specialist companies/contractors. The downhole corrosion monitoring tool electronics section is shown in Fig. 23–6. The sensor head comprises a plurality of mutually insulated electrical sensors (connector electrode array) and is tied to a cable that is extensible so as to raise or lower the sensor head downhole and has a signal conductor (electronic section) to return data from the sensor head to the surface.

23.4.4.3 Field signature methods

Field signature monitoring was created to identify inner corrosion in pipes and pipelines (Strommen, Horn, & Wold, 1994). An array of 3-mm sensing pins is welded to the outside surface of the pipeline (spool) to serve as a measuring sample. The measuring sample is a length of the pipeline (spool) constructed of the same material as the parent pipeline. In order to measure the electric field pattern, the electrodes on the outside surface of the test spool are programmed to detect the very tiny changes in potential between the pairs of pins when a current is passed through them. Variations in electric field patterns are a result of general or localized corrosion on the surface of the test spool. In addition, current variations and temperature impacts are considered. Corrocean (Norway) received the first Norwegian patent for the system in 1984, and it has been licensed globally for commercial use by the company. With a resolution of 0.025 mm, Corrocean claimed that an accuracy of 0.1% of pipe wall thickness could be achieved for a pipe wall thickness of 25 mm. Monitoring large areas (up to several m^2) is possible, and easier corrosion resolution can be achieved by decreasing pin distances. A version of this approach has been described in the literature as

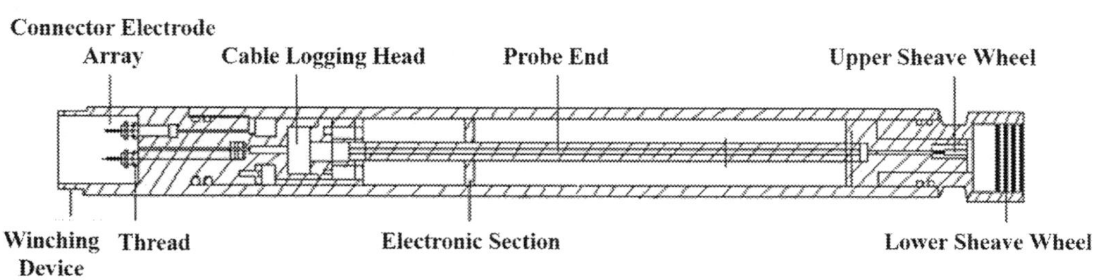

FIGURE 23–6 The downhole corrosion monitoring tool electronics section. *From WO/2006/059133, https://patentscope.wipo.int/search/en/detail.jsf?docId = WO2006059133.*

an ER system. As an alternative, another manufacturer produces a detachable sensor sleeve with two parts. In the oil and gas industry, the technology was used to detect and monitor corrosion in subsea pipes, especially for new systems rather than retrofitting existing lines, which would require extensive shutdowns and additional costs. A decade ago, the technology was extended to include ship topsides. For testing corrosion inhibitors in oilfields, an experimental pressurized pipeline loop was installed with an FSM system in the United Kingdom. The operator stated that the results were good, emphasizing precision. Significant cost savings on corrosion inhibitors were achieved in one onshore oilfield installation when the period between inhibitor treatments was increased by 2 weeks. A technique's noninvasive nature, the large area of actual plant components that can be monitored compared to traditional invasive probes, and the sensitivity that allows it to correlate corrosion rates with process conditions, such as following the corrosion process in crude oil refining due to naphthenic acid, are all advantages (Britton, n.d.). The North Sea has been able to identify pinning and SCC and monitor corrosion across pipeline welds. The technique's high cost limits its use to essential applications, and it requires expert data interpretation, which is only possible with custom-supplied software. The results must be interpreted carefully since corrosion happening more than 1 m on each side of the sensing unit will not be recognized, and extrapolating the test spool performance to the entire system might be deceptive if the corrosion has not been identified. Concerns have been raised about using FSM in high temperatures (60°C in the shade in the Middle East), as well as whether the delicate electronics would withstand sand and dust in desert regions, which applies to any instrumented system. Appropriate and frequent maintenance, on the other hand, may help to alleviate these issues. Overall, the multitude of applications in the oil and gas sector, as well as its increasing use and growth in other industries, make this approach one of the most exciting advancements in corrosion monitoring technology in recent years.

23.4.4.4 Thin layer activation

To monitor engine wear and the performance of various lubricants, thin layer activation (TLA) was developed for monitoring metal wear. It was originally developed to monitor corrosion in 1983 (Asher, Conlon, Tofield, & Wilkins, 1983). Using a radioactive tracer (one part in 1020) injected into a metal surface, the reduction in radioactivity can be used to measure the amount of metal loss caused by corrosion. Successful application of corrosion products requires nonadherent properties. Monitoring can be based on determining whether the corrosion-corroding surface or corroding liquid environment is losing radioactivity or increasing it unless the throughput is too high to distinguish radioactive isotope uptake from background radiation. Invasively exposing irradiated coupons or plant components can be done, for example. An irradiated internal sample surface can be used for measurements outside of equipment, making this approach noninvasive. Although the technique has found use in subsea, water treatment, and paper mill operations, the requirement of exposing samples to a nuclear accelerator, along with the associated logistics, has restricted its widespread use (Britton, n.d.).

23.4.4.5 Chemical analysis

In corrosion monitoring, the need for a well-thought-out and implemented chemical analysis program is often ignored. Changes in corrosivity, as well as changes in process conditions, may be evaluated via liquid analysis. Analysis of corrosion behavior should include oxygen, hydrogen, chloride, trace elements, iron, manganese, corrosion inhibitors, and other treatment chemicals such as oxygen scavengers, biocides, and pH. Hydrogen sulfide and carbon dioxide testing may be performed on-site, and free water and pH monitoring can be done using various test sheets. Temperature measurements should always be taken immediately on-site. In order to ensure adequate coverage throughout the system, a chemical analysis should be included in the process plant design, beginning with the well and continuing through the processing facilities and the export infrastructure. In order to get relevant information, temperature, pressure, flow rates, and stream composition can be interconnected with plant variables. Ideally, automatic analysis systems based on electrodes, that can detect ions, are preferred. Chemical analysis is used to determine the amount of element iron present in product streams in the oil industry. This type of analysis refers to iron counts. If corrosion is measured in a systematic manner, such as starting at an upstream reference point and collecting fluid samples at several downstream reference points, the results may provide an accurate indicator of corrosion. Weighing the amount of iron within a liquid sample is a method of evaluating samples in the lab.

Several recommendations are included in the literature (Asher et al., 1983) for iron analysis in the oil and gas industry. This method does not result in a reduction in wall thickness. If the corrosion distribution is uniform, continuous sampling, on the other hand, provides a good indicator of the amount of corrosion that is happening. A number of flaws in this approach exist, including ineffectiveness against localized corrosion and the fact that it is invalidated by the presence of iron in process fluids such as oil. In this case, a manganese analysis can be used instead, and the results can then be linked.

23.4.4.6 Monitoring hydrogen

Cathodic corrosion creates hydrogen as a result of acidic conditions such as acidizing fluids, sour systems, carbon dioxide, and naphthenates. It is possible to determine the extent of corrosion based on the presence and amount of hydrogen. The presence of hydrogen in the oil sector was previously checked using an empty paint that can be filled with a sample of generated fluids. Blemishes on the outside may indicate the presence of hydrogen due to sour water. Because of their small size, hydrogen atoms can easily travel through a metal lattice. A metal lattice with gaps or flaws allows hydrogen atoms to combine, creating hydrogen molecules, causing internal pressure to rise until it exceeds the material's cohesive strength, producing hydrogen-induced cracking and blistering. As a result of corrosion, the first probes tracked the build-up of hydrogen pressure. However, it does not provide direct evidence of corrosion penetration. In contrast, any increase in gas volume indicates that corrosion is occurring. There are still pressure probes used in oil production and refining activities, allowing for insertion and retraction under system pressure with the same fittings used for other types of intrusive probes and coupons. Continuous bleeding from the probe is required by hydrogen gas, which is inconvenient in vast oilfields. For example, forming teams, particularly for this purpose, has proven successful in the Middle East. Pipelines and piping systems can have an outer sheath welded to them to distribute hydrogen generated by bore corrosion. For noninvasive bore corrosion monitoring, a pressure probe can be used to determine the pressure created. Electrochemical hydrogen monitoring was later introduced by several firms using Devanathan cells mounted on exterior surfaces with sealing mechanisms to contain electrolytes. Portable versions of these cells allow corrosion and inspection specialists to use this technology more widely. More sensitive probes can now be used to measure hydrogen flow through metals. Several versions of noninvasive technologies are available for periodic and continuous monitoring. A variety of sensor configurations are available to accommodate different geometries, so they are capable of detecting hydrogen at extremely low levels, as well as monitoring temperature.

23.4.4.7 Testing of heat exchangers and spool pieces

Test heat exchangers are frequently used in cooling water systems to analyze the efficiency of the corrosion inhibitor and, as a result, the corrosion behavior of the heat exchanger tubes. Corrosion inhibitor suppliers frequently include them as part of their total service package. Multiple tubes on the test equipment can be opened for visual inspection and assessing corrosion rates using destructive metallography. Tubes may be replaced or removed without interrupting the system's operation via valves that isolate the units. Plants can be installed with a small section of pipe (1 m in length) that can provide corrosion data. Removal and replacement of spools can be made easier by inserting them into bypass loops during shutdowns. If data from bypass loops are to be interpreted, caution must be exercised until it is clear that the bypass line is representing mainstream conditions and that any variances are being considered.

23.4.4.8 Monitoring of bacteria

Bacteria may be found in a variety of commercially essential processes, such as cooling water systems and practically every aspect of oil production. Detecting the presence of bacteria is necessary for good maintenance management, which includes biocide treatment. Bacteria acquire tolerance to certain biocides over time, and the kind of biocide must be changed at regular intervals to keep bacterial growth under control. Sulfate-reducing bacteria, which eat sulfate in their metabolism and produce and release hydrogen sulfide, which is highly corrosive and can cause steel breaking, are the most important bacteria in terms of corrosion. Bacteria can live in liquid (planktonic) or beneath scale/deposits (sessile), and bacteria-monitoring methods must take this into account. Bacterial analysis (through filters), side streams, containers, and particularly built corrosion coupons, such as Robbin's instrument, which uses metal studs as detailed in the next paragraph, are all examples of monitoring techniques. It is worth noting that simply monitoring the fluid flow will not reveal the existence of sessile bacteria. Testing for bacteria is expensive, time-consuming, and inconvenient, thus much work has gone into developing spot tests that can detect the presence of bacteria in 10–15 minutes, but without identifying the bacterium kind. Adenosine triphosphate photometry, fluorescence microscopy, and radio respirometry are all quick ways to examine microorganisms. There are standards and best practices to follow (Detecting, Testing, & Evaluation of Microbiological Influenced Corrosion (MIC) of External Surfaces of Buried Pipelines, 2006; Standard Recommended Practice-Monitoring Corrosion in Oil & Gas Production with *Iron Counts*, 1998; TM0194–, 2004; *Microbial* Corrosion & Sequeira, 2000). It is required to utilize a microbiological laboratory to determine whether the sort of bacteria is present while transporting sterile packaging, which might be difficult, especially for offshore operations. Coupons and probes (with flush components) can be used to get information about microorganisms. Any sessile bacteria will deposit on the studs, which may be retrieved and examined, according to 'Robbin's device, which consists of a 25 mm diameter stainless steel pipe holding a series of carbon steel studs. For installation into the plant using conventional access fittings, 40 variations of the 'Robbin's unit are commercially available. Most coupon and corrosion probe suppliers will be able to provide this equipment.

23.4.5 Data management in corrosion inspection and monitoring

At the design stage, the sort of data necessary and how it will be examined must be carefully considered. There has always been a dispute over the utility of continuous data when the end result is massive numbers of records that are never thoroughly reviewed. Integrating corrosion management guidelines into overall plant management provides a structure for what data is needed and to whom it should be communicated. In addition to corrosion monitoring and inspection data, corrosion professionals should consult operating and process data. All data should be included in the appropriate plant and equipment records, with a quarterly summary report given to the relevant departments. Reports should be provided in a computer format. In order to minimize redundancy, duplication, and/or obsolete information being retained in current records, a regular assessment of the equipment being monitored and its accompanying data processing system should be conducted. The efficiency of corrosion-monitoring programs may be substantially harmed if data recording and processing are not standardized. Recently, the author audited corrosion control methods on a major railway bridge in Scotland (Report, 2001). There was a comprehensive coupon monitoring program in place. The report format for describing how the coupons looked was inconsistent, and there were a lot of coupon records. As a result of the absence of one inspector and the substitution of another, there were variations in coupon descriptions that made precise comparisons difficult. Despite the suggestions for systematic documentation of coupon corrosion, the standardization of coupon corrosion was ignored. As an example of data management that was not coordinated, a corrosion breakdown occurred in the saturated gas plant of a UK oil refinery. This led to a large explosion. In order to control plant deterioration, the refinery used materials selection, inspection, corrosive agent sampling, and corrosion monitoring. Corrosion was detected during an ER probe, but no action was taken. The corrosion monitoring and inspection records at the refinery were filled with numerous inaccuracies. The inspection crew at the refinery kept track of corrosion-related occurrences individually, but there was no central repository where it was stored.

Appendix A: Standards and codes of practice

NACE International Standards, Recommended Practices, and Test Methods

RP0102-2002 In-line Inspection of Pipelines

RP0189-2002 Online Monitoring of the Cooling water system

RP0192-98 Monitoring Corrosion in Oil and Gas Production (with Iron Counts)

RP0197-2004 Standard Format for Computerized Electrochemical Polarization Curve Data Files

RP0198-2004 The Control of Corrosion Under Thermal Insulation and Fireproofing Materials – A System Approach

RP0205 Recommended Practice for the Design, Fabrication, and Inspection of Tanks for the Storage of Petroleum Refining Alkylation Unit Spent Sulfuric Acid at Ambient Temperature

RP0296-2004 Guidelines for Detection, Repair, and Mitigation of Cracking of Existing Petroleum Refinery Pressure Vessels in Wet H2S Environments

RP0497-2004 Field Corrosion Evaluation Using Metallic Test Specimens

RP0502–2002 Pipeline External Corrosion Direct Assessment Technique

RP0775-2005 Preparation, Installation, Analysis, and Interpretation of Corrosion Coupons In Oilfield Operations

SP0106-2006 Control of Internal Corrosion in Steel Pipelines and Piping Systems

SP0206-2006 Internal Corrosion Direct Assessment Technique for Pipelines Carrying Normally Dry Natural Gas

SP0294-2006 Design, Fabrication and Inspection of Tanks for the Storage of Concentrated Sulfuric Acid and Oleum at Ambient Temperatures

SP0590-2007 Prevention, Detection, and Correction of Deaerator Cracking

SSPC-VIS2 Standard Technique of Evaluating Degree of Rusting on Painted Steel Surfaces

TM0106-2006 Detecting, Testing, and Evaluation of Microbiological Influenced Corrosion (MIC) of External Surfaces of Buried Pipelines

TM0169-2000 Laboratory Corrosion Testing of Metals

TM0193-2000 Laboratory Corrosion Testing of Metals in Static Chemical Cleaning Solutions at Temperatures below 93C

TM0194-2004 Field Monitoring of Bacterial Growth in Oilfield Systems

TM0286-2001 Cooling Water Test Unit Incorporating Heat Transfer Surfaces

TM0299-99 Corrosion Control and Monitoring in Seawater Injection Surfaces

TM0398-98 Laboratory Corrosion Testing of Metals in Static Chemical Cleaning Solutions at Temperatures Above 100C

TM0498-2006 Evaluation of the Carburization of Alloy Tubes Used for Ethylene Manufacture

TM0499-99 Immersion Corrosion Testing of Ceramic Materials

American Society for Testing Materials (ASTM)

ASTM G4-01 (2008) Standard Guide for Conducting Corrosion Tests in Field Applications
ASTM G96-90 (2008) Standard Guide for Online Monitoring of Corrosion in Plant Equipment (Electrical and Electrochemical Methods)
ASTM G102-89 (2004) Standard Practice for Calculation of Corrosion Rates and Related Information from Electrochemical Measurements
ASTM G1-03 Standard Practice for Preparing, Cleaning, and Evaluating Corrosion Test Specimens

European Federation of Corrosion – EFC (Incorporating ISO) Standards

EN ISO 6509 1995 Corrosion of metals and alloys – Determination of dezincification resistance of brass (ISO 6509:1981)
EN ISO 7441:1995 Corrosion of metals and alloys – determination of metallic corrosion in outdoor exposure corrosion tests (ISO 7441:1984)
EN ISO 7539 Corrosion of metals and alloys – Stress corrosion testing, Parts 1–7 1995: 1989
EN ISO 8565:1995 Metals and Alloys – Atmospheric corrosion testing-general requirements for field tests (ISO 8565:1992)
EN ISO 9400:1995 Nickel-based alloys–Determination of resistance to intergranular corrosion (ISO 9400: 1990)
EN 10229:1998 Evaluation of resistance of steel products to hydrogen cracking (HIC)
EN ISO 11306:1998 Corrosion of metals and alloys – Guidelines for exposing and evaluating metals and alloys in surface seawater (ISO 11306:1998)

Provisional Europeans Standards (As of June 1994)

pr EN ISO 8044 Corrosion of metals and alloys—basic terms and definitions
pr EN 12500 Protection of metallic materials against corrosion—corrosion likelihood in atmosphere environment—classification, determination, and estimation of corrosivity of atmospheric environments

British Standards/ISO

PD 8010 2:2004 Code of practice for pipelines. Subsea pipelines.
PD 8010 1:2004 Code of practice for pipelines. Steel pipelines on land.
BS EN 1473:2007 Gas liquefaction equipment and installation. Onshore installation design.
BS EN ISO 19903:2006 The petroleum industry and the natural gas industry. Offshore fixed-concrete installations.
DD ISO/TS 24817:2006 Natural gas, petroleum, and petrochemical industries. Pipework repairs using composite materials. Design, installation, testing, and inspection of pipework.
BS ISO 12732:2006 Metals and alloys corrosion. Using Cihal's double loop technique, electrochemical potentio-kinetic reactivation is measured.
BS EN ISO 9934-2:2002 Nondestructive testing. Magnetic particle. Detection media

References

Asher, J., Conlon, T. W., Tofield, B. C., & Wilkins, N. J. (1983). *On-line surveillance and monitoring*. Ellis Horwood.

Balakrishnan, K. (1994). Trends in corrosion protection of materials. *Bulletin of Materials Science, 17*(7), 1331–1339. Available from https://doi.org/10.1007/BF02747231.

Britton, C. F. *Britton, Private Communication*.

Bruyn, H. J. D. (1996). Current corrosion monitoring trends in the petrochemical industry. *International Journal of Pressure Vessels and Piping, 66*.

Bullard, S. J., Covino, B. S., Russell, J. H., Holcomb, G. R., Cramer, S. D., Ziomek-Moroz, M., & Eden, D. (2003). 1 2003/01 NACE - International corrosion conference series 03614409 *National Assoc. of corrosion engineers International United States Laboratory evaluation of an electrochemical noise system for detection of localized and general corrosion of natural gas transmission pipelines*. http://www.nace.org.

Corrosion, M., & Sequeira, C. A. C. (2000). Ed.; European Federation of Corrosion.

Detecting, Testing, and Evaluation of Microbiological Influenced Corrosion (MIC) of External Surfaces of Buried Pipelines. NACE International, 2006.106-2006.

Douglas, J. (1995). Measuring the cost of transmission services. *EPRI Journal, 20*(4), 6–15.

Edeleanu, C. (1955). Corrosion monitoring for chemical plant. *Anti-Corrosion Methods and Materials, 2*(7), 204–208. Available from https://doi.org/10.1108/eb019074.

Field Corrosion Evaluation Using Metallic Test Specimens. (2008).

Herts, P. B. (2006). *Method and apparatus detecting corrosion &/or deposit formation in water systems*. EP Patent. 1693664.

Hines, J. (1968). Industrial corrosion monitoring; committee on corrosion.

Hladky, K. & John, D. G. (1984). 1984 *Proceedings of the 2nd international conference on corrosion monitoring and inspection in the oil and petrochemical and process industries using electrochemical noise*.

Iverson, W. P. (1968). J. Electrochem. Soc. 115, pp. 617–618.

Lazzari, S., Ormellese, M., & Pedeferri, P. (2005). *Measuring device, equipment and method for monitoring the onset of corrosion affecting steel reinforcements embodied in reinforced concrete*. WO Patent.

Publication EFC4. Guidelines for Electrochemical Corrosion Measurements, European Federation of Corrosion. 1990.

Rothwell, N., & Tullmin, M. (2000). *The corrosion monitoring handbook*. Kingham.

Salonen, A., & Deleryd, M. (2011). Cost of poor maintenance: A concept for maintenance performance improvement. *Journal of Quality in Maintenance Engineering, 17*(1), 63–73. Available from https://doi.org/10.1108/13552511111116259.

Standard Guide for Conducting Corrosion Tests in Field Applications. ASTM International (2005).

Standard Guide for Examination and Evaluation of Pitting Corrosion. ASTM International (2005).

Standard Recommended Practice-Monitoring Corrosion in Oil and Gas Production with Iron Counts. NACE International (1998).

Standard Test Method for conducting cyclic potentiodynamic polarization measurements for localized corrosion susceptibility of iron-, nickel-, or cobalt-based alloys. ASTM International (2003).

Stern, M., & Geary, A. L. (1957). Electrochemical polarization: I. A theoretical analysis of the shape of polarization curves. *Journal of the Electrochemical Society, 104*(1), 56–63. Available from https://doi.org/10.1149/1.2428496.

Stern, M., & Method, A. (1958). For determining corrosion rates from linear polarization data. *Corrosion, 14*(9), 60–64. Available from https://doi.org/10.5006/0010-9312-14.9.60.

Strommen, R. D., Horn, H., & Wold, K. R. (1994). Mater. Perf, 32, pp. 50–54

TM0194, 2004. TM0194, *Field monitoring of bacterial growth in oil and gas systems*. NACE International, 2004.

PART 8

Corrosion protection in industries

24

High-temperature corrosion

High-temperature corrosion is the oxidation process brought on by the presence of a salt or ash deposit that results in the degradation of metals and alloys. In contrast to gaseous oxidation, hot corrosion subjects metals and alloys to substantially faster rates of deterioration, resulting in the formation of a porous, nonprotective oxide scale on their surface and sulfides in the substrate (John &Young, 2002). High-temperature corrosion, also known as hot corrosion, is a serious problem in energy conversion and chemical process systems such as boilers, internal combustion engines, fluidized bed combustion, and industrial waste incinerators (Rapp, 2002). One type of corrosion that does not need a liquid electrolyte presence is high-temperature corrosion. Instead of reacting with ions in solution, metals in this corrosion mechanism directly interact with gaseous atoms in the environment. The first quantitative analysis of oxidation behavior was made in the early 1920s with the postulation of the parabolic-rate theory of oxidation by Tammann and, independently, by Pilling and Bedworth. This sort of damage is sometimes referred to as "dry corrosion" or "scaling."

High temperatures are sometimes defined as those that exceed 150°C. A durable protective scale that protects alloys from additional corrosion, such as sulfidation, carburization, and other types of high-temperature attacks, is frequently developed via the oxidation reaction (Dove, Messer, & Phillips,2001). Typically, the most prevalent corrosion product determines the names of the corrosion mechanisms(s). For instance, sulfidation implies sulfides, sulfidation/oxidation means a combination of sulfides and oxides, and carburization implies carbides. Oxidation is the general term for a range of reactions (John, 1999).

High-temperature corrosion resistance is determined by the combination of two factors: thermodynamics and kinetics. The former determines if a corrosive reaction will take place, while the latter helps determine the rate at which it will happen. To reduce the rate of the reaction, carefully selected alloying components are used such as the inclusion of a multivalent metal that can react with a greater number of oxidizing atoms. A careful study of the properties and behavior of a heat-resistant material in the environment it will be used in is also important to ensure that the choice is made for the right application. Due to the availability of new materials, it is now possible to make better decisions and establish safe working limits for these materials.

24.1 The various forms of high-temperature corrosion

High-temperature tarnishing, scaling, and high-temperature oxidation are other terms for high-temperature corrosion. In the absence of a liquid electrolyte, high-temperature corrosion develops. Although it may start at about 250°C (480°F), it typically occurs at

Handbook of Corrosion Engineering. DOI: https://doi.org/10.1016/B978-0-323-95185-2.00024-1

temperatures above 450°C (840°F). Scale is typically deposited because of the reaction between gas and metal. The scale is protective when it forms a solid that completely envelops the metal's surface, is adherent to the surface of the metal, has a high melting point so that it can withstand higher temperatures, and has a small amount of species diffusion. High-temperature tarnishing, scaling, and high-temperature oxidation is other terms for high-temperature corrosion. In the absence of a liquid electrolyte, high-temperature corrosion develops.

Depending on the scale type, high-temperature corrosion may be categorized as: (1) carburization, (2) sulfidation, (3) oxidation, (4) chlorination, and (5) nitridation.

24.1.1 Carburization

By facilitating the diffusion of carbon atoms through the surface up to a predetermined depth, the exterior of ferrous workpieces (component parts) is hardened by the process of carburization. It is a technique that is frequently used to make the surface of metallic parts and components harder. By diffusing through the steel's surface, carbon from the carburizing environment weakens the metal and reduces its toughness and creep resistance. As a result, there is a decrease in corrosion resistance and strength as well as an increase in cracking-type failure susceptibility. Carburization frequently affects furnace tubes because it takes place in high-temperature settings. These procedures can cause the carbon content of the carburized surface to increase to 1.2%. In most circumstances, the metal's surface is shielded by an oxide scale, but in hydrocarbon environments, carburization takes place at temperatures rising to 700–850°C (1300–1550°F).

Some of the commonly recognized susceptibilities for carburization include carbon steel, low alloy steels, cast stainless steels, and nickel base alloys. These materials must have been exposed to a carbonaceous or carburizing environment to perform the reactions required for carburization. If atmospheric carbon dioxide is present, then it is possible to assume that high-temperature corrosion is occurring in an environment that is rich in carbon as shown below:

$$CO + H_2 \rightarrow H_2O + C \tag{24-1}$$

$$C + 3Fe \rightarrow Fe_3C \tag{24-2}$$

$$Fe_3C \rightarrow 3Fe + C \tag{24-3}$$

It is important to note that the production of carbides or carburization reduces as H_2S concentration increases. As the concentration of H_2S increases to a critical level for the synthesis of sulfides, sulfidation will take place instead of carburization.

There are several different carburization techniques available. Carburization can be done in five different ways. Each method has advantages and disadvantages of its own. Following are the techniques.

Liquid carburization: Metallic workpieces submerged in a liquid bath containing silicon carbide, sodium chloride, and sodium carbonate undergo a process known as liquid carburization. Around 900°C (1650°F) is the target temperature. Once workpieces have undergone carburization, they are quenched. Items made of metal alloys encounter a collection of carbon molecules while submerged in this mixture. In most cases, these carbon molecules will diffuse into the alloyed materials quickly, enabling a hard casing to form in a short amount of time. Steel that has been liquid-carburized typically has a high carbon content and a low nitrogen content. In this method, soot buildup and oxide production are prevented while the surfaces harden more uniformly.

Vacuum carburization: In this procedure, the workpiece is heated to 1000°C (1830°F) in an enclosed chamber while being exposed to an oxygen-free, low-pressure atmosphere. The atmosphere is then supplied with a carbon-rich gas, like propane, ethylene, or acetylene, allowing carbon molecules to bond to the alloys. The absence of oxygen during the process makes it virtually impossible for steel alloys to oxidize. This makes it possible to rapidly accelerate the carburization process by injecting high heat into the enclosed atmosphere. This procedure is carried out in a furnace that is sealed. The furnace cannot perform the intended process unless oxygen can be completely sealed off.

Plasma carburization: In a plasma carburization system, the workpiece is processed in a vacuum furnace with an associated oil quenching tank that operates at temperatures between 900°C and 950°C (1650°F and 1740°F). A negative voltage between -800 V and -400 V is used to hold the workpiece. A controlled flow of propane or methane gas is delivered into the chamber after being diluted with argon, nitrogen, and oxygen. Due to the existence of the gas and the negative voltage, plasma surrounds the workpiece. Any carbon in the glow discharge that is present in the free state diffuses into the surface of the workpiece. This procedure involves quenching in an oil bath after the required heat-treated depth is reached, which is typically less than 2 mm.

Pack/solid carburization: Pack carburization is the process of putting steel products next to high-carbon products in a furnace. These high-carbon products range from cast iron particles to carbon powder, among other things. These things will be heated with the aid of carbon monoxide once you have entered them. This gas reduces carbon, causing carbon to draw from the surface of the materials that were placed in the furnace that was carbon-dense. These carbon molecules will permeate into the surfaces of the steel products that will be subjected to carburization once they are no longer attached. Execution of this carburization technique is the simplest.You might even be able to complete it in a garage or home workshop. Its reliability and consistency are its main issues. Although it will permit carbon diffusion, this diffusion usually will not take place uniformly across a steel structure.

Gas carburization: Gas carburization is a process that involves the use of natural gas, which is a mixture of methane, ethane, and propane. Parts that are being prepared for this process are heated in a furnace at a rate of 950°C (1740°F). After the gas has been introduced, it takes around 5–10 hours for the process to complete. The heat treatment process used during this process can be used to treat complex parts, such as those with complex geometry. It can also produce accurate surface hardness. This process is very popular in the world and is very useful for mass carburization.

24.1.2 Sulfidation

The process of accelerated corrosion of various metals and alloys in the presence of sulfur is known as sulfidation. This type of process is usually identified by the presence of sulfides within the materials. It is caused by the formation of metal-sulfur scales in the presence of other compounds that are sulfur-containing, such as hydrogen sulfide. Most of the time, the oxide and sulfide scales form together in the presence of oxygen. However, the kinetics of the oxide scale formation is slower than that of the sulfide. Because the sulfide scale melts at a lower temperature than the oxide scale, the production of the sulfide scale increases corrosion. Since sulfur is the most prevalent pollutant in chemical and petrochemical production, it seems sensible that sulfidation-related materials issues will be extremely prevalent in these industrial sectors (Perkins, 1981).

In areas where the temperature is higher than 240°C and sulfur, H_2S, and/or mercaptan can develop, high-temperature sulfidation typically takes place. In the petrochemical sector, high-temperature sulfidation essentially refers to the corrosive attack brought on by active sulfur, such as H_2S. Crude oil's organic sulfides can breakdown at 260°C and release elemental sulfur and H_2S, which can then immediately interact with the metallic parts of the equipment to cause severe corrosion on their surfaces.

The amount of active sulfur and the overall sulfur content in crude oils are the two key determinants of the high-temperature sulfidation rate. The rate at which alloys sulfate tends to increase as H_2S concentration rises. The impact of different alloys commonly used in process equipment is depicted in Fig. 24–1. Solid lines and dashed lines, respectively, reflect interpolations of the data that is currently accessible.

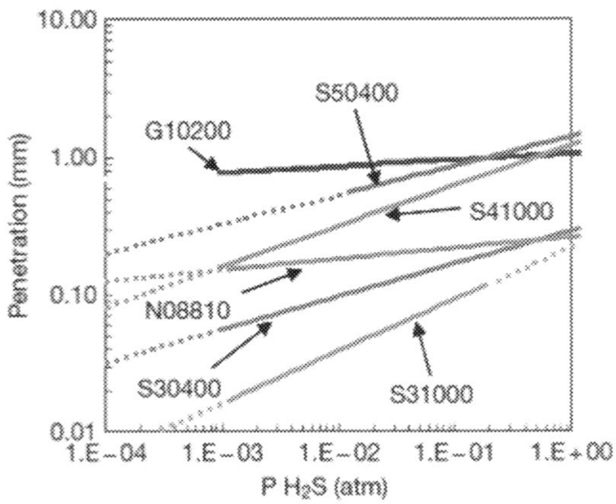

FIGURE 24–1 Effect of H_2S partial pressure upon the sulfidation corrosion after 1 year in H_2-H_2S gases at a total pressure of 1 atm at 723K (John, 2010). *Reproduced with permission from John, R. (2010). Sulfidation and mixed gas corrosion of alloys, shreir's corrosion, 1, pp. 240–271.*

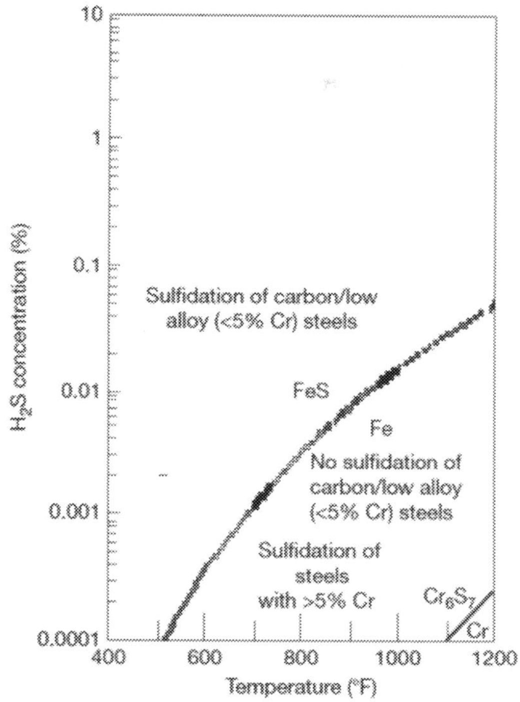

FIGURE 24–2 Conditions for possible sulfidation, based upon H_2S concentrations in H_2–H_2S gases and temperatures above the FeS/Fe line for carbon/low alloy (<5% Cr) steels and above the Cr_6S_7/Cr line for alloys with >5% Cr, 400–1200°F (204°C–648°C, 473–921K). *Reproduced with permission from John, R. (2010). Sulfidation and mixed gas corrosion of alloys, shreir's corrosion, 1, pp. 240–271.*

The line for carbon steel that is used for corrosion is not able to work at low concentrations of hydrogen sulfide because the FeS content of the metal is not stable. This means that it cannot corrode at high temperatures and partial pressures. It is also incorrect to assume that certain types of steel cannot corrode at conditions that are like those indicated by this line. High concentrations of Cr can lead to the formation of CrS in certain types of alloys, such as those with a high temperature and low H_2S concentration. This phenomenon can affect the corrosion properties of low-alloy steels. Fig. 24–2 shows the limits of the H_2S concentration and temperature in relation to the sulfide corrosion products of Cr and Fe. Although the low corrosion rates of low alloy steels are low (such as 0.025–0.25 mm $year^{-1}$), they are still significant for certain conditions, such as those where they are believed to be immune to sulfidation. Figs. 24–1 and 24–2 show the varying ranges of H_2S concentrations used in the design and construction of hydrotreaters and catalytic reformers used in petroleum refineries.

The increase in temperature can lead to accelerated reactions of the sulfides and the degradation of the inactive sulfur. When the temperature reaches 240°C, the sulfidation process becomes more severe (Haycock &Sharp, 1959). At temperatures ranging from 343°C to

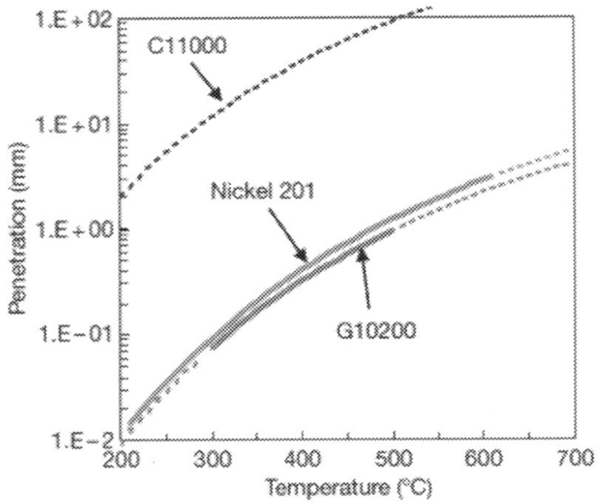

FIGURE 24–3 Effect of temperature upon sulfidation of some simple metals after 1 year in 0.5 atm H_2 and 0.05 atm H_2S (Haycock and Sharp, 1959). *Reproduced with permission from Haycock, E. W., Sharp, W. H. (1959). Corrosion of ferrous alloys by H2S at high temperatures. In* Session from the 24th meeting of the American Petroleum Institute, *New York.*

371°C, the decomposition rate of sulfides into H_2S is the fastest. At 350°C to 400°C, H_2S can further decompose into S and H_2S has a stronger activity than H_2S, leading to a more severe attack on the metal surfaces. The maximum temperature at which the corrosion rate can reach its peak is around 430°C. However, when the temperature exceeds 480°C, the sulfides can almost finish decomposing. Fig. 24–3 shows sulfidation predictions of several simple metals (copper, carbon steel, and nickel).

24.1.3 Oxidation

One of the most common forms of high-temperature corrosion is oxidation. The metals undergo oxidation due to the presence of steam, air, and oxygen. The oxide scale's protectiveness can vary depending on the environment, temperature, and other factors (Wright, 1987).

24.1.4 Chlorination

The process of chlorination involves the formation of metallic chlorides. Due to their high vapor pressures, these materials are unstable and can cause their formation to be extinguished on a continuous oxide scale. In the gaseous phase, the hydrogen produced by the oxygen reacts with the metallic chlorides and forms chlorine gas. This gas then raises localized corrosion. This gas can be extinguished by extinguishing the metal oxide scale as shown in Eq. (24–4):

$$2FeCl_2 + O_2 \rightarrow Cl_2 + 2FeO \qquad (24-4)$$

24.1.5 Nitridation

Nitridation is the process by which the presence of nitrogen or ammonia causes the formation of metal nitrides. The resulting corrosion can be caused by the precipitation of metallic compounds such as titanium, niobium, aluminum, chromium, and chromium. This process can also lead to metal dusting in stainless steel (Dong, 2003).

Like carburizing, the process of applying nitriding can be done in any of the four states of matter, namely, liquid, gas, plasma, and solid. Although pack and gas-nitriding are both relatively safe, they tend to produce brittle and thick compounds. Unlike carburizing, plasma is environmentally safe and can be used to address various problems faced by traditional processes when it comes to the treatment of stainless steel. It can also overcome the surface oxide issue that was encountered in the previous methods. Because of the low temperatures and the absence of a quenching mechanism, the process produces less deformation and distortion. This is very important for certain precision components.

24.2 High-temperature corrosion due to naphthenic acid

A nonaqueous corrosion process called naphthenic acid corrosion (NAC) is brought on by naphthenic acids. NAC primarily takes place in high-velocity sections of crude distillation units in the temperature range of 430°F–750°F (220°C–400°C). The breakdown of naphthenic acids or protection from the coke generated at the metal surface is most likely to blame for the lack of corrosion damage that is typically observed at temperatures above 750°F (400°C). Naphthenic acid-rich crude oils are also known as high total acid number crude oils or high-acid crude oils. The nature of NAC (pitting and impingement) and its intense attack at high velocities in crude distillation units set it apart from sulfidic corrosion.

Sulfur-containing substances that are naturally present in crude oil fractions as soluble sulfides can slow down NAC. Despite the fact that these sulfides can also be corrosive, they can also create iron sulfide scale on metal surfaces, possibly providing some defense against naphthenic acid attack.

24.3 High-temperature corrosion due to the formation of ash, deposits, and molten salts

Molten salt corrosion occurs when Na, S, or V oxides, such as Na_2O, SO_3, and V_2O_5, react with the protective surface oxide coating on a metal substrate at high temperatures, usually around 900°C. Low-grade petroleum in the oil and gas business contains a certain quantity of NaCl, S, and V. Low-grade marine residual fuels, for example, may contain up to 0.2 (wt.%) ash, 5 wt.% S, and 0.06 wt.% V, whereas low-grade marine distillate fuels may contain up to 0.05 wt.% Na, 2 wt.% S, and 0.01 wt.% V. When these elements are exposed to high-temperature O_2-rich gases in the oil and gas industry, they oxidize to their corresponding oxides. These oxides then condense and interact with the metal structures they are exposed to, resulting in a high degree of corrosion.

24.4 Factors affecting prediction of corrosion rates for high-temperature corrosion

Over time, laboratory research has made a substantial contribution to the understanding of high-temperature corrosion processes. The modeling of the kinetics of the governing processes can be used to predict the rates of corrosion for various high-temperature corrosion events. Predictive software tools have been created that can do this function for a variety of high-temperature alloys. Many times, the difficulties in simulating commercial processes can be attributed to extraneous elements, which can be challenging to imitate in the lab. These variables include, but are not limited to, pressure, gas velocity, contaminants, mechanical stress, and thermal stress:

Pressure: The mobility of reacting ions through a surface layer like an oxide is not significantly affected by pressure. Pressure, however, may have a significant impact if the dissociation pressure of a corrosion product is within the specified range of conditions. Usually, the impact of temperature is much more pronounced.

Gas velocity: When combined with potentially corrosive species, very high-velocity gas streams can accelerate erosion, leading to accelerated wasting rates through an erosion–corrosion mechanism. On the other hand, materials rich in molybdenum or vanadium may experience what is known as "catastrophic oxidation" when exposed to such environments. Molybdenum trioxide, or MoO_3, melts at 795°C, while vanadium pentoxide, or V_2O_5, melts at 690°C. These oxides do not volatilize if sufficient gas flow is not provided, which encourages more oxidation.

Impurities: Due to the impossibility of precisely determining the quantity of numerous low-level contaminants, simulating industrial corrosion conditions is frequently exceedingly challenging. This frequently results from ignorance of the particulars of the industrial environment.

Stress: Accelerated grain boundary oxidation has been known to contribute to the development of stress-induced cracking phenomena in various materials. For instance, in the case of Fe-Ni material, the resulting stress-corrosion cracking was observed in a simulation of a water reactor at 360°C (Panter et al., 2006; Soontrapa, 2005). In another case (Le Calvar, Scott, Magnin, & Rieux,1998), the same mechanism was reported to have contributed to the development of cracking in austenitic steels at 610°C. Thermal cycling can result in stresses that are high enough to cause corrosion layers to buckle and crack, possibly eventually separating the corrosion layer entirely off the surface. The likelihood of this happening can vary depending on the expansion coefficient differences between the oxide and substrate as well as the strength of the oxide relative to the substrate. During heat cycling, subsurface oxides can also help to offer mechanical stability.

Short questions

1. Where does high-temperature corrosion occur?
2. Which material is most corrosive for high-temperature corrosion?
3. What is the purpose of the nitriding process? And discuss its types.
4. What is the temperature of carburizing and what different methods of carburizing?
5. Does high-temperature increase oxidation?
6. What is an example of high-temperature oxidation corrosion?

References

Dong, H. (2003). Materials in sports equipment. *Materials in Sports Equipment 7 - Surface Engineering in Sport.*

Dove, D., Messer, B., & Phillips, T. (2001). An austenitic stainless steel, resistant to high temperature creep and naphthenic acid attack in refinery environments, corrosion 2001. In The NACE international annual conference and exposition, Houston, TX, NACE International.

Haycock, E. W., & Sharp, W. H. (1959). Corrosion of ferrous alloys by H2S at high temperatures. In Session from the 24th meeting of the American Petroleum Institute, New York.

John, R. (2010). Sulfidation and mixed gas corrosion of alloys, Shreir's corrosion. Volume 1, pp. 240–271.

John, R. C. (1999). Compilation and use of corrosion data for alloys in various high-temperature gases.In Corrosion/99, paper no. 73 (Houston, TX: NACE).

John, R. C., & Young, A. L. (2002). New understanding on corrosion of alloy in high-temperature sulfidizing gases, corrosion 2002. In The NACE International annual conference and exposition, Houston, TX, NACE International.

Le Calvar, M., Scott, P. M., Magnin, T., & Rieux, P. (1998). *Corrosion, 54*(2), 101–105.

Panter, J., Viguer, B., Cloué, J. M., Foucault, M., Combrade, P., & Andrieu, E. (2006). *Journal of Nuclear Materials, 348*(1–2), 213–221.

Perkins, R. A. (1981). In high temperature corrosion. In R. A. Rapp (Ed.), (pp. 345–353). Houston, TX: National Association of Corrosion Engineers.

Rapp, R. A. (2002). *Corrosion Science, 44*(2), 209.

Soontrapa, C. (2005). *Master's thesis.* Massachusetts Institute of Technology.

Wright, I. G. (1987). High-temperature corrosion. In korb LJ, Olson DL, editors. ASM handbook. Corrosion, vol.13;. Figure 37, page 97, ASM International, ISBN: 0-87170-007-7.

25

Top-of-line corrosion

A gas pipeline's corrosion can occur in different parts of the pipe. These include the lower part, where the moisture accumulates (bottom-of-line corrosion, BOL), or the upper part, where the moisture condenses (top-of-line corrosion, TOL). It can also occur in places where the moisture has accumulated, such as gaps, cracks, and stagnant zones. The process known as top-of-line corrosion occurs on the upper side of the inner wall of the pipelines. It is caused by the accumulation of water vapor that contains harmful gases such as CO_2 and H_2S and organic acids—mostly acetic acid. This occurs due to the heat exchange between the cold ambient medium and the pipe wall (Fig. 25–1). TOL corrosion is commonly encountered in stratified flow systems and cannot be protected with standard corrosion inhibitors. It has been regarded as one of the most challenging types of corrosion in the oil & gas industry. In 1981 the first case of TOL corrosion was reported in an H_2S-containing gas field in France by Paillassa et al. It quickly became apparent that this type of corrosion was prevalent in the

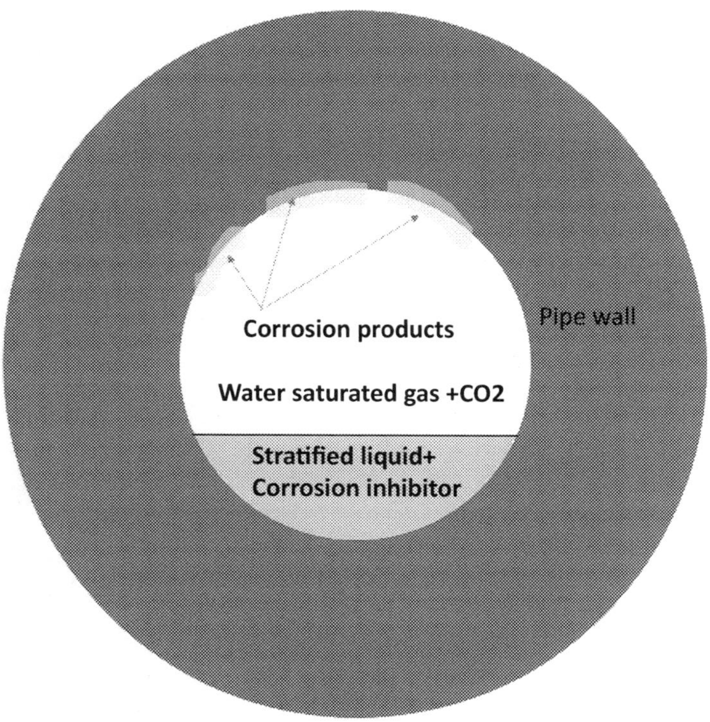

FIGURE 25–1 A typical TOL corrosion in the field. *TOL*, top-of-line.

Handbook of Corrosion Engineering. DOI: https://doi.org/10.1016/B978-0-323-95185-2.00025-3

fields due to its CO_2 environment. Numerous cases of TOL corrosion have been reported in various regions, such as the Asia-Pacific. TOL corrosion is a type of CO_2 corrosion that is mainly confined to the transportation and production of wet gas in pipelines using a stratified flow regime.

In recent decades, TOL has been linked to several pipeline failures. TOL corrosion issues have been documented in a number of gas fields worldwide, including the CrossfieldGasfield (Canada) (Bich & Szklarz, 1988), TunuGasfield (Indonesia) (Supriyatman, Achmad, & Gunaltun, 1999), and LacqGasfield (France) (Estavoyer, 1981).

The corrosive vapor coming from the top of the pipeline can cause this type of corrosion. The temperature difference between the outside and the inside of the pipeline can cause it (Keeper Gate, 2012). The temperature inside the pipeline is usually between 50°C and 105°C (Frenier, 2014), while the outside temperature can vary from the atmosphere to the deep sea temperature and could reach 4°C. The higher the temperature of the gas inside the pipeline, the more likely it is that the corrosion rate will increase.

When the hot, wet natural gas is transported through a poorly insulated pipeline, it can cause TOL corrosion. The high heat transfer coefficient within the pipeline can cause the water vapor to condense at the pipe walls. This type of condensed water is referred to as "hungry water." The lack of dissolved iron and buffering agents can cause the water to become aggressive. CO_2 dissolves in the water, which then turns into carbonic acid, which lowers the pH and causes the water to become more aggressive, and this can induce TOL corrosion. It is believed that a condensation rate of around 0.25 $g/m^2/s$ can induce TOL corrosion, and the necessary rate can be lowered by a magnitude once enough organic acids are present.

Organic acids, specifically acetic acid, are the by-product of natural gas. Usually, the concentration of around a few hundred to thousand ppm is not uncommon. This can lead to a higher corrosion rate, and it can also affect the formation of protective scales in low concentrations.

Based on the corrosive substituents, TOL is divided into the "sweet" and "sour". The former is usually found in areas with high levels of CO_2, while the latter is often found in environments with high levels of H_2S.

25.1 TOL in "sweet" environments

There have been numerous cases of sweet (CO_2) dominated TOL field cases reported in the literature (Gunaltun & Larrey, 2000). These cases are usually associated with severe corrosion and failure of pipelines due to inadequate thermal insulation. Due to the high fluid temperatures (>50°C) and the presence of water vapors, these areas would be subjected to water condensation. Corrosive gases such as carbon dioxide (CO_2) or organic acid vapors would dissolve in the condensed water and lead to steel dissolution. The quick removal of dissolved iron from the steel surface caused by high rates of water condensation would inhibit the growth of a protective iron carbonate ($FeCO_3$) layer on the steel surface. Over the past 10 years, there has been a lot of research done on CO_2 dominated TOL, including experimental (Andersen, Halvorsen, Valle, & Kojen, 2007; Hinkson, Singer, Zhang, & Nesic, 2008) and modeling (Zhang, Hinkson, Singer, Wang, & Nesic, 2007) work. The governing factors of sweet TOL have been determined to be the water condensation rate, fluid temperature, and CO_2/organic acid content (Nyborg & Dugstad, 1999). Field failures frequently occur where the fluid is the warmest, resulting in substantial temperature gradients between the process fluid and the outside environment and, as a result, high rates of water condensation. Additionally, it was shown that the rate of corrosion was significantly accelerated by the presence of high amounts of organic acid. At lower temperatures or when weight loss corrosion rates (WCR) was below a crucial value, there were no TOL problems. A critical WCR of 0.25 $mL/m^2/s$ in the absence of organic acid and 0.025 $mL/m^2/s$ in its presence have been recommended as engineering standards (Gunaltun & Larrey, 2000). These critical values should be used with caution, though, as they logically depend on other field factors including temperature and the presence of corrosive gases.

25.2 TOL in "sour" environments

The mechanisms for sour (H2S-dominated) TOL appear to be more complex. Although it appears to be much less common than its sweet counterpart, sour TOL has been linked to pipe failures. Understanding of corrosion mechanisms is lacking, which makes sense given that little is known about sour corrosion at the bottom of the food chain. This paper's primary objective is to combine pertinent field and laboratory experience that is available in the literature, as well as internal data, to identify any remaining knowledge gaps and to derive some early features of sour TOL (Singer, Camacho, Brown, & Nesic, 2010).

25.3 Factors affecting TOL

The presence of carboxylic acids (acetic acid), temperature increase, and the variability of the steel surface as depicted in Fig. 25–2 can all contribute to the intensification of TOL corrosion. To develop effective TOL mitigation techniques and prevent pipeline failure while in service, a thorough understanding of the various elements that influence the corrosion mechanisms of TOL is essential. Various elements influencing the rate of TOL are discussed in the following section.

25.3.1 High water-condensation rates and low condensation of glycol

In general, it is assumed that the corrosion rate will be low while the condensation rate is low and will rise as the condensation rate increases. This happens because $FeCO_3$ precipitations cannot fully saturate the condensed water due to its rapid renewal at high condensation rates. To reduce the TOL, a thin condensed $FeCO_3$ layer is formed on the surface by the

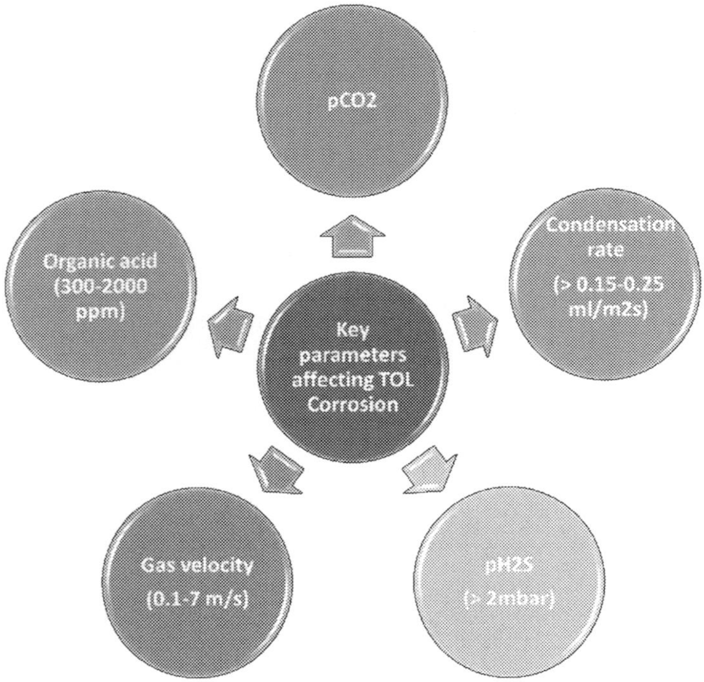

FIGURE 25–2 Factors affecting and critical range for TOL corrosion. *TOL*, top-of-line.

$FeCO_3$-saturated droplets when the condensation rate is low (Singer, Nesic, & Gunaltun, 2004; Vitse, Nesic, Gunaltun, de Torreben, & Duchet-Suchaux, 2003). Water condensation rates have been found to range from 0.001 to 2.25 $mL/m^2/s^1$. Monoethylene glycol has been shown to slow down condensation in the TOL corrosion example by lowering the vapor pressure of the liquid phase of wet gas (Ehsani, 2013).

25.3.2 High pipeline-inlet temperatures of approximately over 80°C

The temperature has been found to be the main factor affecting the TOL rate because of its effect on the rate of condensation, the types of corrosion products that occur, and ultimately the rate of corrosion. The temperature of the gas and the surrounding environment have an impact on the condensate film's thickness, rate of condensation, and temperature (which causes condensation because of extreme cooling) (Chen, Zhang, Qin, Xu, & Lu, 2011). Due to the success rate of condensation throughout the process, there are now a huge number of conflicting studies on the impact of low and high gas temperatures on the rate of corrosion. The rate of condensation is anticipated to be relatively low at low gas temperatures, and the electrochemical response of the condensed water regulates the pace of corrosion. However, higher gas temperature causes thicker films and a faster rate of condensation, both of which ultimately raise corrosion rates.

25.3.3 Gas velocity

Instead of directly affecting the rate of corrosion, gas velocity has a constant impact on the rate of condensation.

25.3.4 Effects of H_2S

Iron sulfide scale properties, which are primarily a function of temperature and H_2S concentration, determine TOL corrosion rates in sour conditions. Condensation rate is a supporting factor in this scenario. At H_2S partial pressures above 2 mbar and as the rate of water condensation increases, TOL corrosion rate tends to increase.

25.3.5 Organic acids

Among the organic acids commonly found in oil and gas wells are propionic, formic, and acetic. The presence of hydrogen at the TOL rate accelerates the rate of corrosion because of its effect on the formation of hydrogen. This phenomenon can be explained by the presence of HAc, which lowers the pH levels and increases the solubility of Fe^{2+}. It is important to note that the total amount of HAc in the condensed liquid is determined by the undissociated HAc.

25.3.6 High CO_2 partial pressures

The presence of pCO_2 in the gas can also have negative and positive effects on the TOL rate. It helps in the formation of a film that is composed of $FeCO_3$ which reduces the rates of corrosion. Additionally, it decreases the condensed liquid's pH and increases the corrosivity of the medium. However, additional environmental factors such as the rate of water condensation at the top of the line have a considerable impact on its greatest impacts. At high rates of condensation over a range of 1–8 bar and at bulk temperatures of 50°C, 70°C, and 90°C (Vitse et al., 2003), the change in pCO_2 has a considerable impact on the rate of corrosion, but not at low condensation rates. At low condensation rates, it may be simpler to supersaturate (or saturate) the condensed liquid with corrosion products, which prevents ions from migrating to and from the metal surface. As a result, the pH level may rise, and the reaction kinetics of CO_2 corrosion may slow down. The buffering effect of Fe ions on the condensed liquid's pH level may be negligible at higher condensation rates, and the pH level may be more sensitive to changes in pCO_2 (Vitse et al., 2003).

25.4 Top-of-the-line corrosion control

Numerous strategies to reduce TOL corrosion have been proposed and put to the test. Some of these initiatives are more fruitful and useful than others. The first, kilometers of a field can be constructed with corrosion-resistant alloys without thermal insulation if it is still being built. The water vapor can then be condensed out, making it disappear further along the line. The pipeline might also be buried further under the ocean's surface to increase heat insulation. These are engineering initiatives that can be used in recently developed fields but not in pipelines that have already been built and are in operation. Inhibition is crucial in preventing TOL corrosion in these situations.

The delivery of the inhibitor to the TOL presents significant challenges. Batch treatment for inhibitors is one of the more widely used methods. Between two Pipeline Inspection Gauges (PIGs), a sticky long-chain organic inhibitor is inserted, and it is then passed through the pipelines. Unfortunately, the operating firms incur a significant loss because of having to stop production of the pipe or even huge portions of a field for this treatment.

Another technique would be to use a spray PIG, which would be novel approach. The PIG sprays a solution from the BOL with a high inhibitor concentration to the TOL using the Venturi effect. The field does not need to be entirely shut off, but the pipeline still requires a location for PIGs to be launched and received in addition to the possibility of a PIG becoming stuck in a pipeline.

Another promising approach is the use of volatile corrosion inhibitors (VCI). These are commonly referred to as small chain amines due to their volatility. They are injected into the BOL and are the same type of corrosion inhibitors that are used in conventional BOL systems. It is not clear how these chemicals work, VCIs are thought to evaporate and co-condense at the TOL with all the other fluids and inhibit the CO_2 corrosion directly where it occurs. Although many of the commercially available VCIs can be used to prevent TOL corrosion, they are not yet effective enough to be considered the sole countermeasure. Instead, they are usually combined with other PIGs to extend the cycle of the treatment.

References

Andersen, T., Halvorsen, A. M. K., Valle, A., & Kojen, G., The influence of condensation rate and acetic acid concentration on TOL corrosion in multiphase pipelines. In *Proc. NACE* International, Houston, USA, Paper 7312.

Bich, N. N., & Szklarz, K. E. (1988). *Crossfield corrosion experience.* USA: NACE/Corrosion 88. Paper No. 196.

Chen, Y., Zhang, L., Qin, H., Xu, L., & Lu, M. (2011). Effects of temperature on CO2 top-of-line corrosion of pipeline steel. In *NACE International, Corrosion/2011*, 13–17 March, Houston, Texas, NACE-11327.

Ehsani, H. Curtin University; Australia (2013). *Influence of monoethylene glycol (MEG) on the corrosion inhibition of wet-gas flow lines.*

Estavoyer, M. NACE; USA (1981). *Corrosion problems at Lacq sour gas field*; p. 905.

Frenier, W. W. Society of Petroleum Engineers (2014). *Multifaceted aproaches for controlling top of the line corrosion in pipelines.*

Gunaltun, Y., & Larrey, D. (2000). Correlation of cases of top-of-the-line corrosion with calculated water condensation rates. In *Proc. NACE International*, Houston, USA, Paper 71.

Hinkson, D., Singer, M., Zhang, Z., & Nesic, S. (2008) A study of the chemical composition and corrosivity of the condensate in top of the line corrosion. In *Proceedings of the NACE International*, Houston, USA, Paper 8466.

Keeper Gate. (2012). *Top of line corrosion.* GAT2004-GKP-2012.

Nyborg, R., & Dugstad, A. (1999). Top of the line corrosion and water condensation rates in wet gas pipelines. In *Proc. NACE International*, Nashville, USA, Paper 7555.

Singer, M., Camacho, A., Brown, B., & Nesic, S. (2010). Sour top of the line corrosion in the presence of acetic acid. In *Proc. NACE International*, Houston, USA, Paper 10100.

Singer, M., Nesic, S., & Gunaltun, Y. (2004). Top of the line corrosion in presence of acetic acid and carbon dioxide. In *NACE Meeting Papers*, Paper No. 04377.

Supriyatman, D., Achmad, J., & Gunaltun, Y. (1999). *Top of line corrosion in multiphase gas line, a case history*. USA: NACE/Corrosion 99. Paper No 036.

Vitse, F., Nesic, S., Gunaltun, Y., de Torreben, D. L., & Duchet-Suchaux, P. (2003). Mechanistic model for the prediction of top-of the- line corrosion risk. *Corrosion*, *59*, 1075–1084.

Zhang, Z., Hinkson, D., Singer, M., Wang, H., & Nesic, S. (2007). A mechanistic model for Top of the line corrosion. In *Proceedings of the NACE International*, Nashville, USA, Paper 7556.

26

Corrosion in CO_2 environments

In the oil and gas industry, high temperatures (16°C–163°C), the presence of several gases (0.2–40 mol.% CO_2 content in the gas phase), the presence of wet hydrocarbons, solutions with high salinity and/or acidity (pH = 3.5–6.5), and flow conditions (liquid velocities of 0.5–5 m/s and gas velocities of 5–20 m/s) are just a few of the operating conditions. As a result, the metals used in all phases of production, including drilling, refining, transportation, and storage, become much more susceptible to corrosion (Kahyarian, Achour, & Nesic, 2017). Steel is susceptible to various types of corrosion under the circumstances, including galvanic corrosion, stress corrosion cracking, and erosion–corrosion (Finšgar & Jackson, 2014). The three most common types of corrosion, however, are those brought on by the presence of CO_2 (sweet corrosion), H_2S (sour corrosion), and oxygen dissolved in injection water. According to studies, more than 25% of failures in the oil and gas sector are linked to corrosion in general, and more than 28% of those failures are attributable to CO_2 corrosion. These failures present serious environmental issues if rusted transportation pipes burst, in addition to their negative economic effects. Therefore it is crucial to lessen the effects of corrosion in this industry.

Natural gas, crude oil, and the surrounding environment are to blame for the CO_2 that is present in the well. To improve oil recovery and lower the viscosity of the pumped fluid, CO_2-saturated water is occasionally pushed into wells (Cui, Wu, Li, Zhu, & Yang, 2004). An example of CO_2 corrosion is the formation of an iron carbonate scale when dissolved CO_2 in condensate reacts with water to create carbonic acid (H_2CO_3). The boiler condensate return systems that are not sufficiently treated with corrosion inhibitors are where CO_2 corrosion is most frequently encountered. The concentration of CO_2 in the environment, the conditions in which the components operate, and the kind of materials they are made of are just a few of the variables that affect how quickly materials exposed to CO_2 corrosion deteriorate. It is also crucial to be careful of the water chemistry near the machinery; for instance, de-aerated alkaline water might accumulate extremely high CO_2 levels.

26.1 Mechanism of sweet corrosion

Understanding the anodic reaction and cathodic reaction occurrence of Fe in strong acids is fundamental to comprehending the mechanism of sweet corrosion in carbon steel. According to Eqs. (26.1–26.3), the oxidation of iron is frequently associated with the anodic reaction in CO_2 corrosion.

$$Fe + H_2O \rightarrow FeOH + H^+ + e^- \qquad (26-1)$$

Handbook of Corrosion Engineering. DOI: https://doi.org/10.1016/B978-0-323-95185-2.00026-5

$$FeOH \rightarrow FeOH^{+} + e^{-} \quad (26-2)$$

$$FeOH^{+} + H^{+} \rightarrow Fe^{2+} + H_2O \quad (26-3)$$

The initial step of sweet corrosion with the presence of CO_2 is:

$$Fe + [\![CO]\!]_2 \rightarrow Fe - CO_2(\text{Adsorbed surface complex}) \quad (26-4)$$

The resultant surface complex that has been adsorbed then interacts with water (H_2O) to produce $Fe^{2+}{}_{(aq)}$ and a weak acid (H_2CO_3) that can serve as the cathodic reactant H+ during dissociation. Consequently, the cathodic reaction's mechanism is:

$$2H^{+}{}_{(aq)} + 2e^{-} \rightarrow H_{2(g)}(pH = < 4) \quad (26-5)$$

$$H_2CO_3 + 2e^{-} \rightarrow H_2 + 2HCO_3^{-} \quad (26-6)$$

When H_2CO_3 is dissociated in the solution, hydrogen ions are produced for the cathodic process, or H_2CO_3 itself may be adsorbed at the cathode and reduced directly.

$$H_2CO_{3_{(aq)}} + e^{-} \rightarrow H^{+}{}_{(aq)} + HCO_{3_{(aq)}}{}^{-}(pH = 4 - 6) \quad (26-7)$$

Overall, a generalized equation for CO_2 corrosion of carbon steel can be represented as follows:

$$Fe_{(s)} + CO_{2_{(aq)}} + H_2O_{(l)} \rightarrow FeCO_{3_{(s)}} + H_{2_{(g)}} \quad (26-8)$$

26.2 Factors affecting sweet corrosion

Even though several studies on CO_2 corrosion have been conducted under various environmental conditions, including pressure, temperature, and pH levels. Below are these characteristics' specifics:

26.2.1 The effect of oxygen concentration in the system

Because $FeCO_3$ is unstable in aerobic conditions, the production of the $FeCO_3$ protective film is more assured in the absence of oxygen. The introduction of oxygen into the production machinery occurs because of procedures like the injection of produced water or inhibitors. Iron ions (Fe^{2+}) oxidize into ferric ions (Fe^{3+}) at oxygen concentrations above 40 ppb, undermining the protective layer that has formed on the steel surface. Furthermore, as oxygen is a potent oxidant, Eq. (26–8) suggests that the presence of oxygen may quicken the cathodic process. In essence, iron's oxidation produces the electrons needed for the reduction reaction in Eq. (26–8). The corrosion rate is accelerated by both phenomena.

$$O_{2_{(g)}} + 2H_2O_{(l)} + 4e^{-} \rightarrow 4OH^{-}{}_{(aq)} \quad (26-9)$$

26.2.2 Effect of pH

The concentration of hydrogen ions, that is, pH in a solution is known as one of the most important factors that can affect the rate of sweet corrosion of steel. At pH <4, the dominant type of reaction is hydrogen evolution. However, between pH 4 and 6, the reduction of carbonic acid is the main component of the cathodic reaction. On the other hand, at pH >6, the research has shown that the reduction of the bicarbonate is the main component of the reaction. The change in the pH level of the solution can affect the rate of sweet corrosion of steel. In either direction, it can lead to a reduction in the rate of catalytic reactions. The reduction in the rate of sweet corrosion caused by the increase in pH can also be achieved by reducing the rate of cathodic reactions due to the limited concentration of H + . On the other hand, the increase in pH lowers the solubility of protective layer of $FeCO_3$ on the surface of the steel and promotes the precipitation of $FeCO_3$ on the surface of the steel, which results in an indirectly reduced rate of corrosion. The change in the pH level shown in Fig. 26–1 revealed that when the pH is lower than 4, the reduction of hydrogen ion concentration is dominant. On the other hand, when the pH is between 4 and 7, the reduction of H_2CO_3 is important.

26.2.3 Effect of the fluid composition

The amount of corrosion caused by CO_2 is also influenced by the fluid composition, or the ratio of water, oil, and gas, and their flow rate. For instance, at lower water cut ratios, there

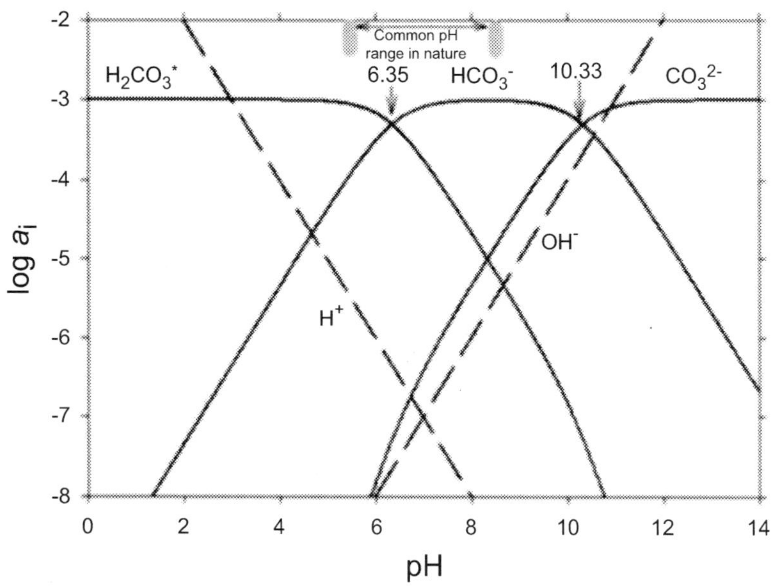

FIGURE 26–1 Showing Bjerrum plot of the activity (concentration) of the various ions in the CO_2-H_2O system as a function of the pH value. H_2CO_3 is sum of $CO_{2(aq)}$ and $H_2CO_{3(aq)}$ (Nesic, 2012). *From Nesic, S. (2012). Effects of multiphase flow on internal CO2 corrosion of mild steel pipelines. In* Energy and Fuels. *26(7), pp. 4098–4111. https://doi.org/10.1021/ef3002795.*

is no corrosion, or an oil layer stops water from coming into touch with the carbon steel surface. However, with high water cut ratios, corrosion rates increase due to increased flow rates, increased reactant flow, and increased protective film breakdown (Han, Nesic, Yang, & Bruce, 2011; Nesic & Lee, 2003).

26.2.4 Effect of CO_2 partial pressure

If the partial pressure of CO_2 (P_{CO2}) increases, the rate of corrosion increases due to increasing the amount of the $H_2CO_{3(aq)}$ and $CO_{2(aq)}$.

$$CO_{2_{(g)}} \leftrightarrow CO_{2(aq)} + H_2O_{(l)} \leftrightarrow H_2CO_{2(aq)} \quad (26-10)$$

The corrosion rate is categorized as serious corrosion when $P_{CO2} > 0.2$ MPa, mild corrosion when $P_{CO2} = 0.02-0.2$ MPa, and no corrosion when $P_{CO2} < 0.02$ MPa (Renpu, 2011).

26.2.5 Effect of solid corrosion products

The presence of siderite can also affect the sweet corrosion rate in mild steel. Due to its nonconductive nature, a film protective layer made of $FeCO_3$ can help delay the corrosion rate by acting as a barrier against the diffusion of harmful substances. The degree of protection depends on the environment and the carbon steel's characteristics (Han et al., 2011).

26.2.6 Effect of temperature

The temperature affects the mass transport between the bulk solution and the steel surface. The effect of the temperature on the sweet corrosion rate is dependent on the solution's chemistry. The temperature can affect the rate of CO_2 corrosion. For instance, at low temperatures (<70°C–90°C), the rate of degradation increases due to the mass transfer and the reduction of the protective scale. On the other hand, at high temperatures (>80°C –100°C), the rate of degradation decreases due to the loss of the iron carbonate's solubility that led to form the $FeCO_3$ protective scale. The increase in temperature during nonfilm forming conditions can lead to the development of H^+ ions, which can destabilize the precipitation and formation of $FeCO_3$ and thus increasing the rate of reaction.

26.2.7 The effect of flow

The flow conditions can affect the sweet corrosion rate. In nonfilm forming conditions, the increased flow can help transport the corrosive agents to the surface of the steel, which can increase the attack rate. On the other hand, in flow conditions, the high wall shear stress can negatively affect the growth rate of the protective passive layer or prevent it from forming. This can also lead to the removal of the existing one (Nesic, Lee, & Ruzic, 2002). Previously, it was believed that the high temperature of around 50°C–80°C could not trigger a flow sensitivity in the corrosion rate due to the mass transfer barrier effect and the formation of corrosion products on the steel surface. However, under certain film-forming conditions, the supply of CO_3^{2-} ions increases, which results in the precipitation of $FeCO_3$ and decreases sweet corrosion rate.

26.2.8 The effect of steel composition

The composition of steel and its properties are two of the most important factors that are considered when it comes to the selection of materials for the oil and gas industry. These materials are usually used to endure harsh working conditions, such as exposure to a combination of gases and corrosive solutions at high temperature and pressures, under dynamic conditions. Before the 1980s, carbon steels such as API N80, L80, and J55 were commonly used in upstream production operations. Due to the increasing number of wells being drilled deeper, the materials used for the construction of these structures were exposed to harsher environments. Fortunately, there are now various types of corrosion-resistant alloys that are commonly used in the oil and gas industry. These include Super-13 and Duplex 22% Cr steel. Despite the higher cost, carbon steels (CSs) are still considered the most practical materials for various applications in the industry. It is widely believed that CSs cover around 90% of all metallic materials utilized in the oil and gas industry. The various types of steel can be used in different ways depending on their working conditions. For instance, API 5CT series steel is commonly used for casings and tubing, while API 5L series is utilized for pipelines (Nesic et al., 2002). P110 steel can be utilized in multiphase flow-saturated CO_2 systems. It can also exhibit corrosion resistance in solutions containing concentrated chloride media. The H_2S corrosion resistance of API L80 grade steel tubing has been found to be reasonable. N80 or API 5A N80 CS is commonly used for downhole tubulars (Lin et al., 2015). Due to the presence of certain metal ions, such as Cu, Mo, and Zinc, in a corrosive solution, they can potentially affect the process of corrosion inhibition. It is also important to note that these ions can form insoluble precipitates with certain corrosion inhibitors (Mishra, Tiwari, Singh, & Singh, 2014). Several organizations, such as the API, ASTM, and AISI, have recognized the standard steels that can be used in the oil and natural gas industry. However, due to the composition of the different types of steel, it can be hard to reproduce the results of studies and compare them with those of other research groups. This makes it difficult for the public to understand the application of steel in the field (Xhanari, Wang, Yang, & Finšgar, 2021).

From a metallurgical standpoint, microalloying steel with components that are resistant to corrosion, such as chromium (Cr), may be able to reduce the rate of sweet corrosion. According to published research, adding as little as 3 wt.% of Cr to steel reinforcement could reduce the rate of sweet corrosion by 3–10 times at an increase in cost of no more than 1.5%. The benefits were much larger when the Cr reinforcement was combined with additional elements like titanium (Ti), copper (Cu), silicon (Si), and even vanadium (V).

26.3 Types of sweet corrosion

Sweet corrosion in oil and gas equipment can be categorized into two types: general and localized corrosion (Sun, Nešić, & Papavinasam, 2008). In localized corrosion, the breakdown of the protective film on the metal surface causes a metal dissolution. This process then leads to the formation of a pitting pattern. This characteristic of the pitting occurs in low to medium flow conditions and is attributed to the processing techniques used in the manufacturing of the components. The smooth and tapering side pitting corrosion that can occur in moderate to stagnant flow conditions is referred to as ringworm corrosion. This type of corrosion is caused by the failure of the welding process to fully normalize the tubular after processing.

References

Cui, Z. D., Wu, S. L., Li, C. F., Zhu, S. L., & Yang, X. J. (2004). Corrosion behavior of oil tube steels under conditions of multiphase flow saturated with super-critical carbon dioxide. *Materials Letters, 58*(6), 1035–1040. Available from https://doi.org/10.1016/j.matlet.2003.08.013.

Finšgar, M., & Jackson, J. (2014). Application of corrosion inhibitors for steels in acidic media for the oil and gas industry: A review. *Corrosion Science, 86*, 17–41. Available from https://doi.org/10.1016/j.corsci.2014.04.044.

Han, J., Nesic, S., Yang, Y., & Bruce, N. (2011). Brown spontaneous passivation observations during scale formation on mild steel in CO2 brines. *Electrochimica Acta, 56*.

Kahyarian, A., Achour, M., & Nesic, S. (2017). *CO2 corrosion of mild steel. Trends in oil and gas corrosion research and technologies: Production and transmission* (pp. 149–190). United States: Elsevier Inc. Available from https://doi.org/10.1016/B978-0-08-101105-8.00007-3, http://www.sciencedirect.com/science/book/9780081011058.

Lin, Y., Singh, A., Ebenso, E. E., Quraishi, M. A., Zhou, Y., & Huang, Y. (2015). Use of HPHT autoclave to determine corrosion inhibition by Berberine extract on carbon steels in 3.5% NaCl solution saturated with CO2. *International Journal of Electrochemical Science, 10*(1), 194–208. Available from http://www.electrochemsci.org/papers/vol10/100100194.pdf.

Mishra, M., Tiwari, K., Singh, A. K., & Singh, V. P. (2014). Synthesis, structural and corrosion inhibition studies on Mn(II), Cu(II) and Zn(II) complexes with a Schiff base derived from 2-hydroxypropiophenone. *Polyhedron., 77*, 57–65. Available from https://doi.org/10.1016/j.poly.2014.04.003.

Nesic, S., & Lee, K. J. (2003). A quasi 2-D localized corrosion model. *Corrosion, 59*.

Nesic, S., Lee, K. L. J., & Ruzic, V. (2002). NACE - International corrosion conference series 03614409, In *National Assoc. of Corrosion Engineers International Australia A mechanistic model of iron carbonate film growth and the effect on CO2 corrosion of mild steel*. http://www.nace.org.

Nesic, S. (2012). 7 19 2012/07/19 energy and fuels 10.1021/ef3002795 15205029 7 4098-4111 United States effects of multiphase flow on internal CO2 corrosion of mild steel pipelines 26.

Renpu, W. (2011). Advanced well completion engineering (third edition). *Chapter 11: Oil and gas well corrosion and corrosion prevention*. Gulf professional publishing.

Sun, W., Nešić, S., & Papavinasam, S. (2008). Kinetics of corrosion layer formation. Part 2 - Iron sulfide and mixed iron sulfide/carbonate layers in carbon dioxide/hydrogen sulfide corrosion. *Corrosion, 64*(7), 586–599. Available from https://doi.org/10.5006/1.3278494, http://corrosionjournal.org.

Xhanari, K., Wang, Y., Yang, Z., & Finšgar, M. (2021). A review of recent advances in the inhibition of sweet corrosion. *The Chemical Record, 21*(7), 1845–1875. Available from https://doi.org/10.1002/tcr.202100072.

27 Corrosion in H_2S environments

During the 1950s and 1960s, the oil and gas industry started to develop new fields. In Canada, for instance, the H2S content of the gas phase was higher. Compared to the environmental conditions during the operation of sour-oil fields, the development of these new fields had more severe effects. The development of new fields led to the discovery of metal failures in the surface equipment and wells. The industry then started investigating the cause of these issues. Through the use of laboratory data and field observations, the researchers were able to identify the factors that caused these problems. The exact environmental conditions that can cause a failure were complex to predict. They involved analyzing various factors such as the presence of liquid water, temperature, pressure, and concentration of H_2S. Sour gas is produced by many gas fields. It contains high levels of H_2S and CO_2 (Yépez et al., 2016). It is typically treated at a processing plant. Some upstream compressors in certain applications, such as reinjection, gas lift, and gas gathering, operate with untreated gas. CO_2 and H_2S react with liquid water and form corrosive acids. High levels of these substances can also cause sulfide stress cracking in certain materials.

One of the most common problems with understanding sour corrosion is the presence of various iron sulfides. They can behave in different ways. For instance, they can cause an electromotive force to corrode a metal, which results in a galvanic coupling reaction and an increase in the rate of corrosion. They can also stimulate the hydrogen evolution reaction and prevent further damage. It inhibits corrosion altogether and protect the surface from further attack.

27.1 Sour corrosion rates and electrochemistry

Understanding the mechanisms by which hydrogen sulfide can affect carbon steel corrosion is very important in order to predict its effects. This can be done through the study of the FeS layer's electrochemical development and kinetics. Besides these, other factors such as the presence of corrosion products are also taken into account to determine its effects. The catalytic role of iron sulfide corrosion products is to catalyze the hydrogen evolution reaction. It can also act as a cathodic fuel. Unfortunately, the exact mechanisms by which carbon steel corrosion can be triggered by hydrogen sulfide are not known. Some of the unanswered questions include the role of FeS corrosion film breakdown in the development of localized corrosion, the contribution of galvanic effects to pit growth, and the impact of steel microstructure on the condition. This chapter aims to discuss the various mechanisms by which

Handbook of Corrosion Engineering. DOI: https://doi.org/10.1016/B978-0-323-95185-2.00027-7

hydrogen sulfide can affect metal corrosion. It was based on observations made in the field and laboratory.

27.2 Electrochemical reactions

This section describes the various mechanisms involved in the process of sour corrosion. The main net reaction is the reduction of the protons to elemental hydrogen and the oxidation of iron. The process of sour corrosion involves a series of electrochemical and chemical reactions. The main net reaction is the reduction of the protons to elemental hydrogen and the oxidation of iron (0) to iron (II). The hydrogen can either be absorbed by the metal or formed by an adsorbed state on the surface of the iron sulfide or steel. The latter can lead to the formation of hydrogen embrittlement and the increase of the susceptibility of the metal to cracking such as hydrogen-induced cracks (HIC), Stress Corrosion Cracking (SSC). The fraction of hydrogen that can be absorbed depends on various factors such as the microstructure of the metal, its permeability, and the defect types and concentration.

$$Fe + H_2S \rightarrow FeS + 2H_{ads} \tag{27-1}$$

Bare steel surfaces are not suitable for the measurement of sour corrosion rates due to the presence of iron sulfide layers. This makes it difficult to estimate and measure sour corrosion on gas and oil field setups. In a study conducted on FeS-covered iron, Shoesmith, Taylor, Bailey, and Owen (1980) discovered that the anodic and the cathodic Tafel slopes were greater than 240 mV at pH 4. This was due to the heavy polarization of the reactions generated by the protective mackinawite ($Fe_{1+x}S$) layer on the surface.

27.2.1 Anodic reactions—iron dissolution

The studies on the anodic dissolution of iron in a solution containing H_2S have been carried out by several workers (Cheng et al., 1998; Pound, Wright, & Sharp, 1989). They noted that the bisulfide ions in the solution acted as a complexing agent during the initial phase of the process. They then performed two electron transfer steps to determine the rate at which FeSH complexes would form. The formation of the FeSH complexes leads to the formation of various types of corrosion films, such as mackinawite or aqueous Fe^{2+} ion. Anodic Tafel gradients were reported in the range of 40–80 mV/decade on bare steel surfaces in acidic H_2S solutions (pH $>$ 3). On the other hand, in the presence of iron sulfide deposits, steeper gradients were observed.

27.2.2 Cathodic reactions-hydrogen reduction

Cathodic charge transfer reactions are commonly triggered by the hydrogen evolution during the dissolution of iron in CO_2 and H_2S solutions. In addition to H^+ ions other species such as H_2CO_3, H_2S, HS, and HCO_3 are also candidates for electron acceptors. The reduction of carbonic acid through the dissolution of H_2S can contribute to the rate of the reaction. The dissolution of H_2S can also contribute to the rate of the reaction through mechanisms similar to that of carbonic acid reduction. The mechanisms and experimental evidence for the direct reduction of H_2S in acidic solutions were presented by Naumann and Carius (1959). This was well before the theory of carbonic acid's reduction was established and has later been confirmed by other workers (Morris, Sampaleanu, Veysey, Sampaleanu, & Veysey, 1980; Zheng, Brown, & Nešić, 2014). The cathodic Tafel slopes can be measured between 110 and 120 mV/decade in solutions with pH less than five. However, steeper slopes can also be observed in the presence of sulfide layers. The cathodic reaction rates can be affected by the concentration of H + and H_2S. In areas with high rates of corrosion, the mass transport limitations may apply. The reaction can also occur on the surfaces of deposits of iron sulfide and on the electrical contact between the steel and the surface. Due to the presence of deposits and porous layers, the iron sulfide can potentially have a larger area of Cathodic protection than the adjacent steel surface hydrogen electrode (Bolmer, 1965; Kawashima, Hashimoto, & Shimodaira, 1976). This can lead to a significant corrosion attack. In order to study the electrochemical properties of different iron sulfides, Tjelta and Kvarekvål (2016) have conducted studies. In a study conducted on the dissolution of iron in aqueous solutions containing H_2S based on the formation of a mackinawite film, Sun (2006) proposed a different mechanism that could be utilized (Fig. 27–1).

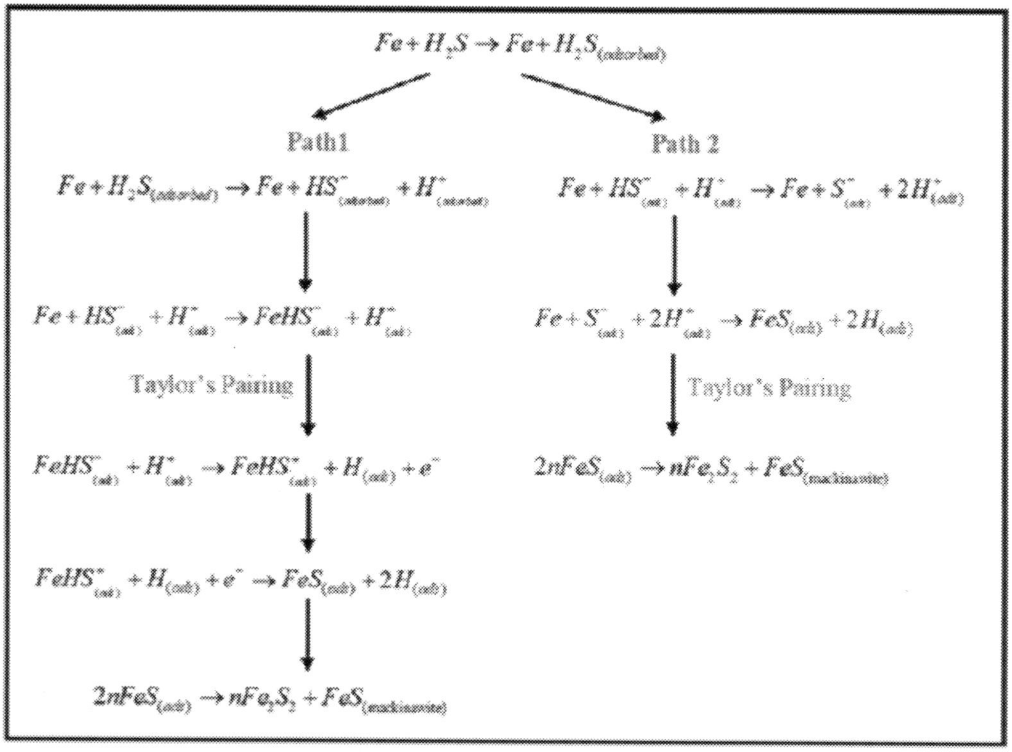

FIGURE 27–1 Mechanism for iron dissolution in aqueous solutions of H_2S.

27.3 Various iron sulfides formed during sour corrosion

It has been reported that the most common type of corrosion product on iron with a CO_2 + 13% H_2S gas mixture is mackinawite (Sardisco & Pitts, 1965). The presence of both troilite and pyrite was also detected in the experiment. The films were subjected to varying pressure and pH values at 24°C (75°F), 1 atm bar (14.5 psi), and total pressure and pH values between 3.1 and 11. At pH between 6.5 and 8.8, the films mainly consisted of mackinawite. The properties of these films were not good compared to those made with either troilite or pyrite. Bisulfide (HS) ions predominate in the pH range of maximum mackinawite yield and were thus regarded as unfavorable reactants in terms of protective iron sulfide formation. Sulfide films formed at pH 3.2 and 11.0 provided the best corrosion protection, with half being mackinawite and the other half troilite, pyrite, or marcasite.

Shoesmith et al. (1980) stated that the corrosion product from H_2S exposure at room temperature is composed of three phases: mackinawite, troilite, and cubic iron sulfide. The first phase (mackinawite), which is formed by cracking, leads to high local corrosion rates and the precipitation of troilite and cubic iron sulfide. Troilite thermodynamically considered to be the most favorable of the compounds. However, due to its low nucleation rate, it only forms small amounts at room temperature. At higher temperatures (>50°C [>122°F]), the nucleation rate increases due to the presence of supersaturated Fe^{2+} ions and higher corrosion rates. In solutions with pH below 5, the H_2S partial pressures can reach up to 20 bar. The higher sulfides can be produced by mackinawite when subjected to temperatures ranging from 80°C to 180°C to the following sequence (Wikjord, Rummery, Doern, & Owen, 1980): Mackinawite/cubic FeS/troilite/pyrrhotite/pyrite. According to Rhodes (1976), the main corrosion product is pyrrhotite ($FeS_{1.15}$), which is produced at 1 bar (14.5 psi) H_2S and 90°C (194°F) in 1 M NaCl brines. In his experiments, he found that a layer of $FeCl_2$ between the steel surface and the film of corrosion maintained the anodic reaction rate. Milliams and Kroese (1979) also noted that the presence of chloride ions prevented the formation of protective iron sulfide films at 25°C (77°F).

The transformation sequence ends with the formation of troilite (Shoesmith et al., 1980), which occurs at atmospheric H_2S partial pressure and room temperature. The dissolution of iron sulfides triggers the formation of modified iron sulfide, which then forms on the top of the metal surface or under the corrosion film. Numerous modifications of iron sulfide have coexisted. Smith and colleagues (Milliams & Kroese, 1979; Smith & Joosten, 2006; Smith & Brown, 2006; Smith, Brown, & Sun, 2011) have reviewed and analyzed numerous publications on the various iron sulfides' stability, occurrence, sequence, and corrosion products under different conditions. They found that pyrrhotite and mackinawite are the most common corrosion products during the sour conditions. They also created conceptual phase diagrams to show the transition between the polymorphs. The researchers found that some types of iron sulfides, like the chloride-intolerant stoichiometric cubic iron sulfide, were less stable.

27.4 Factors affecting sour corrosion

There are various factors that can affect the behavior and rate of sour corrosion in pipelines (Dougherty, 1988; Li, Ling, Cao, Pacheco, & Desai, 2013). Some of these include the degree of pressure, temperature, concentration, and dissociation of hydrogen sulfide. Also, the steel type, the flow velocity, the chemistry of the fluid, and the chemistry of fluid (water cut, pH, oil wettability, phase ratios, and organic acid) are some of the factors that can influence the corrosion rate. Although these factors can influence the rate and behavior of sour corrosion, it is not possible to determine the exact effects of these on the condition. This is why it is important to thoroughly discuss the effects of testing parameters on the corrosion process.

27.4.1 Gas composition

The composition of the gas is thought to be an important factor influencing the rate of steel corrosion observed in oil and gas production. The partial pressures of H_2S and CO_2 in particular had been used to demonstrate the effects of gas composition on the corrosion rate.

27.4.1.1 The partial pressure of H_2S

Several studies have been conducted on the effects of H_2S and CO_2 partial pressures on the rate of corrosion in different types of environments, such as N_2/H_2S and CO_2/H_2S. The results of these studies revealed that the P_{H2S} varied, while all other parameters were kept constant.

27.4.1.2 The CO_2 partial pressure in CO_2/H_2S environments

The CO_2 partial pressure has been established to be one of the key factors that affect the corrosion rate in pure sweet, CO_2 corrosion. However, the role of CO_2 in CO_2/H_2S corrosion is less well established, so the effect of CO_2 partial pressure on sour corrosion is discussed.

27.4.2 Iron sulfides scale

Scale formation is another factor that can affect the rate of corrosion in pure CO_2 and CO_2/H_2S. In contrast to the simple form of corrosion, wherein only one type of iron carbonate can form, there are several types of iron sulfides that can form in this type of corrosion. The characteristics and thermodynamics of different types of iron sulfides can be affected by the scale formation. For instance, the growth rate of FeS can be determined by the kinetics of the scale formation. On the other hand, the degree to which a particular species can be stable in a given environment can be determined by the thermodynamics of the scale. Besides the formation of iron sulfide during the corrosion process, mill scale is also present on the surface of steel pipe before it is installed in gas and oil production fields. This presence or absence of mill scale can have a minor impact under certain conditions like thin, less stable mill scales are formed. In-house experiments were conducted on the surface of mill-scaled and acid cleansed steel pipes. It was observed that the former's presence did not significantly affect the rate of corrosion in both liquid and gas phases. Some of the experiments also indicated that mill-scaled steel pipes have a slightly higher corrosion rate than those of bare steel. Since mill scale can affect the corrosion rate of a pipeline, one should be careful when it comes to predicting its condition using a sour corrosion model.

27.4.3 Effect of temperature

The effects of temperature on the corrosion rate are not apparent in the experiments conducted with low H_2S concentrations. High concentrations of H_2S can cause alterations in the protective and type of scales (Kvarekvål & Svenningsen, 2015).

27.4.4 Test duration

The initial corrosion rate in CO_2/H_2S corrosion was high, but it drops off significantly at longer periods due to the formation of sulfide deposits. In short-term tests, the researchers (Sun, Nešić, & Papavinasam, 2008) found that the rate of corrosion decreased with increasing reaction time. The initial rate of corrosion can be as high as 2.5 mm/h, while it can drop to less than 0.5 mm/h after 24 hours. In a series of experiments and in-house tests, the researchers (Kapusta, Raghu, & Richard, 2008) were able to show that the initial rate of corrosion decreased with increasing reaction time. This effect was mainly due to the formation of iron sulfide scale deposits. It is important to conduct studies comparing the data collected in the laboratory with the data gathered in the field to understand the effects of test duration.

27.4.5 Flow effects

Some studies have been conducted to determine the effects of flow velocity on the sour corrosion rate. For instance, studies of Tewari and Campbell (1979) showed that the dissolution rates of mackinawite were accelerated by increasing flow velocity.

27.4.6 Solution composition

The composition of the solution can also affect the rate of corrosion in acid gas environments. Some of the parameters that are commonly used to determine the solution composition include the pH level, Fe^{2+} concentration, total ionic content, brine composition, and chloride concentration. In-house data revealed that the rate of corrosion decreased significantly when the solution's pH level increased at trace amounts of H_2S. This was because at trace amounts of H_2S and low pH (such as pH below 4–4.5), the solution was undersaturated with respect to iron sulfide, which caused the FeS scale to dissolve. The presence of other corrosive substances, such as organic acids, H^+, and CO_2, can also affect the rate of corrosion. When the solution's pH level is constant and the partial pressure of H_2S is increased, it can become supersaturated, which can lead to the formation of an iron sulfide scale. However, if the pH level is maintained at a lower level, the rate of corrosion can be influenced by the presence of H_2S.

27.4.7 Organic acid

When organic acids are used in CO_2 systems, they can accelerate the process of corrosion by decreasing the pH, exposing the steel to a corrosive environment, and destabilizing the $FeCO_3$ scale. According to studies conducted by Pugh et al. (2009), organic acid can slightly accelerate the rate of corrosion in sour systems. A flow loop study conducted by Singer, Brown, Camacho, and Nesic (2007, 2011) revealed that a combination of H_2S (0.004–0.13 bar) and acetic acid (1000 ppm in the aqueous phase) significantly increased the rate of corrosion compared to that of H_2S alone. They also observed that the films had cracks at the higher H_2S partial pressures. This suggests that the internal stress in the films could be the reason for the increased corrosion rate. In the study, Sun, Blumer, and Davis (2010) found that organic acids, when used in combination with CO_2, increased the pitting rate in tests at 50°C with 1–3 bar H_2S and 2.5–4.5 bar CO_2. However, the results of the tests revealed that the pitting rates were below 1 mm/year and decreased with increasing test duration. In the tests conducted on X70 and X80 steels with 10%–100% H_2S (Omweg, Frankel, Bruce, Ramirez, & Koch, 2003), the presence of 0.5% acetic acid caused pitting corrosion at ambient pressure. Since the presence of this acid affected the SSC results, it was difficult to interpret them. Each pit was tested to determine if it had terminal crack features.

27.4.8 Fe^{2+} concentration

In order to understand the effects of Fe^{2+} concentration on the scale formation and corrosion, Sun et al. (2008) conducted experiments in H_2S environments. The results show that Fe^{2+} concentration has minimal effect on the rate of corrosion. It is believed that the iron sulfide scale can be easily dissolved in water. During the experiments, the increasing Fe^{2+} concentration in the solution accelerated the formation of iron sulfides which had been observed during experiments as black powder suspensions in the solution. However, this effect did not affect the rate of corrosion. As a result, the precipitation of iron sulfide on the steel surface did not contribute to the corrosion rate. It is believed that the presence of Fe^{2+} can influence the iron sulfide scale's formation at low pH. In addition, other dissolved Fe ions, such as FeS_{aq} and $FeHS^+$, can exist as a function of pH (Rickard & Luther, 2007).

27.4.9 Chloride concentration

In-house studies revealed that the presence of chloride can affect the pitting corrosion rate in H_2S environments. They found that when the concentration of chloride is between 0% and 3%, the weight loss rate is constant, while when it exceeds 20%, the rate of corrosion is doubled. In-house investigations also suggest that the presence of chloride can cause a reduction in the consistency of the iron sulfide scale (Sun, Pugh, Stephen, Ling, & Pacheco, 2014).

27.4.10 Effects of elemental sulfur, polysulfides, and oxygen

Rhodes and Schutt (Gregg & Lerbscher, 2005; Schutt & Rhodes, 1996) discovered that the presence of H_2S and O_2 in neutral aqueous solutions significantly increases the rate of corrosion possibly a result of oxidation of sulfide to elemental sulfur. The researchers measured the corrosion rates in samples up to 15 mm/year (600 mpy) with pH 7, 8 ppm O_2 and 1 mM H_2S. The pH was adjusted by adding ammonia to the solution. In another study about the effects of oxygen on sour corrosion, Song and colleagues (Song, Palencsár, Svenningsen, Kvarekvål, & Hemmingsen, 2011; Song et al. 2011) discovered that when H_2S and O_2 react, oxysulfur compounds, such as thiosulfate, sulfate, and sulfite are formed with the formation of elemental sulfur. These compounds, which are stronger acids, cause localized corrosion within H_2S solutions and increase the rate of corrosion in the solution. Contrary to popular belief, the presence of oxygen in H_2S systems does not lead to the development of elemental sulfur as the only reaction product.

27.5 Mitigation of localized corrosion of oil and gas pipeline

27.5.1 Corrosion inhibitors

While being transported or stored, corrosion inhibitors can protect against the effects of aggressive species. These chemicals are usually cheap and can be used in small concentrations to provide effective protection. Despite the technological advancements that have occurred in the field of corrosion inhibitor technology over the past decade, the dose rates needed for sour systems are still higher than those under sweet conditions. In regions where deposited sulfur corrosion is occurring, the use of corrosion inhibitors can only be performed if there is a sulfur-based solvent. As a result, the application of these chemicals is limited. Due to the presence of other chemicals, such as scale inhibitors, biocides, and hydrate inhibitors, the use of corrosion inhibitors in areas where there are still deposits of sulfur can be more challenging.

27.5.2 Environmental conditions (scavengers)

The use of environmental conditioners can help remove aggressive species from the metal's environment. They can also reduce the medium's corrosivity by scavenging various types of corrosive substances. For example, in alkaline or near-neutral solutions, oxygen is reduced in the cathodic reaction as follows:

$$O_2 + 2H_2O + 2e^- \rightarrow 4OH^- \qquad (27-2)$$

In this case, the aggressive species oxygen is scavenged by the addition of a scavenger such as hydrazine, which controls corrosion via the reaction described below (Singh, 2009):

$$5O_2 + 2(NH_2 - NH_2) \rightleftharpoons 2H_2O + 4H^+ + 4NO_2^- \qquad (27-3)$$

Commercial scavengers, such as those based on 1,3,5-tris (2-hydroxyethyl) -1,3,5-triazinane, can also be used to remove corrosive hydrogen sulfide (H_2S) during oil and gas production and transportation through pipelines (Rajagopal, Lacerda, Slobodcicov, & Campagnolo, 2009; Rasmussen, Wright, Schofield, & Lund, 2004).

27.5.3 Interface Inhibitors

Through various mechanisms, interface inhibitors are used to reduce corrosion in oil and gas pipelines. Interface inhibitors prevent corrosion by forming a protective film between the metal and the corrosive medium. Interface inhibitors are divided into two types: vapor phase and liquid phase. Liquid interface inhibitors are classified as anodic, cathodic, or mixed depending on the type of electrochemical reaction inhibited.

27.5.4 Cladding

A treatment column that has a thermal spray-resistant coating on its interior can prevent the formation of hydrogen atoms in the steel and prevent the sour process fluid from flowing into it.

27.5.5 Coatings

A sacrificial thermal spray coating is usually designed to protect the substrate from the corrosive constituents of the fluid. However, a barrier thermal spray coating has to completely block the corrosive elements from rusting the metal. The density and porosity of the applied coating are also important factors to consider when it comes to preventing corrosion.

References

Bolmer, P. W. (1965). Polarization of iron in H2S-NaHS buffers. *Corrosion.*, *21*(3), 69–75. Available from https://doi.org/10.5006/0010-9312-21.3.69.

Cheng, X. L., Ma, H. Y., Zhang, J. P., Chen, X., Chen, S. H., & Yang, H. Q. (1998). Corrosion of iron in acid solutions with hydrogen sulfide. *Corrosion.*, *54*(5), 369–376. Available from https://doi.org/10.5006/1.3284864.

Dougherty, J. (1988). *Factors effecting H2S and H2S/CO2 attack on carbon steels under deep hot well corrosion.* 88.

Gregg M. R. Lerbscher J. (2005). Inhibitor developments providing mitigating benefits against pitting corrosion to carbon steel constructed assets used to process wet sulfur contaminated sour gas production. In *NACE - international corrosion conference series 03614409*, National Assoc. of Corrosion Engineers International Canada, http://www.nace.org.

Kapusta, S., Raghu, D., & Richard, J. (2008). Managing corrosion in sour gas systems: Testing, design, implementation and field experience. In *NACE - International Corrosion Conference Series 03614409*, Netherlands, 086411-0864112.

Kawashima, A., Hashimoto, K., & Shimodaira, S. (1976). Hydrogen electrode reaction and hydrogen embrittlement of mild steel in hydrogen sulfide solutions. *Corrosion.*, *32*(8), 321–331. Available from https://doi.org/10.5006/0010-9312-32.8.321.

Kvarekvål, J. & Svenningsen, G. (2015). Norway effect of high H2S partial pressures on localized corrosion of carbon steel. In *NACE - International corrosion conference series 03614409*, National Assoc. of Corrosion Engineers International, http://www.nace.org.

Li, C., Ling, S., Cao, F., Pacheco, J., & Desai, S. (2013). Effect of sodium chloride concentration on carbon steel sour corrosion. In *NACE - International corrosion conference series 03614409*, United States.

Milliams, D. E. & Kroese, C. J. (1979). Aqueous corrosion of steel by H2S and H2S/CO2 mixtures. In *Third international conference on the internal and external protection of pipes.*

Morris, D. R., Sampaleanu, L. P., Veysey, D. N., Sampaleanu, L. P., & Veysey, D. N. (1980). The corrosion of steel by aqueous solutions of hydrogen sulfide. *Journal of the Electrochemical Society, 127*(6), 1228–1235. Available from https://doi.org/10.1149/1.2129861.

Naumann, F. K. & Carius, W. (1959). Die Bedeutung der Korrosionsvorg€ange in w€assrigen Schwefelwasserstoff e L€osungen f€ur die Bruchbildung an St€ahlen. 30

Omweg, G. M., Frankel, G. S., Bruce, W. A., Ramirez, J. E., & Koch, G. (2003). Performance of welded high-strength low-alloy steels in sour environments. *Corrosion, 59*(7), 640–653. Available from https://doi.org/10.5006/1.3277595. Available from http://corrosionjournal.org/.

Pound, B. G., Wright, G. A., & Sharp, R. M. (1989). The anodic behavior of iron in hydrogen sulfide solutions. *Corrosion, 45*(5), 386–392. Available from https://doi.org/10.5006/1.3582033.

Pugh, D. V., Asher, S. L., Cai, J., Sisak, W. J., Pacheco, J. L., Ibrahim Che, F., . . . Robson, D. (2009). Top-of-line corrosion mechanism for sour wet gas pipelines. In *NACE - International corrosion conference series 03614409*, United States.

Rajagopal, K., Lacerda, R., Slobodcicov, I., & Campagnolo, E. (2009). Modeling and simulation of hydrogen sulfide removal from petroleum production lines by chemical scavengers. *Chemical Engineering Communications, 196*(10), 1237–1248. Available from https://doi.org/10.1080/00986440902832100.

Rasmussen, K., Wright, A., Schofield, M., & Lund, H. (2004). *NACE Meeting Papers Denmark Investigation of a corrosion failure in an offshore gas compression system: The role of H2S scavenger and organic acids.*

Rhodes, P. R. (1976). Corrosion mechanism of carbon steel in aqueous H2S solution. *The Electrochemical Society.*

Rickard, D., & Luther, G. W. (2007). Chemistry of iron sulfides. *Chemical Reviews, 107*(2), 514–562. Available from https://doi.org/10.1021/cr0503658.

Sardisco, J. B., & Pitts, R. E. (1965). Corrosion of iron in an H2S-CO2-H2O system mechanism of sulfide film formation and kinetics of corrosion reaction. *Corrosion., 21*(8), 245–253. Available from https://doi.org/10.5006/0010-9312-21.8.245.

Schutt, H. U., & Rhodes, P. R. (1996). Corrosion in an aqueous hydrogen sulfide, ammonia, and oxygen system. *Corrosion, 52*(12), 947–952. Available from https://doi.org/10.5006/1.3292088. Available from http://corrosionjournal.org/.

Shoesmith, D. W., Taylor, P., Bailey, M. G., & Owen, D. G. (1980). The formation of ferrous monosulfide polymorphs during the corrosion of iron by aqueous hydrogen sulfide at 21°C. *Journal of the Electrochemical Society, 127*(5), 1007–1015. Available from https://doi.org/10.1149/1.2129808.

Singer, M., Brown, B., Camacho, A., & Nesic, S. (2007). Combined effect of CO2, H2S and acetic acid on bottom of the line corrosion. In *NACE - International corrosion conference series 03614409*, United States.

Singer, M., Brown, B., Camacho, A., & Nešić, S. (2011). Combined effect of carbon dioxide, hydrogen sulfide, and acetic acid on bottom-of-the-line corrosion. *Corrosion., 67*(1). Available from https://doi.org/10.5006/1.3546849.

Singh, G. (2009). Corrosion inhibitors. *Corrosion Reviews, 27*(Supplement), 367–416. Available from https://doi.org/10.1515/corrrev.2009.27.s1.367.

Smith, S. N., Brown, B., & Sun, W. (2011). Corrosion at higher H2S concentrations and moderate temperatures. In *NACE - International corrosion conference series 03614409*, United States.

Smith, S. N. & Joosten, M. W. (2006). Corrosion of carbon steel by H2S in CO2 containing oilfield environments. In *NACE - International corrosion conference series 03614409*, United States, 061151–0611526.

Song, Y., Palencsár, A., Svenningsen, G., Kvarekvål, J., & Hemmingsen, T. (2011), Norway Effect of O2 and temperature on sour corrosion. In *NACE - International corrosion conference* series 03614409.

Sun, W. (2006). *Kinetics of iron carbonate and iron sulfide scale formation in CO2/H2S corrosion.*

Sun, H., Blumer, D., & Davis, J. (2010). Pit propagation of carbon steel in sour conditions. In *NACE - International corrosion conference series 03614409*, United States.

Sun, W., Nešić, S., & Papavinasam, S. (2008). Kinetics of corrosion layer formation. Part 2 - Iron sulfide and mixed iron sulfide/carbonate layers in carbon dioxide/hydrogen sulfide corrosion. *Corrosion, 64*(7), 586–599. Available from https://doi.org/10.5006/1.3278494. Available from http://corrosionjournal.org/.

Sun, W., Pugh, D., Stephen, S. N. S., Ling, S., & Pacheco, J. (2014). A parametric study of sour corrosion of carbon steel.

Tewari, P. H., & Campbell, A. B. (1979). Dissolution of iron during the initial corrosion of carbon steel in aqueous H 2 S solutions. *Canadian Journal of Chemistry, 57*(2), 188–196. Available from https://doi.org/10.1139/v79-031.

Tjelta, M., & Kvarekvål, J. (2016). Norway Electrochemistry of iron sulfide and its galvanic coupling to carbon steel in sour aqueous solutions. In *NACE - International corrosion conference series 9781510821354*, National Assoc. of Corrosion Engineers International, 4, 2531–2544. http://www.nace.org.

Wikjord, A. G., Rummery, T. E., Doern, F. E., & Owen, D. G. (1980). Corrosion and deposition during the exposure of carbon steel to hydrogen sulphide-water solutions. *Corrosion Science, 20*(5), 651–671. Available from https://doi.org/10.1016/0010-938X(80)90101-8.

Yépez, O., Obeyesekere, N., & Wylde, J. (2016). Study of sour corrosion mechanism under controlled pH. In *NACE - International corrosion conference series 9781510821354*, National Assoc. of Corrosion Engineers International, United States, 6, 4326–4337. http://www.nace.org.

Zheng, Y., Brown, B., & Nešić, S. (2014). Electrochemical study and modeling of H2S corrosion of mild steel. *Corrosion, 70*(4), 351–365. Available from https://doi.org/10.5006/0937.

PART 9

Sustainable corrosion management by computational modelling

28

Corrosion management using computational simulations

28.1 Introduction

28.1.1 Corrosion inhibition: modern developments and limitations

Through their interactions with the elements in their environment, almost all metals and their alloys are susceptible to corrosion (Böhni, Suter, & Schreyer, 1995; Singh & Dahotre, 2007). One of the most innovative, beneficial, cost-effective, and simple techniques for slowing down metal corrosion is the use of synthetic (organic and inorganic) substances (Xhanari & Finšgar, 2016; Verma, Verma, Ebenso, & Quraishi, 2018a; Ju, Kai, & Li, 2008). By creating a barrier of protection at the point where the metal meets the harsh environment, organic compounds inhibit corrosion. Inorganic oxides, such as nitrates, chromates, molybdates and phosphates, among others, are known to be efficient inhibitors of metal corrosion. Nevertheless, owing to their significant toxicity and bioaccumulation potential, there have been campaigns against using these very powerful chemicals (Peter, Obot, & Sharma, 2015). As a result, somewhat less harmful organic molecules are taking the place of inorganic inhibitors. In particular, heteroatoms as polar substituents are researched as potential corrosion inhibitors in several types of organic compounds. Organic compounds can bond with metallic surfaces by adsorbing to their electron-rich centers, which can include polar substituents like −OMe (methoxy), OH (hydroxyl), >NH (2-amine), $-NH_2$ (1-amine), $-NO_2$ (nitro), >N- (3-amine), −COOH (carboxylic acid), −CN (nitrile), $-COOC_2H_5$ (ester), $-SO_3H$ (sulfonic acid),−COX (acid halide), $CONH_2$ (amide), >C=S (thiocarbonyl), C=O (carbonyl), >C = N− (imine), and −N=N− (azo) (Verma et al. 2017; Verma, Olasunkanmi, Ebenso, Quraishi, & Obot, 2016). Additionally, side chains and many pi-bonds in aromatic compounds facilitate their adsorption on metallic surfaces.

Considering several kinds of organic molecules have undergone testing as corrosion inhibitors for metals. Although these substances may have varying degrees of environmental effect when used or introduced through synthetic workups, they cannot be regarded as "green" or environmentally beneficial. Multistep reactions (MSRs) are used to synthesize the majority of efficient organic inhibitors. These processes include the use of pricey and sometimes harmful compounds, catalysts, and solvents. After purification and workup operations, MSRs often involve the usage and discharge of massive volumes of hazardous wastes and solvents. MSRs are also incredibly time-consuming and challenging to manage. MCRs, which combine three or more reactants in a single step, may be used to synthesize the chemicals and get around the MSR problem (Verma, Haque, Quraishi, & Ebenso, 2019). MCRs are

Handbook of Corrosion Engineering. DOI: https://doi.org/10.1016/B978-0-323-95185-2.00028-9

associated with low waste and by-product levels due to the decrease in the number of reactions, purification, and workup stages. MCRs are particularly simple to conduct and manage since they are one-step processes. MCRs are thus acknowledged as one of the greenest synthesis methods for organic molecules that might be used as green corrosion inhibitors, particularly in conjunction with MW and US irradiations (Verma et al., 2019; Verma Quraishi & Ebenso, 2018d).

Due to their relatively tiny molecular size and area, the bulk of organic molecules utilized as corrosion inhibitors only provide little surface coverage, which results in minimal metal protection even at moderately high concentrations (Saji & Cook, 2012; Saji & Thomas, 2007). This problem may be resolved by using polymeric materials and nanoparticles as corrosion inhibitors. Numerous artificial and organic polymers have been successfully used as corrosive damage inhibitors. According to a literature review, G and graphene oxide and carbon nanotubes (Deyab, 2014; Montemor & Ferreira, 2008) are often used to prevent corrosion. Even though polymers and nanoparticles provide high surface coverage, their commercial applicability is limited by the low degree to which they can dissolve in polar electrolytes. In coating formulations, however, inhibitors may be added to polymers and nanoparticles (Chauhan, Quraishi, Ansari, & Saleh, 2020). As they often include pricey solvents, catalysts and chemical precursors, the synthesis, functionalization, and purification of polymers and nanomaterials are likewise very costly processes. This is in addition to the less eco-friendly synthetic processes.

As a result, R&D is centered on the creation of substances from natural and biological sources that have minimal to no negative environmental consequences. In this context, corrosion inhibitors based on AAs (amino acids), carbohydrates and their derivatives are favored as metal/alloy defenders in various electrolyte media (Umoren & Eduok, 2016; Hamadi, Mansouri, Oulmi, & Kareche, 2018). Corrosion prevention techniques often use carbohydrate-based polymers and their derivatives, including chitosan, cellulose, chitin, starch, etc. It is generally known that derivatization increases the number of active (donor or binding) sites on polymeric chains of carbohydrates and increases the amount of solubility of such chains in polar electrolytes, both of which increase the protective power of those chains. Additionally, extensively investigated in Corrosion Inhibitors (CIs) are AAs and their heterocyclic derivatives. Along with carbohydrates, recognized ecologically acceptable substitutes that are extensively used in CIs include AAs and their derivatives, plant extracts, and chemical medications (Gece, 2011; Pathak & Mishra, 2016; Popoola, 2019). A review of the literature revealed that extracts from various plant components are often employed as corrosion inhibitors. The results of the literature review demonstrate that, among the various extracts, leaf extracts exhibit considerably greater efficacy and are the subject of the majority of studies. Several literature reviews are reporting the anticorrosive properties of plant extracts (Verma, Ebenso, Bahadur, & Quraishi, 2018b). Another approach for preventing corrosion in the environment is chemical medications or pharmacological molecules (Gece, 2011; Pathak & Mishra, 2016). Oleochemicals (Shehata, Korshed, & Attia, 2018), biosurfactants (Malik, Hashim, Nabi, Al-Thabaiti, & Khan, 2011; Zhu, Free, Woollam,

& Durnie, 2017), and natural gums (Peter et al., 2015; Arthur, Jonathan, Ameh, & Anya, 2013) are a few other natural substances that are being studied as corrosion inhibitors.

28.1.2 Corrosion inhibition: role of computational modeling

The use of organic chemicals, particularly heterocyclics, is one of the most common, cost-efficient, and successful techniques. To illustrate the effectiveness of these chemicals as protectors, several experimental and computational techniques are created (Fig. 28–1).

A few instances of experimental techniques for monitoring corrosion include hydrogen evolution (HE), open circuit potential (COCP), linear polarization (LPR), potentiodynamic polarization (PDP), electrochemical impedance spectroscope (EIS), scanning electron microscope (SEM), energy dispersive X-ray (EDX), atomic force microscope (AFM), X-ray difraction (XRD) and X-ray photoelectron spectroscopy (XPS). The most widely used computationally based corrosion monitoring approaches are DFT, MDS, MCS, artificial neural network (ANN), and quantitative structure–activity relationship (QSAR). The results of experimental and computational demonstration point to the fact that organic corrosion inhibitors work by adhering to the boundary between metal and a corrosive environment (Adachi, Osako, Tanaka, Hojo, & Hollister, 2006; Edmonds & Hales, 2005; Obot, Macdonald, & Gasem, 2015). These substances provide a surface coating that prevents corrosion by their adsorption.

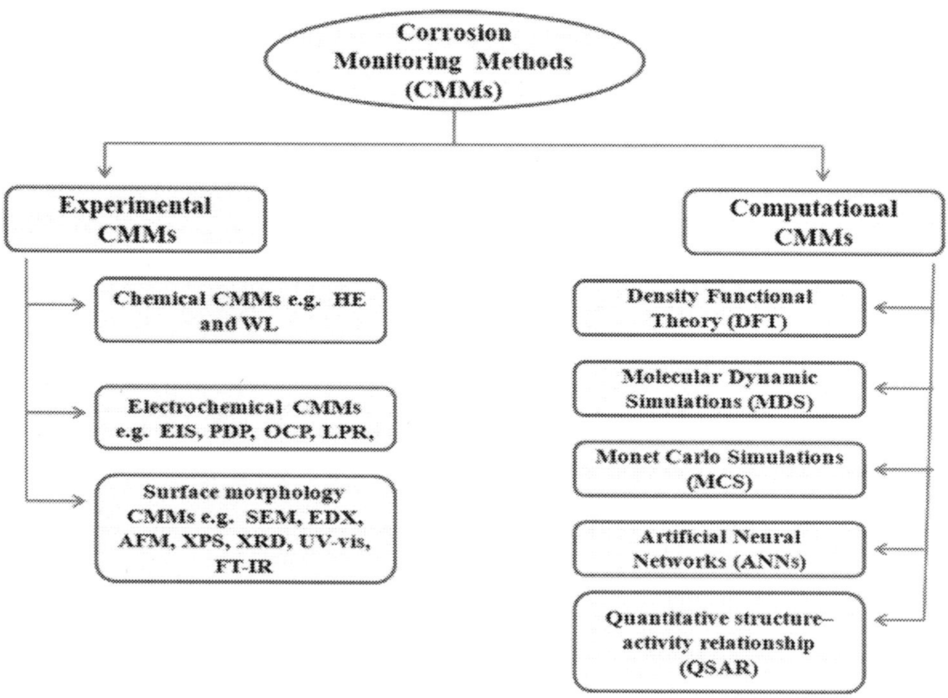

FIGURE 28–1 Experimental and computational methods of corrosion inhibition testing.

Authors from several groups have shown that organic corrosion inhibitors adhere to metallic surfaces via a donor–acceptor (charge-sharing) process before physicochemical adsorption. Electrostatic interactions between a charged metallic surface and a charged inhibitor molecule cause physisorption to take place. It is crucial to note that in aggressive electrolytes, heteroatom-containing organic molecules and metallic surfaces both readily charge. Chemisorption takes place when an inhibitor molecule and a charged metallic surface share charges. Authors found that the buildup of counterions (chloride ions) on the metallic surface causes the Schiff base (SB) of pyridyl substituted triazoles to become negatively charged, whereas the protonation of the heteroatoms in the SBs causes them to become positively charged (Ansari, Quraishi, & Singh, 2014). Through the physisorption process, these oppositely charged species attracted one another. These SB molecules' heteroatoms also interact with metallic surfaces by donating (transferring) their spare electron pairs into the metal's d-orbitals. These molecules use a retro-(back-donation) mechanism to acquire electrons from metallic surfaces. Fig. 28–2 depicts a schematic representation of an SB molecule on a surface made of mild steel. Similar findings have also been observed elsewhere (Verma, Quraishi, & Singh, 2015a).

To illustrate the donor–acceptor mode, DFT-based computational simulations are often performed. When assessing the corrosion inhibition capacity of organic corrosion inhibitors, the presence and type of substituents are key factors. In general, donating groups like -OH, -OMe, $-NH_2$, and $-NMe_2$ boost the performance, but electron-withdrawing substituents like -COOH,$-NO_2$, and -CN have been shown to have the opposite effect. In an acidic solution containing 1 M HCl, our study team examined the inhibitory potential of three AACs with different types of substituents (AAC-1: R = NO2, AAC-2: R = -H, and AAC-3: R = -OH) (Verma, Quraishi, Olasunkanmi, & Ebenso, 2015b). The results of experimental and DFT experiments demonstrated that AACs functioned as powerful inhibitors and that changing the substituents significantly changed the effectiveness of their protection. Because of its

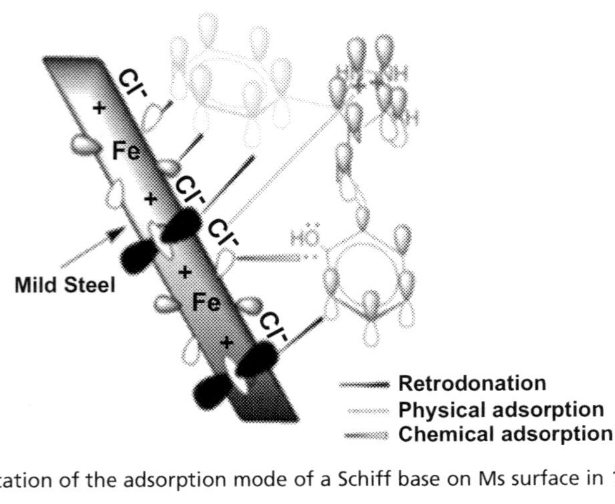

FIGURE 28–2 The presentation of the adsorption mode of a Schiff base on Ms surface in 1 M HCl.

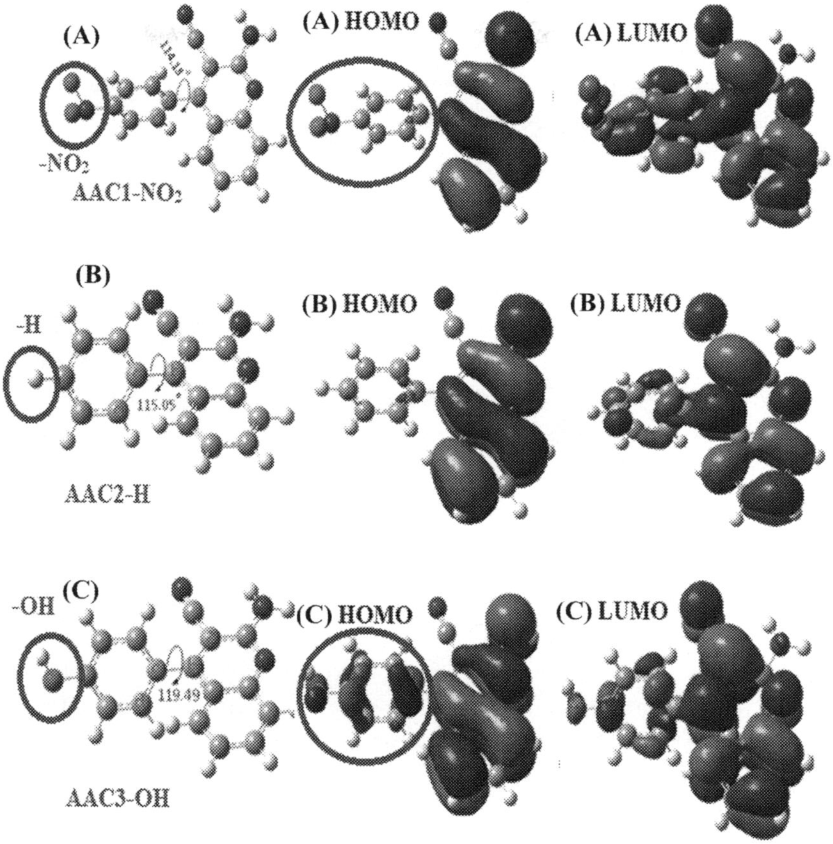

FIGURE 28–3 FMOs of (A) AAC-1-NO_2, (B) AAC-2-H and (C) AAC-3-OH (Verma et al., 2015b).

electron-withdrawing properties, the -NO_2 substitution was shown to reduce the HOMO contribution, which led to reduced protection efficacy for the compounds when compared to the unsubstituted inhibitor molecule (AAC-2). Due to its electron-donor activity, the presence of the hydroxyl group, on the other hand, boosts HOMO contribution. The frontier molecular orbitals (FMOs; optimized, HOMO, and LUMO) of AACs are shown in Fig. 28–3, and replacements are shown in red circles.

28.2 Computational modeling in corrosion inhibition studies

28.2.1 DFT, MDS, and MCS modeling

28.2.1.1 MDS modeling for iron alloys

In studies on corrosion inhibition, MDS may provide some extremely pertinent and reliable insights into how chemical compounds employed as corrosion inhibitors interact with metals

(Quraishi, Chauhan, & Saji, 2020; Verma et al., 2018c). Since these insights are challenging to glean from laboratory-based studies, molecular dynamics (MD) simulations are often utilized in conjunction with other methods in corrosion inhibition research. Combining experiments with MD simulations has several benefits, including a thorough atomic-level description of the chemical system that aids in the development of adsorption mechanism theories. The locations of the particles are determined by resolving Newton's equations of motion in this simulation method, which is based on the rules of classical physics (Berendsen, 1999; Leimkuhler & Matthews, 2016). Atomistic models (empirical potential or force field), as well as interatomic potential functions, characterise many components of the simulated system as well as the forces that exist between them (Monticelli & Tieleman, 2013). Since well-defined (fixed) thermodynamic states may be modified, the main benefit of MD simulations may be the capacity to alter various system characteristics.

Different computational methods have been used by researchers to simulate how inhibitors prevent the corrosion of iron alloys, but they have all been focused on building simulation systems that could produce insightful data about the relationships between experimental inhibitor performance and inhibitor molecular structure (Quraishi et al., 2020; Verma et al., 2018c). Molecular dynamics simulations are being used more and more to study how to prevent corrosion in iron alloys. The statistics unambiguously show that this field is expanding without a lot of articles being produced annually.

Numerous for-profit and free computer applications enable computations of molecular dynamics (Berendsen, 1999). But as several studies have shown, using commercial software drastically cuts down on the time needed to prepare computer files and scripts. There has been a lot of interest in selecting appropriate iron planes for simulations. To investigate the adsorption of pyrrole, furan, and thiophene on iron surfaces, Guo et al. (2017) used the CASTEP module integrated into the Materials Studio programme to calculate the morphological characteristics and the surface energy values of three iron planes, Fe(110), Fe(100), and Fe(111). Fe (110) has a high packed density (111). In addition to these key variables, the force fields, ensemble, and boundary conditions have all been taken into account while doing MD simulations to prevent corrosion in iron alloys.

Ensemble: MD simulations use an ensemble to depict the thermodynamic states that are used to replicate the experimental circumstances. It is an assortment of every feasible system that has the same thermodynamic characteristic or macroscopic property but distinct microscopic states. One parameter variable is left unfixed while the others are fixed by an ensemble (Berendsen, 1999). The NVE, NPT, NVT, and VT are the most popular ensembles. These ensembles allow for dynamic energy during simulation. It is common practice to employ canonical and microcanonical ensembles in MD simulations of corrosion inhibition (Verma et al., 2018c; Obot, Haruna, & Saleh, 2019).

Force fields: The force field is the first parameter considered in molecular dynamics simulations. The difficult part in any simulation is selecting or developing the right force field for a given system since it is by far the most important in gathering pertinent data (Jensen, 2016). The collection of formulae and constants characterising all the aspects of potential energy operating on atomic nuclei constitutes the force field. Both experimental data and those from ab initio computations are used to parameterize force fields. The most used force field for corrosion inhibition studies is COMPASS. In addition to ab initio calculations, COMPASS was parameterized taking into account a variety of experimental data, such as organic compounds created with halogens and other atoms and metals (Obot et al., 2019). Other force fields in use include the CVFF, GVFF, and UFF, an all-atom potential with parameters for each atom (Hagler & Lifson, 1974; Hagler, Huler, & Lifson, 1974; Lifson, Hagler, & Dauber, 1979).

Periodic boundary conditions: A primitive cell in MD simulations is a unit cell that depicts a box with every atom and bond represented much like a crystal structure. However, since it includes hundreds of atoms and molecules, the real corrosion inhibition process cannot be replicated using an infinite system (Obot et al., 2019). Researchers created many boundary conditions, including the periodic boundary conditions, which may make the corrosion inhibition simulation almost an infinite system, to provide more realistic simulations at a lower computing cost.

28.2.1.2 DFT-based molecular modeling

Aluminum is mostly employed in construction materials and aerospace transportation due to its low weight and resistance to corrosion. Aluminum typically exhibits less corrosion than iron, mostly because the surface of aluminum develops a protective oxide coating that shields it from additional corrosion (Guo et al., 2017; Oguzie, 2007; Xhanari et al., 2017). Nevertheless, the corrosion rate of aluminum surfaces rises when they are exposed to air and moisture environments that include a variety of corrosive agents, such as aggressive anions, salts, and other elements. As a result, there are financial losses and a decline in mental efficiency. There are several kinds of corrosion inhibitors, and plant extracts, polymers, nonmaterials, and organic molecular medicines are often employed to stop this corrosion in metals. All of these organic compounds have functional groups and unsaturated Pi electrons, which make them electron-rich, and are thus thought to be the most efficient. Organic molecules often include heteroatoms like S, O, P, and N, which have spare electrons that they may readily share with the metal's unoccupied d orbital and deposit as a protective layer on the metal surface. Between inhibitor compounds and metal surfaces, there are two different forms of interactions: chemisorption and physisorption.To comprehend the corrosion prevention process in metal, a range of experimental approaches are used, with gravimetric analysis and electrochemical characterization standing out. However, in the last several decades, computer-based quantum chemical calculations have become the predominant method. The computer-based calculations are fully eco-friendly since they do not need any expensive lab supplies or reagents, and they also provide more reliable and precise findings than experimental calculations. A useful method for explaining the chemical behavior, reactivity, and adsorption behavior of inhibitor compounds is quantum chemical calculation. The link between molecular surface and inhibition efficiency may be readily explored with the use of computational tools, which are employed for research on corrosion inhibition at the molecular level.

Theoretically demonstrating the anticorrosive activity of organic compounds has been done extensively using the newly developed computational calculation-based DFT modeling (Verma et al., 2020; BinSabt, Nazeer, Madkour, & Al-Sagheer, 2016). The primary benefit of DFT is that it allows for the theoretical calculation of the reactivity or inhibition efficiency of organic compounds. It also aids in the development of corrosion inhibitors, eliminating the need for complex labs and the manufacture of organic chemicals (Khojasteh, Kazerooni, Salarian, & Kamali, 2016). The interaction with the metallic surface is often caused by molecule size and substituents. DFT calculations based on quantum chemistry may be used to explain several theoretical calculations on the effectiveness of preventing corrosion and the molecular reactivity of organic molecules. The highest occupied molecular orbital (HOMO) and lowest unoccupied molecular orbital (LUMO) of any inhibitor molecule, respectively, are known as the FMOs, and they are what determine an inhibitor's ability to inhibit (Vashisht et al., 2016; Mandour, Nazeer, Al-Hetlani, Madkour, & Abdel-Monem, 2018). In addition to this, a few additional local and global characteristics also affect how reactive organic molecules are. The HSAB paradigm is often followed in the interaction between the metal surface and inhibitor molecule, where the inhibitor acts as a nucleophile and the metal surface behaves as an electrophile owing to the unoccupied d orbital. DFT-based computational calculations may also be used to represent the HSAB notion mentioned above. Therefore it can be argued that several conventional and empirical concepts in corrosion investigation may be linked with a quantum mechanism using DFT-based modeling. The most essential and relevant parameter among all the numbers shown by DFT is electron density (P(r)), which allows for the computation of other valuable parameters (Lai et al., 2017; Parr, Donnelly, Levy, & Palke, 1978). Nowadays, theoretical calculations based on DFT that include certain helpful factors are often used for studies on corrosion inhibition to evaluate the interaction between the metal surface and inhibitor molecules.

28.2.2 Concept of chemical reactivity

Quantum chemical calculations include descriptors that relate to the chemical processes and the electronic structure of the inhibitor molecules that are involved in the establishment of a strong interaction between the inhibitor molecules and the metal surface. Here are some examples of chemical reactivity indicators based on DFT.

Frontier molecular orbital global reactivity descriptor: The fundamental rules controlling stereochemistry and chemical reactions that are derived from border molecular orbitals were initially researched and effectively implemented by Fukui (1982a). FMOs, such as the LUMO, HOMO, and energy gap (ΔE), are crucial in determining the chemical behavior and reactivity of organic molecules. The E_{HOMO} values and organic molecules' capacity to donate electrons are in excellent accord with the inhibitor's molecule's ability to inhibit. The donor−acceptor and acid−base concepts serve as the foundation for corrosion inhibition. When an inhibitor molecule comes into contact with an electron-reaching species, it donates an electron; yet, when an empty d orbital is present on a metal surface, it may readily absorb an electron. In this approach, the organic molecules are deposited above the metal surface by their reactive centers, depositing it in the shape of a protective layer, blocking the active site of the metal surface, and guarding it against further corrosion thanks to the donor–acceptor idea. This leads to the conclusion that the more electrons that reach the molecule and may readily donate their electron to the metal surface, the greater the value of E_{HOMO} (Khaled & Amin, 2008). E_{LUMO} will take electrons from metal high-lying orbitals through a base donation to a greater extent when its value decreases. Corresponding to this, the energy gap $E = E_{LUMO}$-E_{HOMO} of organic molecules is another crucial characteristic for the Frontier molecular orbitals. Typically, the molecule's reactivity decreases with a larger value of *E*, whereas its ability to donate an electron increases with a lower value of ΔE (Bentiss, Traisnel, Vezin, Hildebrand, & Lagrenee, 2004). Furthermore, the estimated values of E_{LUMO}, E_{LUMO}, and ΔE in their study in which they performed DFT calculation of *n*-alkyl carboxylic acids on Al (lll), and they showed a high correlation with inhibitory efficiency discovered by weight loss measurement (Poberžnik et al., 2020; Verma et al., 2015b).

Global hardness and softness: The primary feature results from the understanding of the chemical hardness (η) idea with the energy second derivative about the number of electrons (Parr & Pearson, 1983):

$$\eta = \left(\frac{\delta^2 E}{\delta N^2}\right)_v = \left(\frac{\delta \mu}{\delta N}\right)_v \tag{28-1}$$

Iczkowski and Margrave (1961) state that the above equation may be reduced as follows:

$$\chi = -\mu = \left(\frac{I + A}{2}\right) \tag{28-2}$$

where *I* and *A* stand for the ionization potential and the electron affinity, respectively. A more straightforward version of this equation is represented by the energies of FMOs and hardness (η) (Sharma, Peter, & Obot, 2015):

$$\eta = \left(\frac{I - A}{2}\right), \eta = -\frac{E_{LUMO} - E_{HOMO}}{2} \tag{28-3}$$

Furthermore, Sanderson's notion of electronegativity equalization has a theoretical justification (Sanderson, 1983). The global softness (*S*), which is the opposite of the global hardness (η), may be shown as follows (Yang & Parr, 1985):

$$S = \frac{1}{2\eta} = \left(\frac{\delta N}{\delta \mu}\right)_{v(r)} \tag{28-4}$$

Additionally, the physical correlation between hardness, softness, and energy gape (ΔE) provides a fundamental understanding of the reactivity and electron-donor nature of organic molecules. The molecule will typically be equally isolated and stiff in an electrical system with a greater energy gap between E_{HOMO}-E_{LUMO} (ΔE), which will result in a lower propensity for electron donation, compared to a molecule with a higher degree of softness. It may quickly and effectively donate low-lying orbitals of the electron to the metal, which would prevent the metal's surface from further corrosion. According to N. Ammouchi, Allal, Belhocine, Bettaz, and Zouaoui (2020), a molecule may be easily deformed when its chemical hardness is high, but an inhibitor moiety's chemical softness is low. They listed the three pyrazinamide derivatives in order of having the lowest energy gap (E) and highest softness (SIGMA). Meo-PZA> ME-PZA> PZA, which is also described as Meo-PZA (82.95)> ME-PZA (75.25)> PZA in terms of efficiency (67.64) (Ammouchi et al., 2020).

Electronegativity and chemical potential: According to Parr et al. (1978), chemical potential (μ) is equivalent to the negative magnitude of electronegativity (χ), which is the fundamental connection of DFT of chemical reactivity:

$$\mu = \left(\frac{\delta E}{\delta N}\right)_{v} = -\chi \tag{28-5}$$

Sanderson's electronegativity: According to the Sanderson electronegativity equation principle, the higher the reactivity of a molecule is, the lower its value of electronegativity and chemical potential is; in other words, the stronger the interaction will be if there is less electronegativity difference between the metal and the molecule (Sanderson, 1976). In their study, Abdallah, Gad, Sobhi, Al-Fahemi, and Alfakeer (2019) used the drug tramadol to prevent aluminum from corroding, and they reported electronegativity (χ) 0.1001 and chemical potential (eV) 0.2361, both of which are relatively low values. As a result, they were able to donate their electrons to the Al (lll) surface quite easily. Chemical potential and electronegativity values were in line with experimentally observed inhibitory efficiency. The drug compounds in Fig. 28–4 are shown in HOMO, LUMO, and MEP images, which indicate the drug molecules' electron-donating properties and where the electron density is positioned about the benzene ring and methoxy group.

Dipole moment (μ):A molecule's polarizability or polarity may be described by the physical quantity known as the dipole moment. Typically, the greater value of the dipole moment reveals the better inhibitory effect of the molecule, which may be explained by the stronger connection between the inhibitor molecule and the metal surface (Obot & Obi-Egbedi, 2010). According to a certain study, when the value of the dipole moment decreases, innovation efficiency will rise (Gao & Liang, 2007). Li et al. (2014) applied oxime compounds, namely acetone oxime (AO) and 2-butanone oxime (BO), for aluminum in HCl solution and reported the noticeable, =5.305 (AO)

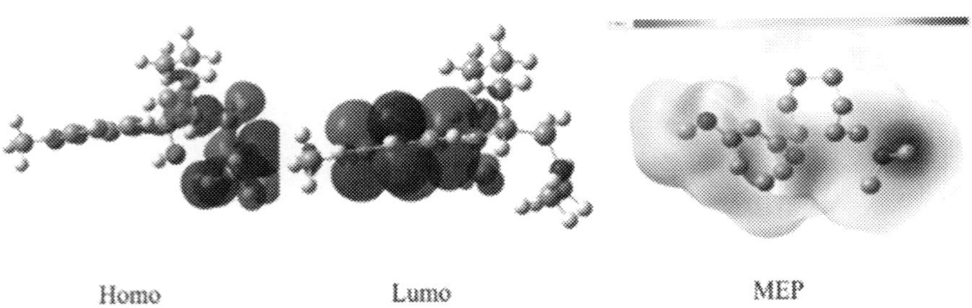

FIGURE 28–4 HOMO, LUMO and MEP of tramadol drug. *Copyright 2017@ Elsevier.*

$>$=5.247 (BO). This suggests that the better inhibitory performance of BO might not be caused by the intermolecular electrostatic force (Li, Deng, & Xie, 2014).

Electrophilicity and nucleophilicity: Through the work of Parr, Szentpaly, and Liu (1999), the global electrophilicity index has been introduced:

$$\omega = \frac{\mu^2}{2\eta} \tag{28-6}$$

It has been shown using the parabolic model that may also be stated in terms of I and A as follows (Geerlings, De Proft, & Langenaeker, 2003):

$$\omega = \frac{(I+A)^2}{8(I+A)} \tag{28-7}$$

wBoth demonstrate the propensity of any organic species or moiety to accept electrons, that is, the stronger the inhibitor's electrophilicity value, the greater the inhibitor's propensity to accept electrons; consequently, the good electrophile will be the one whose value is higher while the relative propensity will be lower and the nucleophilicity value will be higher. As a result, it can be deduced that a molecule's nucleophilicity value will grow along with its inhibitory characteristic, causing it to interact with metal surfaces more strongly (Chattaraj, Maiti, & Sarkar, 2003).

28.2.3 Electron transfer function

The relationship between ΔN (electron transfer charge) and ΔE (energy change) (Kovačević & Kokalj, 2011):

$$\Delta N = \frac{\mu_B - \mu_A}{2(\eta_A + \eta_B)} \tag{28-8}$$

And

$$\Delta E = \frac{(\mu_B - \mu_A)^2}{2(\eta_A + \eta_B)} \tag{28-9}$$

The replacement of the metal and mol by the subscripts A and B, respectively, and the substitution of -μ with - χ results in:

$$\Delta N = \frac{\chi_{metal} - \chi_{mol}}{2(\eta_{metal} + \eta_{mol})} = \frac{\phi - \chi_{mol}}{2\eta_{mol}} \tag{28-10}$$

$$\Delta E = -\frac{(\chi_{metal} - \chi_{mol})^2}{4(\eta_{metal} + \eta_{mol})} = \frac{(\phi - \chi_{mol})^2}{4\eta_{mol}} \tag{28-11}$$

In the more complicated system, the electron transfer function term Aries often results in electron following. When calculating the value of electron transfer under the assumption that bulk metal $I = A$, the values of global hardness ηFe$=0$ and bulk iron XFe$=7$ eV are often used. According to Lukovists, if ΔN is more than 3.6, the metal surface's improved ability to track electrons typically results in a rise in the inhibitor efficiency (Lukovits, Kalman, & Zucchi, 2001).

The electronic Fukui function and lower softness: In recent times, metals' wave functions have been employed instead of their electronegativities. Fukui indices estimate may be used to produce various DFT-based local parameters (Fukui, 1982b). A solid foundation for FET is provided by the Fukui function density functional definition (frontier-electron theory) (Saha & Banerjee, 2015). The following is a derivation of the molecular sites for nucleophilic, electrophilic, and radical assaults utilizing finite difference approximations of the Mulliken population study:

$$f_k^+ = q_k(N+1) - q_k(N) \text{ (for nucleophilic attack)} \tag{28-12}$$

$$f_k^- = q_k(N) - q_k(N-1) \text{ (for electrophilic attack)} \tag{28-13}$$

$$f_k^0 = \frac{q_k(N+1) - q_k(N-1)}{2} \text{ (for radical attack)} \tag{28-14}$$

where r is a point in space close to the molecule, and q_k, and k, are the atom's and molecule's respective gross charges, or the electron density. A singly charged cation with an electron withdrawn from the neutral molecule HOMO is represented by N-1, whereas a singly charged anion with an electron added to the neutral molecule LUMO is represented by $N+1$. The local S is defined as:

$$s(r) = \left[\frac{\delta n(r)}{\delta \mu}\right]_{\upsilon(r)} \tag{28-15}$$

Upon integration, Eq. (28–8) yields the global S:

$$S = \int s(r)dr \tag{28-16}$$

The following is a presentation of the connection between the local S and the Fukui function:

$$s(r) = Sf(r) \tag{28-17}$$

To calculate theoretical descriptors, many computing strategies are created. The Fukui indices may be used to identify any organic molecule's low reactivity (Fukui, 1982b). Along with the inhibitor's low reactivity as measured by the Fukui indices, electrophilicity, nucleophilicity, and a reactive area are also found. A section of an organic molecule with a big Fukui function may be thought of as having a hander, while a region with a Fukui function that is less than the molecule can be thought of as having a softer chemical nature. Fukui function is a crucial index that may be used to identify the nucleophilic attack Fukui function and the electrophilic attack Fukui function, which both help to explain the local site selectivity of an inhibitor molecule. The high value of the attack nucleophilic and electrophilic Fukui functions often denotes an atom's propensity to re-electrify or lose an electron.

28.2.3.1 Quantitative structure–activity relationship-based modeling

28.2.3.1.1 General introduction and a brief history of QSAR

A statistical concept known as the QSAR uses mathematical models to depict the range of potential correlations between molecule structure (or related properties) and chemical (or biological) activity. It is an adaptable technique for descriptive investigation of the relationship between a compound's molecular structure and its physical, chemical, or biological characteristics or activities. Polanski defined QSAR as the mapping of chemical space onto property space that is created by simulating the mathematical function that connects qualities to chemical structure or structural-related characteristics, which are commonly referred to as molecular descriptors. The resilience of a good QSAR model resides in its capacity to operate as a "two-way door" between the chemical and property/activity spaces, allowing, in theory, the prediction of a new molecule's properties or the design of a chemical compound with a potential activity (Polanski, 2019).

Since Crum-Brown and Fraser described what appeared to be the first connection between chemical structure and activity in 1868 based on their research on the reliance of physiological activities of certain quaternary ammonium salts on chemical structure, QSAR has undergone numerous developments to become a modern tool that has discovered implementations in many fields of science and technology. Crum-Brown and Fraser's first QSAR equation was straightforward yet generic (Crum-Brown and Fraser, 1868–1869).

$$\Phi = f(C) \tag{28-18}$$

The discovery of a connection between an organic molecule's cytotoxicity and its solubility in water, as reported by Richet, is one of the earliest advances in the history of QSAR (Richet, 1893), the formulation of the Meyer–Overton hypothesis of anesthesia, which focused on the connection between lipid's anesthetic properties and hydrophobicity or lipophilicity (Meyer, 1899; Overton, 1901), the Hammett formula, which links equilibrium constant to electronic characteristics and effects of substituents (Hammett, 1935), and so on. An important step toward the creation of the updated QSAR approach was the 1964 development of the LFER method in pharmaceutical research, which described the QSAR between biological activities and chemical structures based on the paper of Hansch and Fujita (1964). It had been thought that the Hansch correlation was arguably the earliest and most effective semi-empirical QSAR model (Growcock, 1989). The "inclusion of solubility/dispersibility terms and 2 in the LFER" was a special contribution of Hansch's QSAR. This was done to ensure accurate modeling of the surface activity of the solute in an aqueous solution, in line with Traube's rule, where a less soluble solute (in an aqueous solution) has a greater surface activity (Whalen, 1983). The Hansch model could be mathematically represented as (Growcock, 1989):

$$\log A = \alpha_1 \rho\sigma + \alpha_2 E_s + \alpha_3 \psi + \alpha_4 \psi^2 \qquad (28-19)$$

where A is the characteristic being observed, E_s is a separate parameter relating to the effective size and shape of the molecule, and it is often coupled with the electronic terms in parameters like E_{LUMO} or E_{HOMO}. The octanol/water partition coefficient might be used to determine the words for solubility and dispersibility. The mathematics of the structure–activity connection proposed by Free and Wilson (1964) brought up another step. The review paper by Cherkasov et al. (2014), which examines the past and the future of QSAR, has more details on the history of QSAR. Dearden's "Whither QSAR?" review essay (Dearden, 2017a), review article of Lukovits, Shaban, and Kálmán (2003) on the methodologies employed in QSAR studies, study done by Dearden (2017b) on the history and development of QSAR, and a plethora of additional materials. There have been hundreds of scientific articles that concentrate on various uses of the "tool" from the era of QSAR's covert introduction and subsequent acceptance as a contemporary approach for characterising relationships between activities and structural factors.

28.2.3.1.2 Development of QSAR models

Statistics, experimental research, and computer science all come together in QSAR modeling (Fig. 28–5). Even though many writers have described the processes involved in creating QSAR models in a variety of ways, for the sake of this study, we will take a three-step approach viz (OECD):

1. A data collection of chemical/biological activity for a group of chemicals/compounds that are being used to create a QSAR model, ideally associated with some kind of criterion. For the sake of developing a model, these data—which are empirically derived—are regarded as the dependent variables.

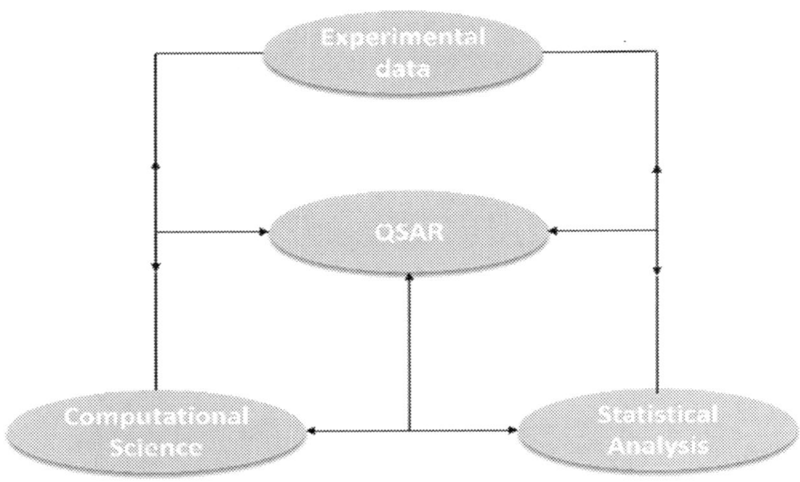

FIGURE 28–5 QSAR as a cross-link in experimental–statistical–computational science relationship.

2. A collection of information on the structural, molecular, or electronic characteristics of a particular group of molecules (as in I above), which is sometimes referred to as the descriptors and would be regarded as the independent factors for the building of a model.

The data sets in I and II above have some kind of mathematical–statistical correlation function that establishes that connection, which is supported by statistical analysis using goodness-of-fit.

A successful model must be able to forecast the property of a molecule that, although it may belong to the class of compounds used in model formation, was not only used in the formulation. This is why QSAR model validation is crucial. At the 37th Joint Meeting of the Chemicals Committee and Working Party on Chemicals, Pesticides, and Biotechnology in November 2004, countries that are members of the Organization for Economic Co-operation and Development (OECD) adopted principles for the validation of QSAR models (for regulatory purposes). As stated in the OECD publication, these guidelines, which are now known as the OECD principles, are as follows: "To assist the consideration of a (Q)SAR model for regulatory reasons, it should be coupled with the following information:

1. a certain endpoint
2. a clear algorithm
3. a specific area of application
4. suitable goodness-of-fit, durability, and predictability metrics
5. If feasible, a mechanistic interpretation" (O.S.O)

An additional definition of "an unambiguous algorithm" (point 2 of the OECD principles) describes an "algorithm" in this context as the structure of the relationship between the chemical behavior (biological or chemical activity), also known as the "endpoint" (point 1 of the OECD principles), and descriptors of the chemical structure. The three components of each algorithm are the chemicals dataset, endpoint values, and descriptor values. Univariate regression, a linear or nonlinear (less often employed) technique, and multiple linear regression are two examples of typical algorithms used in the creation of QSAR/QSPR models (MLR). Ordinary least squares are the MLR technique that is most often used (O.S.O). Many descriptors necessary for MLR must be carefully derived to prevent overfitting and descriptor redundancy, which might complicate and invalidate the model. Principal component analysis, factor analysis, and other approaches might be used to profile a collection of descriptors for MLR. Grover, Grover, and Sharma (2016) go into further depth on the traditional methods used to create QSAR/QSPR models.

28.2.3.1.3 QSAR as a tool in corrosion inhibition studies

The previous research of Bergman (1954) and Hoijtink (1955), which functioned as expositions of QSAR into electrochemical processes, may be linked to the use of QSAR in corrosion inhibition investigations. Before this, Bergman had described certain relationships between the wave numbers of the p-bands in various series of polycyclic aromatic hydrocarbons' absorption spectra and their half-wave potentials (Bergman, 1954). Building on Bergman's findings, Hoijtink proposed Maccoll's relationship between the lowest unoccupied molecular orbital energy of a sequence of alternant and non-alternant hydrocarbons and the

Table 28–1 Selected QSAR equations from previous studies over the last 40 years.

Family of compounds	Alloy/Corrosive medium	QSAR model	*R*	Refs.
p-substituted anilines	Fe/5% HCl	I_{corr} = -5.597 (± 0.7) FN (C1) + 40.67 (± 4.8) FN (N11) + 0.1507 (+0.02)	0.973	Abdul-Ahad and Al-Madfai (1989)
		K = -109.9 (± 16.0) FN (C1) + 389.1 (± 106.9) FN (N11) + 2.157 (+0.6)	0.951	
TCAs	API J55 steel/ 15% HCl	$\log R = 10.93 - 12.5\log P + 2.66 (\log P)^2 - 1.43\ E_{LUMO}$[a]	0.98	Growcock, Frenier, and Andreozzi (1989)
		$\log R = -1.23 - 1.07\log P - 1.78\ E_{LUMO}$[a]	0.93	
		$R = 0.49 - 0.342\log P + 0.061 (\log P)^{2}$[a]	0.93	
		$\log R = -2.27 - 0.645\log P - 2.27\ E_{LUMO}$[b]	0.91	
		$\log R = -6.14 + 2.99\log P - 0.85 (\log P)^2 - 2.38\ E_{LUMO}$[b]	0.91	
		$R = 1.04 - 0.667\log P + 0.109 (\log P)^{2}$[b]	0.93	
Thioureas	CS/1 N H_2SO_4	$E_i = \frac{(7123\pi + 3276)c_i}{1 + (7123 + 3276)c_i}$ [c]	0.95	Lukovits, Palfi, Bako, and Kalman (1997)
		$E_i = \frac{(-36053\mu + 206079)c_i}{1 + (-36053\mu + 206079)c_i}$ [d]	0.93	
		$E_i = \frac{(209056E_{HOMO} - 27869E_{LUMO} + 27870\mu + 1638720)c_i}{1 + (209056E_{HOMO} - 27869E_{LUMO} + 27870\mu + 1638720)c_i}$ [e]	0.95	
Schiff Bases	Cu/1.0 M NaCl	$P = 431.82 + 45.26\ E_{HOMO} - 15.60\ E_{LUMO}$	1.0	Li et al. (1999)
		$P = 584.8 + 69.52\ E_{HOMO} - 11.47\ E_{LUMO}$	1.0	
Thioureas	CS/1 N H_2SO_4	$E = \frac{(43678E_{LUMO} + 250S - 65671)c}{[1 + (43678E_{LUMO} + 250S - 65671)c]}$ [f]	0.97	Lukovits, Bakó, Shaban, amd Kálmán (1998)
		$E = 1.65\Lambda - 6.20\Lambda^2 + 11.50\Lambda^3 - 5.95\Lambda^4$ for $\Lambda = \frac{(58136E_{LUMO} + 335S - 88168)c}{[1 + (58136E_{LUMO} + 335S - 88168)c]}$	0.982	
Polymethylene amines	Fe/Acid	$E_i = \frac{(80.1E_{HOMO} - 673.8E_{LUMO} + 60.1\mu + 2971.3)c_i}{[1 + (80.1E_{HOMO} - 673.8E_{LUMO} + 60.1\mu + 2971.3)c_i]} \times 100$	NR	
		$E_i = \frac{(-748.4\Delta - 2816.4\mu + 12660.2)c_i}{[1 + (-748.4\Delta - 2816.4\mu + 12660.2)c_i]} \times 100$ [g]	0.93	
		$E_i = \frac{(532.6E_{HOMO} + 778.87E_{LUMO} - 63.531\mu + 5031)c_i}{[1 + (532.6E_{HOMO} + 778.87E_{LUMO} - 63.531\mu + 5031)c_i]} \times 100$	0.95	
N-containing compounds	A3 steel/5% HCl	$E_i = 140.2454 + 462.3744E_{HOMO} + 6.4262\mu + 4.9455^{r}f^{1N}_{HOMO} + 0.4609^{r}f^{1N}_{HOMO} + 9.2514^{2}X^{\nu}$	0.8145	Zhang, Lei, Xia, and Wang (2005)
		$E_i = 151.0746 + 530.52286E_{HOMO} + 6.7669\mu - 4.4722Don^{1N} + 2.3781Acc^{1N} + 1.3329Polar$[h]	0.8184	
Schiff bases	CS/1 M HCl	$E_i = 2.084E_{HOMO} - 3.041E_{LUMO} + 115.772$	1	Ashassi-Sorkhabi, Shaabani, and Seifzadeh (2005a)
		$E_i = \frac{(-604.90E_{HOMO} + 5864.86E_{LUMO} + 1990.06\mu - 642.67)c_i}{[1 + (-604.90E_{HOMO} + 5864.86E_{LUMO} + 1990.06\mu - 642.67)]c_i} \times 100$	NR	
Schiff bases	Ms/1 M HCl	$E_i = 3.362E_{HOMO} - 7.094E_{LUMO} + 122.761$	0.9377	Ashassi-Sorkhabi, Shaabani, and Seifzadeh (2005b)
		$E_i = \frac{(-526.325E_{HOMO} + 977.031E_{LUMO} + 107.937\mu - 2150.147)c_i}{[1 + (-526.325E_{HOMO} + 977.031E_{LUMO} + 107.937\mu - 2150.147)]c_i}$		

Thiosemicarbazides	Ms/H_2SO_4	$E_i = \frac{(-6.7E_{HOMO} - 5.9E_{LUMO} - 3.5\mu - 43.7)c_i}{[1 + (-60 - 6.7E_{HOMO} - 5.9E_{LUMO} - 3.5\mu - 43.7)]c_i} \times 100$	0.915	Khalil (2003)
		$E_i = \frac{(227.2 + 23.5E_{HOMO} - 3.8\mu)c_i}{[1 + (227.2 + 23.5E_{HOMO} - 3.8\mu)]c_i} \times 100$	0.962	
Oxadiazoles	C38 CS/1 M HCl	$R_{ct} = 30.67 + (-3.05 \times 10^5 E_{HOMO} + 4.29 \times 10^5 E_{LUMO} - 2.42 \times 10^4 \mu)c_i$	0.99	Outirite et al. (2010)
		$R_{ct} = 30.67 + (-3.15 \times 10^5 E_{HOMO} + 4.50 \times 10^5 E_{LUMO})c_i$	0.93	
		$R_{ct} = 30.92 + (-1.02 \times 10^5 E_{HOMO} + 1.78 \times 10^5 E_{LUMO} - 2.46 \times 10^4 \mu)c_i$	0.98	
		$R_{ct} = 36.84 + (-8.01 \times 10^5 E_{HOMO} + 1.58 \times 10^5 E_{LUMO})c_i$	0.93	
Thiosemicarbazides	Ms/1 M H_2SO_4	$E_i = \frac{(1.0175E_{HOMO} + E_{LUMO} + 0.997\Delta + \mu + \log P + Pol + 58.61)c_i}{[1 + (1.0175E_{HOMO} + E_{LUMO} + 0.997\Delta + \mu + \log P + Pol + 58.61)]c_i} \times 100$	0.8589	Ebenso, Isabirye, & Eddy (2010)
Triazoles and oxadiazoles	Ms/1 M HCl	$\frac{1}{R_{ct}} = 0.00282 + (10.08\mu + 8.42E_{HOMO} - 35.24E_{LUMO})c_i$	0.9576	Bentiss, Traisnel, Vezin, and Lagrenée (2003)
		$\frac{1}{R_{ct}} = 0.00381 + (1.70\mu + 5.03E_{HOMO} - 32.13E_{LUMO})c_i$	0.91	
Carbazoles	Ms/HCl	$E_i = \frac{(1.0176E_{HOMO} + 0.9743E_{LUMO} + 1.0351\Delta E + CosA + CosV + 428.6731)c_i}{[1 + (1.0176E_{HOMO} + 0.9743E_{LUMO} + 1.0351\Delta E + CosA + CosV + 428.6731)c_i]}$	0.9124	Eddy and Ita (2011)
Phthalocyanins	Al/1 M HCl	$E_i = 0.553 \times E_{HOMO} \times c_i - 0.457 \times E_{LUMO} \times c_i + 194.786$	0.938	Dibetsoe et al. (2015)
Naphthalocyanins		$E_i = -0.332 \times \Delta E \times c_i - 0.197 \times \Delta N \times c_i + 91.578$	0.934	
Sulphonamides	Ms/1 M HCl	$E_i = \frac{(2.47\omega + 0.0856E_{LUMO} - 0.0527\mu + 6.20\eta - 0.0411\log P - 21.81) \times 5000}{[1 + (2.47\omega + 0.0856E_{LUMO} - 0.0527\mu + 6.20\eta - 0.0411\log P - 21.81) \times 50]}$	1.000	Murulana, Kabanda, and Ebenso (2016)
Thio-based compounds	Ms/1 M HCl	$E_i = \frac{(-3432.53 - 273.13E_{HOMO} - 156.77E_{LUMO} + 2.77\lambda + 3370.15\Delta N - 28.17c_i) \times 100c_i}{[1 + (-3432.53 - 273.13E_{HOMO} - 156.77E_{LUMO} + 2.77\lambda + 3370.15\Delta N - 28.17c_i)c_i]}$	0.8894	Musa, Mohamad, Kadhum, Takriff, and Ahmoda (2012)
Triazoles	Ms/1 M HCl	$E_i = \frac{(5.31E_{HOMO} - 5.83E_{LUMO} + 1.35\mu + 17.09) \times 100c_i}{[1 + (5.31E_{HOMO} - 5.83E_{LUMO} + 1.35\mu + 17.09)c_i]}$	0.9738	Guo, Zhu, Zhang, He, and Li (2014)

[a] R is the correlation coefficient.
[b] For TCA + n-dodecylpyridinium bromide.
[c] For TCA + adduct of trimethyl-1-heptanol with 7 moles of ethylene oxide.
[d] π denotes the hydrophobic substituent constant.
[e] μ denotes the dipole moment.
[f] S denotes the molecular surface area.
[g] Δ denotes the HOMO–LUMO energy gap.
[h] Not reported.
[i] Don^{1N} and Acc^{1N} denote indices for donor and acceptor characteristics, respectively, for the inhibitor.

experimental half-wave potentials of those molecules. This relationship was derived as the root of the molecular orbital secular equation (Hoijtink, 1955). Donahue and Nobe (1966) found a link between organic compound adsorption propensity and Hammett-like substituent constants in 1965, around 10 years after Hoijtink's study, by applying the LFER method. Based on the presumption that chemisorption is the mechanism of adsorption, Donahue and Nobe's work was expanded to include organic corrosion inhibitors. Since what seemed to be the first research—that is, Donahue and Nobe's work—featured QSAR application in corrosion inhibition investigation both in title and content (Donahue & Nobe, 1966). On the topic, many comprehensive and review studies have been written. Table 28−1 represents the major reports on the use of QSAR in corrosion inhibition reported in the previous studies over the last 40 years. This chapter of the present perspective review aims to compile the important information presented in prior articles and provides some insights into the topic's future viewpoint.

28.2.4 Artificial neural network-based modeling

28.2.4.1 Historical perspectives of ANN

To describe the nonlinear connection between the input variables (molecular descriptors) and their output (chemical activities) in a particular dataset, models such as ANNs, support vector algorithms, and genetic algorithms are used recently (Khaled, Al-Nofai, Abdel-Shafi, & Mater, 2016). A high-tech technique for correlating, predicting, and categorizing complicated and nonlinear correlations between input and output variables has been the use ANN, a subclass of machine learning (Abiodun et al., 2018). Without any prior understanding of the underlying mathematical procedure, these networks are finding increasing use in modeling complicated nonlinear input/output interactions across a range of human achievements (Khaled & Al-Mobarak, 2012). An ANN is a representation of analysis that applies artificial intelligence (AI) to mathematical modeling and is designed to resemble how the human brain works. The debut of ANN might be referred to as the previous work of McCullouch and Pitts in 1943. Together, the first author—a neurophysiologist—and the second author—a mathematician—wrote a scholarly article endorsing the use of a simple artificial neuron (Cohen & Feigenbaum, 1982). In a joint scientific study, the first author—a neurophysiologist —and the second author—a mathematician—recommended the use of a simple artificial neuron. To address character recognition issues, the oldest neural network in use today, the perceptron, was developed. Due to several limitations and critiques of a basic neural network, ANN research endured a significant downturn and obstruction from the middle of the 1970s to the early 1980s.However, studies by LeCun, Parker, and Hopfield, among others, gave ANN a new lease of life and sparked a surge in interest in AI (Hopfield, 1982). Smets and Bogaerts (1992) conducted the first research on the use of neural networks in corrosion science in 1992, using three backpropagation networks to predict the stress corrosion cracking of austenitic stainless steels in near-neutral fluids. In an effort to comprehend corrosion events and forecast corrosion inhibition efficiencies, research on ANN has so continued to draw increasing interest in the field of corrosion studies.

28.2.4.2 Representation of an ANN

An ANN is made up of a number of discrete, highly linked processors known as neurons or nodes. Three key characteristics—node character, network architecture, and learning rules—define each neural network. The characteristics of a node, such as its number of inputs and outputs, the weights (strengths) assigned to each input and output, and the type of activation function, all dictate how it processes signals. The organization and connectivity of the neurons are controlled by the network architecture. Learning strategies specify how an algorithm initializes and modifies its strengths (Livingstone, 2008). A neutral network's fundamental architecture, shown in Fig. 28–6, is made up of synthetic neurons that are similar to the nerve cells in the human brain.

Through connections with connected weights, each artificial neuron (node) gets various inputs from other neurons. Positive weights stimulate the neuron, while negative weights inhibit it and are associated with every connection of this kind. A simple neural network construction that matches the equation is shown in Eq. (28.20):

$$O_j = \sum w_{nj} x_n + \theta_j \qquad (28-20)$$

where θj is the bias or threshold limit of the jth node, and the inputs (x_1, x_2..., x_n) are linked to neuron j with weights ($_{w1j}$, w_{2j}..., w_{nj}) on each connection. Each signal is multiplied by the connection strength it is linked with before the node does a summing of all the input signals it receives. The output is then sent via a transfer function, φ which is often nonlinear, to produce the node's ultimate output, *Oj*. The S-shaped sigmoid function is the best activation function because, when used with a back-propagation neural network, its functions are limited, continuous, and readily differentiable everywhere (Livingstone, 2008; Roy, Kar, & Das, 2015b). The design of a neural network reveals that similar to a multilayer perceptron, the nodes are often grouped in three main levels: the input, hidden, and output layers. The input

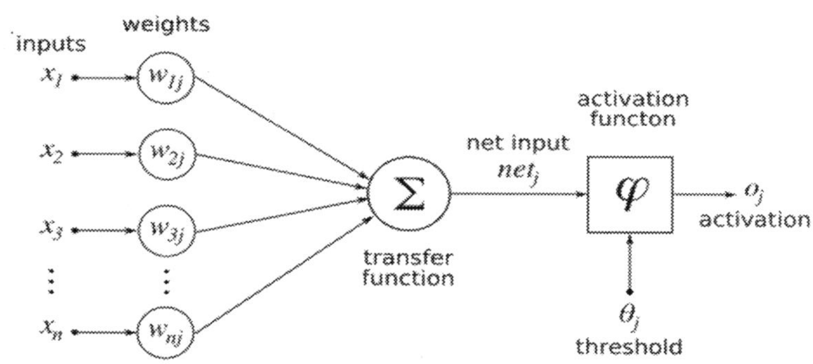

FIGURE 28–6 A representative structure of an artificial neuron (Roy, Kar, & Das, 2015a).

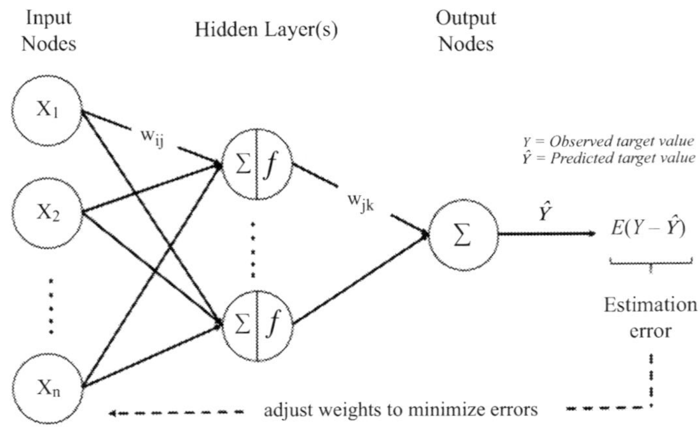

FIGURE 28–7 A representation of multilayer feed-forward back propagation neural network (Abdel-Ilah, Veljović, & Badnjević, 2017).

layer receives linear input data that are often scaled or normalized to 0–1 to avoid hidden nodes from being prematurely saturated and huge values from predominating (Basheer & Hajmeer, 2000). The multilayer perceptron, shown in Fig. 28–7, is the neural network that is most often utilized. It is often referred to as the back propagation feed forward architecture, and it looks to be the most effective method for resolving issues involving pattern recognition, classification, and forecasting (Devillers, 1996; Rumelhart, Hinton, & Williams, 1988; Maltarollo, Honório, & da Silva, 2013). Supervised learning, which gets its name from the fact that data information moves forward, is a key component of feed-forward network topology.

In light of this, it is difficult to analyze and comprehend the complex and nonlinear processes involved in metal corrosion using analytical methods or empirical laws. As a result, ANNs have emerged as the best method for analyzing corrosion events (Luciano, Traverso, & Letardi, 2010). Several inputs are received by the network, which then multiplies them by the strength of its connections, adds the products (squared errors of prediction) with any bias present in the hidden layer, and processes the result using an activation function to produce an output. Backpropagation spreads the mistake back into the network to reduce the difference between the experimental and anticipated output. Depending on the size of the initial error, the strengths of each link between the nodes are changed.

28.2.4.3 Applications of ANN in corrosion inhibition studies

ANNs have been used as a model for forecasting the effects of chemicals in inhibiting corrosion since the early 1990s (Smets & Bogaerts, 1992). In research published by Rosen and Silverman (1992), corrosion type was predicted from Tafel curves using two AI techniques (ANN and an expert system). Some of the input factors were repassivation potential, hysteresis and passive current density, and pitting potential, and the data outputs included general, crevice, and pitting corrosion hazards.

Short questions

1. What are the advantages and disadvantages of computational modeling and simulations in corrosion management?
2. Write the name of some common descriptors derived from DFT study used to describe the interactions between metal surface.
3. What do you mean by adsorption energy? Is it different from binding energy?
4. Describe the role of adsorption energy on mode and effectiveness of inhibitor adsorption.
5. What is the significance of molecular orbital pictures in identifying the sites responsible for binding?
6. What do you mean by energy band gap (delta E) and what is its significance?
7. Write the full form of QSAR, ANN, HOMO, LUMO, and ANN.

References

Abdallah, M., Gad, E., Sobhi, M., Al-Fahemi, J. H., & Alfakeer, M. (2019). *Egyptian Journal of Petroleum, 28,* 173–181.

Abdel-Ilah, L., Veljović, E., Gurbeta, L., & Badnjević, A. (2017). *International Journal of Engineering Research & Technology, 6,* 582–587.

Abdul-Ahad, P., & Al-Madfai, S. (1989). *Corrosion, 45,* 978–980.

Abiodun, O. I., Jantan, A., Omolara, A. E., Dada, K. V., Mohamed, N. A., & Arshad, H. (2018). *Heliyon, 4,* e00938.

Adachi, T., Osako, Y., Tanaka, M., Hojo, M., & Hollister, S. J. (2006). *Biomaterials, 27,* 3964–3972.

Ammouchi, N., Allal, H., Belhocine, Y., Bettaz, S., & Zouaoui, E. (2020). *Journal of Molecular Liquids, 300,* 112309.

Ansari, K., Quraishi, M., & Singh, A. (2014). *Corrosion Science, 79,* 5–15.

Arthur, D. E., Jonathan, A., Ameh, P. O., & Anya, C. (2013). *International Journal of Industrial Chemistry, 4,* 2.

Ashassi-Sorkhabi, H., Shaabani, B., & Seifzadeh, D. (2005a). *Applied Surface Science, 239,* 154–164.

Ashassi-Sorkhabi, H., Shaabani, B., & Seifzadeh, D. (2005b). *Electrochimica Acta, 50,* 3446–3452.

Basheer, I. A., & Hajmeer, M. (2000). *Journal of Microbiological Methods, 43,* 3–31.

Bentiss, F., Traisnel, M., Vezin, H., Hildebrand, H., & Lagrenee, M. (2004). *Corrosion Science, 46,* 2781–2792.

Bentiss, F., Traisnel, M., Vezin, H., & Lagrenée, M. (2003). *Corrosion science, 45,* 371–380.

Berendsen, H. J. (1999). Molecular dynamics simulations: The limits and beyond. *Computational molecular dynamics: challenges, methods, ideas* (pp. 3–36). Springer.

Bergman, I. (1954). *Transactions of the Faraday Society, 50,* 829–838.

BinSabt, M., Nazeer, A. A., Madkour, M., & Al-Sagheer, F. (2016). *RSC Advances, 6,* 6888–6895.

Böhni, H., Suter, T., & Schreyer, A. (1995). *Electrochimica Acta, 40,* 1361–1368.

Chattaraj, P. K., Maiti, B., & Sarkar, U. (2003). *The Journal of Physical Chemistry. A, 107,* 4973–4975.

Chauhan, D. S., Quraishi, M., Ansari, K., & Saleh, T. A. (2020). *Progress in Organic Coatings, 147,* 105741.

Cherkasov, A., Muratov, E. N., Fourches, D., Varnek, A., Baskin, I. I., Cronin, M., ... Todeschini, R. (2014). *Journal of Medicinal Chemistry, 57,* 4977–5010.

Cohen, P. R., & Feigenbaum, E. A. (1982). *The handbook of artificial intelligence*, Stanford, Calif: HeurisTech Press; Los Altos, Calif: William Kaufmann, Inc., Stanford, Calif.

Crum-Brown, A., & Fraser, T. R. (1868–1869). *Transactions of the Royal Society of Edinburgh, 25*, 151–203, 693–739.

Dearden, J. C. (2017a). *Pharma Science, 23*, 82–83.

Dearden, J. C. (2017b). *The history and development of quantitative structure-activity relationships (QSARs). Oncology: breakthroughs in research and practice* (pp. 67–117). IGI Global.

Devillers, J. (1996). *Neural networks in QSAR and drug design*. Academic Press.

Deyab, M. (2014). *Journal of Power Sources, 268*, 50–55.

Dibetsoe, M., Olasunkanmi, L. O., Fayemi, O. E., Yesudass, S., Ramaganthan, B., Bahadur, I., ... Ebenso, E. E. (2015). *Molecules (Basel, Switzerland), 20*, 15701–15734.

Donahue, F. M., & Nobe, K. (1966). *Journal of the Electrochemical Society, 112*, 886–891.

Ebenso, E. E., Isabirye, D. A., & Eddy, N. O. (2010). *International Journal of Molecular Sciences, 11*, 2473–2498.

Eddy, N. O., & Ita, B. I. (2011). *Journal of Molecular Modeling, 17*, 359–376.

Edmonds, B., & Hales, D. (2005). *Journal of Mathematical Sociology, 29*, 209–232.

Free, S. M., & Wilson, J. W. (1964). *Journal of Medicinal Chemistry, 7*, 395–399.

Fukui, K. (1982a). *Science (New York, N.Y.), 218*, 747–754.

Fukui, K. (1982b). *Angewandte Chemie International Edition in English, 21*, 801–809.

Gao, G., & Liang, C. (2007). *Electrochimica Acta, 52*, 4554–4559.

Gece, G. (2011). *Corrosion Science, 53*, 3873–3898.

Geerlings, P., De Proft, F., & Langenaeker, W. (2003). *Chemical Reviews, 103*, 1793–1874.

Grover, A., Grover, M., & Sharma, K. (2016). *World Journal of Pharmaceutical Sciences, 5*, 427–437.

Growcock, F. (1989). *Corrosion, 45*, 1003–1007.

Growcock, F., Frenier, W., & Andreozzi, P. (1989). *Corrosion, 45*, 1007–1015.

Guo, L., Obot, I. B., Zheng, X., Shen, X., Qiang, Y., Kaya, S., & Kaya, C. (2017). *Applied Surface Science, 406*, 301–306.

Guo, L., Zhu, S., Zhang, S., He, Q., & Li, W. (2014). *Corrosion Science, 87*, 366–375.

Hagler, A. T., Huler, E., & Lifson, S. (1974). *Journal of the American Chemical Society, 96*, 5319–5327.

Hagler, A. T., & Lifson, S. (1974). *Journal of the American Chemical Society, 96*, 5327–5335.

Hamadi, L., Mansouri, S., Oulmi, K., & Kareche, A. (2018). *Egyptian Journal of Petroleum, 27*, 1157–1165.

Hammett, L. P. (1935). *Chemical Reviews, 17*, 125–136.

Hansch, C., & Fujita, T. (1964). *Journal of the American Chemical Society, 86*, 1616–1626.

Hoijtink, G. J. (1955). *Recueil des Travaux Chimiques des Pays-Bas, 74*, 1525–1539.

Hopfield, J. J. (1982). *Proceedings of the National Academy of Sciences of the United States of America, 79*, 2554–2558.

Iczkowski, R. P., & Margrave, J. L. (1961). *Journal of the American Chemical Society, 83*, 3547–3551.

Jensen, B. (2016).

Ju, H., Kai, Z.-P., & Li, Y. (2008). *Corrosion Science, 50*, 865–871.

Khaled, K., & Amin, M. A. (2008). *Journal of Applied Electrochemistry, 38*, 1609–1621.

Khaled, K. F., Al-Nofai, N., Abdel-Shafi, N., & Mater, J. (2016). *Environmental Sciences, 7*, 2121–2136.

Khaled, K. F., & Al-Mobarak, N. A. (2012). *International Journal of Electrochemical Science, 7,* 1045–1059.

Khalil, N. (2003). *Electrochimica Acta, 48,* 2635–2640.

Khojasteh, D., Kazerooni, M., Salarian, S., & Kamali, R. (2016). *Journal of Industrial and Engineering Chemistry, 42,* 1–14.

Kovačević, N., & Kokalj, A. (2011). *The Journal of Physical Chemistry C, 115,* 24189–24197.

Lai, C., Guo, X., Wei, J., Xie, B., Zou, L., Li, X., . . . Wang, C. (2017). *Open Chemistry, 15,* 263–271.

Leimkuhler, B., & Matthews, C. (2016). *Molecular dynamics*. Springer.

Li, S., Wang, Y., Chen, S., Yu, R., Lei, S., Ma, H., & Liu, D. X. (1999). *Corrosion Science, 41,* 1769–1782.

Li, X., Deng, S., & Xie, X. (2014). *Corrosion Science, 81,* 162–175.

Lifson, S., Hagler, A. T., & Dauber, P. (1979). *Journal of the American Chemical Society, 101,* 5111–5121.

Livingstone, D. J. (2008). *Artificial neural networks: methods and applications*. Springer.

Luciano, G., Traverso, P., & Letardi, P. (2010). *Corrosion Science, 52,* 2750–2757.

Lukovits, I., Shaban, A., & Kálmán, E. (2003). *Russian Journal of Electrochemistry, 39,* 177–181.

Lukovits, I., Kalman, E., & Zucchi, F. (2001). *Corrosion, 57,* 3–8.

Lukovits, I., Bakó, I., Shaban, A., & Kálmán, E. (1998). *Electrochimica Acta, 43,* 131–136.

Lukovits, I., Palfi, K., Bako, I., & Kalman, E. (1997). *Corrosion, 53,* 915–919.

Malik, M. A., Hashim, M. A., Nabi, F., Al-Thabaiti, S. A., & Khan, Z. (2011). *International Journal of Electrochemical Science, 6,* 1927–1948.

Maltarollo, V. G., Honório, K. M., & da Silva, A. r. B.F. (2013). *Artificial neural networks-architectures and applications,* 203–223.

Mandour, H. S., Nazeer, A. A., Al-Hetlani, E., Madkour, M., & Abdel-Monem, Y. K. (2018). *New Journal of Chemistry, 42,* 5914–5922.

Meyer, H. (1899). *Naunyn-Schmiedeberg's Arch Exp Pathol Pharmakol, 42,* 109–118.

Montemor, M., & Ferreira, M. (2008). *Surface and Coatings Technology, 202,* 4766–4774.

Monticelli, L., & Tieleman, D. P. (2013). *Force fields for classical molecular dynamics. Biomolecular simulations* (pp. 197–213). Springer.

Murulana, L. C., Kabanda, M. M., & Ebenso, E. E. (2016). *Journal of Molecular Liquids, 215,* 763–779.

Musa, A. Y., Mohamad, A. B., Kadhum, A. A. H., Takriff, M. S., & Ahmoda, W. (2012). *Journal of Industrial and Engineering Chemistry, 18,* 551–555.

O.S.O. Pesticides.

Obot, I., Macdonald, D., & Gasem, Z. (2015). *Corrosion Science, 99,* 1–30.

Obot, I., Haruna, K., & Saleh, T. J. A. J. f. S. (2019). *Engineering, 44,* 1–32.

Obot, I., & Obi-Egbedi, N. (2010). *Corrosion Science, 52,* 657–660.

OECD, in.

Oguzie, E. E. (2007). *Corrosion science, 49,* 1527–1539.

Outirite, M., Lagrenée, M., Lebrini, M., Traisnel, M., Jama, C., Vezin, H., & Bentiss, F. (2010). *Electrochimica Acta, 55,* 1670–1681.

Overton, C. E. (1901). Studien über die Narkose: zugleich ein Beitrag zur allgemeinen Pharmakologie, G. Fischer.

Parr, R. G., Szentpaly, L. v., & Liu, S. (1999). *Journal of the American Chemical Society, 121,* 1922–1924.

Parr, R. G., Donnelly, R. A., Levy, M., & Palke, W. E. (1978). *The Journal of Chemical Physics, 68,* 3801–3807.

Parr, R. G., & Pearson, R. G. (1983). *Journal of the American Chemical Society, 105*, 7512–7516.

Pathak, R., & Mishra, P. (2016). *International Journal of Science and Research, 5*, 671–677.

Peter, A., Obot, I., & Sharma, S. K. (2015). *International Journal of Industrial Chemistry, 6*, 153–164.

Poberžnik, M., Chiter, F., Milošev, I., Marcus, P., Costa, D., & Kokalj, A. (2020). *Applied Surface Science*, 146156.

Polanski, J. (2019).

Popoola, L. T. (2019). *Corrosion Reviews, 37*, 71–102.

Quraishi, M. A., Chauhan, D. S., & Saji, V. S. (2020). 3 - Computational methods of inhibitor evaluation. In M. A. Quraishi, D. S. Chauhan, & V. S. Saji (Eds.), *Heterocyclic organic corrosion inhibitors* (pp. 59–86). Elsevier.

Richet, M. (1893). *Compt. Rend. Soc. Biol., 45*, 775–776.

Rosen, E. M., & Silverman, D. C. (1992). *Corrosion, 48*, 734–745.

Roy, K., Kar, S., & Das, R. N. (2015a). *Statistical methods in QSAR/QSPR. A primer on QSAR/QSPR modeling: fundamental concepts* (pp. 37–59). Cham: Springer International Publishing.

Roy, K., Kar, S., & Das, R. N. (2015b). *Understanding the basics of QSAR for applications in pharmaceutical sciences and risk assessment.* Academic press.

Rumelhart, D. E., Hinton, G. E., & Williams, R. J. (1988). ch. Learning representations by back-propagating errors, 696–699.

Saha, S. K., & Banerjee, P. (2015). *RSC Advances, 5*, 71120–71130.

Saji, V. S., & Thomas, J. (2007). *Current Science*, 51–55.

Saji, V. S., & Cook, R. (2012). *Corrosion protection and control using nanomaterials.* Elsevier.

Sanderson, R. (1976). *New York, 77.*

Sanderson, R. T. (1983). *Journal of the American Chemical Society, 105*, 2259–2261.

Sharma, S. K., Peter, A., & Obot, I. B. (2015). *Journal of Analytical Science and Technology, 6*, 26.

Shehata, O. S., Korshed, L. A., & Attia, A. (2018). *Corrosion Inhibitors, Principles and Recent Applications*, 121.

Singh, R., & Dahotre, N. B. (2007). *Journal of Materials Science: Materials in Medicine, 18*, 725–751.

Smets, H. M. G., & Bogaerts, W. F. L. (1992). *Corrosion, 48*, 618–623.

Umoren, S. A., & Eduok, U. M. (2016). *Carbohydrate Polymers, 140*, 314–341.

Vashisht, H., Bahadur, I., Kumar, S., Goyal, M., Kaur, G., Singh, G., . . . Ebenso, E. (2016). *Journal of Molecular Liquids, 224*, 19–29.

Verma, C., Verma, D. K., Ebenso, E. E., & Quraishi, M. A. (2018a). *Heteroatom Chemistry, 29*, e21437.

Verma, C., Ebenso, E. E., Bahadur, I., & Quraishi, M. (2018b). *Journal of Molecular Liquids, 266*, 577–590.

Verma, C., Lgaz, H., Verma, D. K., Ebenso, E. E., Bahadur, I., & Quraishi, M. A. (2018c). *Journal of Molecular Liquids, 260*, 99–120.

Verma, C., Quraishi, M., & Ebenso, E. E. (2018d). *Sustainable Chemistry and Pharmacy, 10*, 134–147.

Verma, C., Haque, J., Quraishi, M., & Ebenso, E. E. (2019). *Journal of Molecular Liquids, 275*, 18–40.

Verma, C., Olasunkanmi, L. O., Ebenso, E. E., Quraishi, M. A., & Obot, I. B. (2016). *The Journal of Physical Chemistry C, 120*, 11598–11611.

Verma, C. B., Quraishi, M., & Singh, A. (2015a). *Journal of the Taiwan Institute of Chemical Engineers, 49*, 229–239.

Verma, C., Quraishi, M., Olasunkanmi, L., & Ebenso, E. E. (2015b). *RSC Advances, 5*, 85417–85430.

Verma, C., Quraishi, M. A., Kluza, K., Makowska-Janusik, M., Olasunkanmi, L. O., & Ebenso, E. E. (2017). *Scientific Reports, 7*, 44432.

Verma, D. K., Al Fantazi, A., Verma, C., Khan, F., Asatkar, A., Hussain, C. M., & Ebenso, E. E. (2020). *Journal of Molecular Liquids, 314,* 113651.

Whalen, J. W. (1983). *Journal of Chemical Education, 60,* A322.

Xhanari, K., Finšgar, M., Hrnčič, M. K., Maver, U., Knez, Ž., & Seiti, B. (2017). *RSC Advances, 7,* 27299–27330.

Xhanari, K., & Finšgar, M. (2016). *RSC Advances, 6,* 62833–62857.

Yang, W., & Parr, R. G. (1985). Proceedings of the National Academy of Sciences, 82, 6723–6726.

Zhang, S., Lei, W., Xia, M., & Wang, F. (2005). *Journal of Molecular Structure: THEOCHEM, 732,* 173–182.

Zhu, Y., Free, M. L., Woollam, R., & Durnie, W. (2017). *Progress in Materials Science, 90,* 159–223.

PART

10

Emerging trends in corrosion protection

29 Recent trends in corrosion protection

Besides being an integral part of everyday life, fighting corrosion is also a challenge that is constantly being faced by various industries. The rapid emergence and evolution of new materials have made it possible to improve the performance and reliability of industrial equipment. The market for corrosion protection is incredibly important and valuable. With the continuous emergence of new technologies and trends, it is important that we keep up with the latest in order to ensure that our products are performing at their best. This chapter will talk about some of the recent trends that have been affecting the industry.

29.1 Cutting edge technology and innovative products

The rapid emergence and evolution of new scientific methods and techniques related to corrosion control have raised the bar for the future of this industry. According to NACE's International Measures of Prevention, Application, and Economics of Corrosion Technology study, the global cost of corrosion is expected to reach over $2.5 trillion (International Measures of Prevention, Application, and Economics of Corrosion Technologies Study, 2016 by NACE International, 15835 Park Ten Place, n.d.). As the industry continues to grow, new research and methods are needed to address the increasing demand for corrosion-resistant materials. Researchers are currently studying various materials such as nickel to develop new and effective ways to improve the analyses of corrosion. They are also exploring the applications of these complex alloys in reinforced concrete structures. Among the other materials that can help structures endure corrosion are steel alternatives. The development of new scientific methods and techniques will allow the industry to improve the efficiency and effectiveness of its corrosion protection processes.

29.2 The environment consideration

Besides affecting the industrial environment; corrosion can also affect the natural environment around us. In addition to protecting the asset, experts are also suggesting ways to minimize the environmental damage caused by corrosion. This includes the use of materials that are designed to minimize the impact of corrosion. It is widely known that the degradation of materials can cause severe environmental damage, such as the pollution of bodies of water. This relationship between environmental acceleration and protection should be considered when it comes to addressing this issue. Due to the increasing importance of industrial

Handbook of Corrosion Engineering. DOI: https://doi.org/10.1016/B978-0-323-95185-2.00029-0

toxicity and environmental sustainability, the design and engineering construction industry is now more focused on the protection of its workers and the environment. This is why it is important that the various factors that affect the environment are considered when it comes to the selection of protective systems. Some of the factors that can affect the environment include global warming, acid rain, and waste management. New trends related to sustainability are emerging. These include the use of products that are both sustainable and long-lasting. In addition, they can help prevent corrosion and keep surfaces looking their best. One of the most important factors that can be considered when it comes to protecting the economy and the environment is the use of protective systems. In order to improve the management of corrosion, it is important that the various technical and nontechnical aspects of an industrial facility are considered. This can be done through preventive strategies. However, environmental protection legislation has been restricted due to the carcinogenic and toxic effects of chromium. The use of rare earth elements as corrosion inhibitors has been attributed to their ability to prevent the harmful effects of chromium salts. In addition to protecting the environment, the process of coating also involves using environmentally friendly compounds.

29.3 Planning protection at stage 1 of design

In order to lower the costs associated with equipment repairs and maintenance, design engineers must be able to identify the effects of corrosion on the structure and function of the equipment. This step should be carried out in the design stage. Having the necessary expertise in both material science and maintenance will ensure that the parts work seamlessly throughout the life of the equipment.

29.4 Using earth's natural materials

In addition to take better care of our assets, we also consider the development of new materials that can be used to protect them. For instance, a Japanese research team has been studying the effects of combining certain bacteria namely *Bacillus subtilis* natto with reinforced concrete. Advancements in materials science and technology that have the potential to improve the lifespan of equipment are some of the factors that contribute to the increasing value of assets. The coating inspector is constantly looking for ways to protect our biggest assets.

29.5 Using smart coatings

Around 786,000 workers in the European Union are currently exposed to hexavalent chromium, a carcinogenic substance (3IARC IARC Monographs on the Evaluation of Carcinogenic Risks to Humans. Chromium VI Compounds. Volume 100C Lyon, France: International Agency for Research on Cancer, 2012). Due to the complexity of the situation, the development of effective and safe alternatives to chromate has been a major challenge for the industry. The extension of the regulations on the registration, evaluation, and approval of chemicals was made to allow the industry to develop new and safer alternatives to chromate. During this time, an innovative new product has been introduced to the market that can replace toxic chemicals. This new innovation, which is a microreservoir inhibitor, can create long-lasting coatings.

29.6 Oil and gas corrosion protection market continues to thrive

The global oil and gas corrosion prevention market is expected to reach a market size of around USD 12.2 billion by 2025, according to a report by Grand View Research. The increasing demand for corrosion protection in the North American region is expected to drive the market's growth. The oil and gas corrosion prevention market is mainly driven by the increasing demand from the upstream and downstream sectors. The increasing need for transportation and supply infrastructure is expected to fuel the demand for advanced products that can prevent atmospheric corrosion. North America is expected to grow at a higher rate due to the presence of an abundant supply of natural resources and the presence of an increasing number of rigs. The different types of protection available in the market are coatings, inhibitors, and paints (Oil & Gas Corrosion Protection Market Size Worth $12.2 Billion By 2025, 2017). The coating type held the largest share of the global oil and gas corrosion prevention market in 2016. North America is expected to be one of the fastest-growing regions in the market during the next few years. The region is mainly driven by the increasing production of shale gas in the United States and Canada. The rapid emergence and growth of the oil and gas industry in the Asia Pacific and Africa are some of the factors that are expected to drive the market's growth. In addition, the increasing number of discoveries in these regions is also expected to fuel the need for corrosion protection.

29.7 Recent trends in corrosion protection

The *Materials Performance Journal* published an issue about the emerging trends in corrosion over the next 25 years. Three finalists were selected: John Scully, FNACE; the Charles Henderson Chaired Professor of Materials Science and Engineering and the co-director of the Centre for Electrochemical Science and Engineering in the School of Engineering and Applied Science at the University of Virginia, USA; Neil G. Thompson, PhD, FNACE, senior vice president and head of Pipeline Services at DNV GL USA, Inc., Dublin, Ohio, USA; and Jack Tinnea, who is the technical editor of the journal. They discussed their predictions about the future of the industry (*Roundtable on the Future of Corrosion Control: Part 3*, 2020). They believe that the future of corrosion-control technologies lies in the development of new and sophisticated innovations that will allow companies to continue to improve their operations over the next couple of decades. One of the most important factors that will allow them to continue to improve their operations is the emergence of cyber-physical systems.

This concept will allow them to monitor and analyze the health of their systems and structures in real time. The concept of corrosion control will likely follow the same path as the development of the internet of things. In real time, various sensors will be able to monitor and analyze the health of a structure or system. They will then be able to make decisions based on the data collected. In the future, many of the sensors that will be used will be powered by the environment, without the need for a connection to the grid. This will allow them to collect huge amounts of data. According to experts, the cyber-physical world will continue to expand and eventually corrode. According to experts, the amount of data collected by these sensors will continue to increase. In the last 2 years, over 90% of the world's data has been collected. Unfortunately, the data collected by these sensors is too large to manage. Several strategies, such as data sciences, materials informatics, and machine learning, will need to be developed to handle and interpret this data. However, they will be able to establish relationships and trends that can help prevent corrosion. Data sciences can also reveal the relationships between various factors that can affect the development and maintenance of corrosion. This strategy can help them develop effective strategies to prevent corrosion. In addition, nanotechnology has been discussed in engineering and science publications (X-Ray Microscopy Platform Probes Material Composition, 2018; Valdez et al., 2018). Understanding the nanoscale is not just about the manufacture of new products. It also includes our understanding of how to construct materials that can endure the effects of corrosion. This knowledge will allow us to develop incredibly strong and flexible materials that can be used in various applications. Through the reactive side, this knowledge can help us understand the causes of corrosion in different locations. For instance, it can be used to monitor the condition of major infrastructure elements such as bridges and roads by placing sensors in concrete. For monitoring major infrastructure elements, such as bridges and roads, sensors typically need to be embedded in concrete for up to 80 years. Since the literature on embedded sensors has been published, there have been a few discussions about failures of these devices (Andrade & Martinez, 2009; Larsen, 2016; Lee et al., 2018). Cross-pollination can be used to improve the reliability of embedded sensors that are designed to fit the demands of long-term infrastructure service lives. After treating them with a vapor phase inhibitor, it is important to apply a material with good wetting characteristics. This process can be performed by either treating the embedded sensors with a vapor phase inhibitor followed by a potting material that has good wetting characteristics or by coating them with a hydrophobic or an alkaline-resistant coating.

All of these possibilities, and more, point to a promising future in the corrosion control industry.

References

3IARC IARC Monographs on the Evaluation of Carcinogenic Risks to Humans. Chromium (VI) Compounds. Volume 100C Lyon, France: International Agency for Research on Cancer. (2012).

Andrade, C. & Martinez, I. (2009). *Non-destructive testing in civil engineering conference proceedings embedded sensors for the monitoring of corrosion parameters in concrete structures.*

International Measures of Prevention, Application, and Economics of Corrosion Technologies Study, 2016 by NACE International, 15835 Park Ten Place. (n.d.).

Larsen, K. R. (2016). Evaluating sensors to monitor steel corrosion in concrete structures. *MP, 55.*

Lee, S. K., Jang, B. S., Ahn, J. H., & Kim, K. C. (2018). Corrosion monitoring of world's largest tidal power plant. *Materials Performance, 57*(11), 48–52. Available from https://mp.epubxp.com/i/1043615-nov-2018.

Oil & Gas Corrosion Protection Market Size Worth $12.2 Billion By 2025. (2017). https://www.grandviewresearch.com/press-release/global-oil-gas-corrosion-protection-market#.

Valdez, B., Schorr, M., Salinas, R., & Eliezer, A. (2018). Food industry: Equipment, materials, and corrosion. *Materials Performance, 57*(5), 40–43. Available from https://mp.epubxp.com/i/973683-may-2018.

X-Ray Microscopy Platform Probes Material Composition. MP. 57 (2018).

Index

Note: Page numbers followed by "*f*" and "*t*" refer to figures and tables, respectively.

H

T

X

Z

Printed in the United States
by Baker & Taylor Publisher Services